Urban Land Systems: An Ecosystems Perspective

Special Issue Editors

Andrew Millington
Harini Nagendra
Monika Kopecká

MDPI • Basel • Beijing • Wuhan • Barcelona • Belgrade

MDPI

Special Issue Editors
Andrew Millington
Flinders University
Australia

Harini Nagendra
Azim Premji University
India

Monika Kopecká
Institute of Geography, Slovak Academy of Sciences
Slovakia

Editorial Office
MDPI
St. Alban-Anlage 66
Basel, Switzerland

This edition is a reprint of the Special Issue published online in the open access journal *Land* (ISSN 2073-445X) from 2017–2018 (available at: http://www.mdpi.com/journal/land/special_issues/ urbanlandsystems).

For citation purposes, cite each article independently as indicated on the article page online and as indicated below:

Lastname, F.M.; Lastname, F.M. Article title. *Journal Name* **Year**, *Article number*, page range.

First Edition 2018

ISBN 978-3-03842-917-3 (Pbk)
ISBN 978-3-03842-918-0 (PDF)

Cover photo courtesy of Monika Kopecká.

Table of Contents

About the Special Issue Editors

Andrew Millington is a Professor in the College of Science and Engineering, Flinders University. His research interests are in remote sensing, land-use dynamics, biogeography and human impacts on the environment. Most of his work lies in the area described as coupled human-natural systems in which he examine the influences of human systems on land use and its impacts on things like vegetation change, landscape fragmentation and biodiversity. In 1978 he joined one of Africa's oldest universities (Fourah Bay College, University of Sierra Leone) as a lecturer in the Geography Department. While teaching there he did his doctoral degree at the University of Sussex.

Harini Nagendra is a Professor of Sustainability at Azim Premji University. She is an ecologist who uses satellite remote sensing coupled with field studies of biodiversity, archival research, institutional analysis, and community interviews to examine the factors shaping the social-ecological sustainability of forests and cities in the south Asian context. She completed her PhD from the Centre for Ecological Sciences in the Indian Institute of Science in 1998.

Monika Kopecká, Ph.D., is a senior research scientist at the Institute of Geography, Slovak Academy of Sciences and the Head of the Department of Geoinformatics. Her research is focused on land use and land cover mapping, landscape changes, and landscape indicators. She participated in several projects related to spatial analysis and assessment of landscape structure and its changes. Currently, her research activities are oriented to monitoring of urban landscape dynamics and agricultural landscape abandonment.

"By far the greatest and most admirable form of wisdom is that needed to plan and beautify cities and human communities."

Socrates

land

MDPI

Editorial

Urban Land Systems: An Ecosystems Perspective

Monika Kopecká [1],* [ID], Harini Nagendra [2] [ID] and Andrew Millington [3] [ID]

[1]   Department of Geoinformatics, Institute of Geography, Slovak Academy of Sciences, Stefanikova 49, 814 73 Bratislava, Slovakia

[2]   Azim Premji University, PES Institute of Technology Campus, Pixel Park, B Block, Electronics City, Hosur Road, Bangalore 560100, India; harini.nagendra@apu.edu.in

[3]   College of Science and Engineering, Flinders University, GPO Box 2100, Adelaide, SA 5001, Australia; andrew.millington@flinders.edu.au

*   Correspondence: Monika.Kopecka@savba.sk; Tel.: +421-905-345-556

Received: 8 January 2018; Accepted: 9 January 2018; Published: 11 January 2018

1. Introduction

We live in an urbanizing world. Since 2008, more than half of humanity lives in cities, both large and small, and old and new. We also live in a world that is becoming even more urbanized, it is expected that by 2050, 66 per cent of the world's population will live in cities [1]. The process of urbanization, accompanied by the rapid expansion of cities and the sprawling growth of metropolitan regions over the world, is one of the most important transformations of a natural landscape.

In the context of land systems science, contemporary urbanisation is a set of land-use change processes and the various contemporary cityscapes are the resulting land systems. Population growth increases urban footprints with consequences on biodiversity and climate. Much of the explosive urban growth has been unplanned and conflicting land-use demands often arise as land is a limited resource. Increased requirements for living space and intensive landscape utilization constitute two of the principal reasons for the environmental change, with significant impacts on quality of life and ecosystems.

This special issue of LAND explores urban land dynamics with particular regard to ecosystem structure, and discusses consequent environmental changes and their impacts. The studies cover a wide range of countries and contexts, and draw on a number of disciplinary methods and interdisciplinary approaches from the social and natural sciences. The papers have been authored by 41 researchers from 29 institutions in countries worldwide: from Australia, Bangladesh, China, India, Iraq, Italy, Japan, Nigeria, Philippines, Saudi Arabia, Slovakia, Spain, Thailand, the United Kingdom, and the USA.

2. Dynamics of Urban Land Systems

Earth observation data can contribute considerably in monitoring complex urban land cover patterns for various applications in different environments. Several papers focus on the detection of urban growth based on spaceborne remote sensing data at multiple scales, spatial and temporal resolutions, and the evaluation of environmental impacts through well-established concepts of landscape metrics.

MacLachlan et al. [2] evaluate multi-temporal urban expansion for Perth, Western Australia, derived from Landsat imagery and the related decrease of natural resources. Their results indicate that the spatial extent of the Perth Metropolitan Region has increased considerably in the period of 1990–2015. Irreversible and unsustainable agricultural landscape changes related to urban growth in peri-urban areas of Adelaide city are assessed by Wadduwage et al. [3]. Fragmentation of agro-ecosystems due to urban expansion is analyzed using several landscape metrics indicators: percentage of land, mean parcel size, patch density, and modified Simpson´s Diversity.

Rapidly growing cities in the global south, particularly in Asia and Africa, represent the dominant urban footprint of the present and future. A great deal of attention is currently being focused on these cities, about which we know relatively little in comparison to northern cities. In this special issue, Zhang et al. [4] use Landsat imagery to detect the changes in urban land use in the Yellow River Delta to document systematic changes in natural wetlands in 1976, 1984, 1995, 2006 and 2014. Their cartographic outputs document systematic wetland degradation, wetland conversion to salt pans and aquafarms, and significant urbanization.

Similarly, Landsat data are used by Rimal et al. [5] to analyze the spatiotemporal patterns of urbanization and LULC changes in Kathmandu Valley, Nepal. Results show that the urban coverage of Kathmandu Valley has increased tremendously from 20.19 km^2 in 1976 to 139.57 km^2 in 2015, at the cost of cultivated lands and forests. This study also reports the impacts of the recent disastrous earthquake in the valley on the urban areas and discusses the high risks associated with different geological formations.

The third example illustrating rapidly developing countries in Asia is a land use/land cover change assessment of the Laguna de Bay area in the Philippines. Iizuka et al. [6] present a visual interpretation of the future changes in LULC classes of built-up, crop-grass, trees, and water up to the year 2030. The probability of changes occurring in different years in the future is calculated using three different scenarios: business-as-usual, compact development, and high sprawl. In total, a large proportion of the study area is modeled to be converted to urban built-up land cover classes by 2030, varying in extent depending on the development scenario.

3. Perception of Urban Green Spaces (UGS) and Ecosystem Services

Urban vegetation is essential for urban ecosystems and ecosystem services and can be determined by well-established methods in remote sensing. In the light of past and future urbanization trends, accurate information on the state, accessibility, distribution, and supply of UGS plays an increasingly important role in sustainable urban development, human well-being, and also for conservation of ecosystem functionality.

Kopecká et al. [7] demonstrate the potential for UGS extraction from newly available Sentinel-2A satellite imagery, provided within the frame of the European Copernicus program. UGS classes are described by the proportion of tree canopy and their ecosystem services. A comparative analysis of three cities in Slovakia indicates the relatively high importance of urban greenery in family housing areas, represented mainly by privately owned gardens.

Cultivated parks and urban gardens play an important role as providers of aesthetic and psychological benefits that enrich human life, reduce stress, and increase physical and mental health. Paul and Nagendra [8] present the results of a survey of UGS perception by park visitors in the megalopolis of Delhi that aimed to understand the importance of parks for them. For example, large parks tend to attract more visitors from further distances, despite their having small neighborhood parks in the vicinity of their homes.

On the other hand, Rupprecht [9] points attention to residents' perceptions of informal UGS—vacant lots, street verges, brownfields, power line corridors, and waterside spaces—in four shrinking cities in Japan. He proposes eight major planning principles derived from the findings as a potential basis for managing non-traditional green spaces to urban planners.

4. Urban Landscape Structure and Urban Heat Island (UHI) Effect

Urban areas influence the local microclimate in several ways, e.g., by air pollution, altered wind speeds and directions, heat stress, or changes in surface ozone concentrations. UHI describes the phenomenon that atmospheric and surface temperatures are higher in urban areas than in surrounding rural areas. Among the long-term consequences on microclimate, the UHI effect has received wide attention from geographers, urban planners, and climatologists over recent years. Rasul et al. [10] review the current research on this topic, methods, data, and techniques used in UHI detection.

They conclude with recommendations for conducting further research on surface urban cool islands that especially occur in arid and semi-arid climates.

Rahman et al. [11] investigate the increase of land surface temperature in Dammam city, the capital of Saudi Arabia's Eastern Province, due to urban expansion in the period 1990–2014. Based on land use/land cover changes and predictive modeling, this study projects a dramatic increase of land surface temperatures for the year 2026.

5. Conclusions

Global and local urbanization is creating very significant challenges for sustainability and human well-being. This special issue highlights some important aspects related to urban sprawl dynamics and urban ecosystem management. Observations and studies presented in 10 papers show that urbanization affects essential ecological, economic, and social landscape functions, whose importance are often undervalued in cities across the globe.

The special issue arises from a session convened by the editors at the GLP Third Open Science Meeting in Beijing, October 2016. We believe that the results presented in these studies will provide useful information for decision and policymakers involved in urban and spatial planning at local, regional, and national levels and can help better plan and design UGS, responding to the needs and preferences of urban communities.

Acknowledgments: We are grateful to the many anonymous reviewers whose comments greatly improved the quality of the special issue. This work was supported by the project: "Effect of impermeable soil cover on urban climate in the context of climate change" (Slovak Research and Development Agency—Grant Agency No. APVV-15-0136). Harini Nagendra acknowledges funding from Azim Premji University's Centre for Urban Sustainability. Andrew Millington acknowledges support from Flinders University College of Science and Engineering.

Author Contributions: Monika Kopecká, Harini Nagendra, and Andrew Millington conceptualized and wrote this paper.

Conflicts of Interest: The authors declare no conflict of interest.

References

1. United Nations, Department of Economic and Social Affairs, Population Division. World Urbanization Prospects: The 2014 Revision, Highlights (ST/ESA/SER.A/352). 2014. Available online: https://esa.un.org/unpd/wup/Publications/Files/WUP2014-Highlights.pdf (accessed on 11 December 17).
2. MacLachlan, A.; Biggs, E.; Roberts, G.; Boruff, B. Urban Growth Dynamics in Perth, Western Australia: Using Applied Remote Sensing for Sustainable Future Planning. *Land* **2017**, *6*, 9. [CrossRef]
3. Wadduwage, S.; Millington, A.; Crossman, N.; Sandhu, H. Agricultural Land Fragmentation at Urban Fringes: An Application of Urban-To-Rural Gradient Analysis in Adelaide. *Land* **2017**, *6*, 28. [CrossRef]
4. Zhang, B.; Zhang, Q.; Feng, C.; Feng, Q.; Zhang, S. Understanding Land Use and Land Cover Dynamics from 1976 to 2014 in Yellow River Delta. *Land* **2017**, *6*, 20. [CrossRef]
5. Rimal, B.; Zhang, L.; Fu, D.; Kunwar, R.; Zhai, Y. Monitoring Urban Growth and the Nepal Earthquake 2015 for Sustainability of Kathmandu Valley, Nepal. *Land* **2017**, *6*, 42. [CrossRef]
6. Iizuka, K.; Johnson, B.; Onishi, A.; Magcale-Macandog, D.; Endo, I.; Bragais, M. Modeling Future Urban Sprawl and Landscape Change in the Laguna de Bay Area, Philippines. *Land* **2017**, *6*, 26. [CrossRef]
7. Kopecká, M.; Szatmári, D.; Rosina, K. Analysis of Urban Green Spaces Based on Sentinel-2A: Case Studies from Slovakia. *Land* **2017**, *6*, 25. [CrossRef]
8. Paul, S.; Nagendra, H. Factors Influencing Perceptions and Use of Urban Nature: Surveys of Park Visitors in Delhi. *Land* **2017**, *6*, 27. [CrossRef]
9. Rupprecht, C. Informal Urban Green Space: Residents' Perception, Use, and Management Preferences across Four Major Japanese Shrinking Cities. *Land* **2017**, *6*, 59. [CrossRef]

10. Rasul, A.; Balzter, H.; Smith, C.; Remedios, J.; Adamu, B.; Sobrino, J.; Srivanit, M.; Weng, Q. A Review on Remote Sensing of Urban Heat and Cool Islands. *Land* **2017**, *6*, 38. [CrossRef]
11. Rahman, M.; Aldosary, A.; Mortoja, M. Modeling Future Land Cover Changes and Their Effects on the Land Surface Temperatures in the Saudi Arabian Eastern Coastal City of Dammam. *Land* **2017**, *6*, 36. [CrossRef]

land

MDPI

Letter

Urban Growth Dynamics in Perth, Western Australia: Using Applied Remote Sensing for Sustainable Future Planning

Andrew MacLachlan [1,*], **Eloise Biggs** [2], **Gareth Roberts** [1] **and Bryan Boruff** [2]

[1]   Geography and Environment Department, The University of Southampton, University Road,
      Southampton SO17 1BJ, UK; G.J.Roberts@soton.ac.uk
[2]   School of Agriculture and Environment, The University of Western Australia, Crawley WA 6009, Australia;
      eloise.biggs@uwa.edu.au (E.B.); bryan.boruff@uwa.edu.au (B.B.)
*   Correspondence: A.C.MacLachlan@soton.ac.uk; Tel.: +44-023-8059-9586

Academic Editors: Andrew Millington, Harini Nagendra and Monika Kopecka
Received: 16 December 2016; Accepted: 18 January 2017; Published: 24 January 2017

Abstract: Earth observation data can provide valuable assessments for monitoring the spatial extent of (un)sustainable urban growth of the world's cities to better inform planning policy in reducing associated economic, social and environmental costs. Western Australia has witnessed rapid economic expansion since the turn of the century founded upon extensive natural resource extraction. Thus, Perth, the state capital of Western Australia, has encountered significant population and urban growth in response to the booming state economy. However, the recent economic slowdown resulted in the largest decrease in natural resource values that Western Australia has ever experienced. Here, we present multi-temporal urban expansion statistics from 1990 to 2015 for Perth, derived from Landsat imagery. Current urban estimates used for future development plans and progress monitoring of infill and density targets are based upon aggregated census data and metrics unrepresentative of actual land cover change, underestimating overall urban area. Earth observation provides a temporally consistent methodology, identifying areal urban area at higher spatial and temporal resolution than current estimates. Our results indicate that the spatial extent of the Perth Metropolitan Region has increased 45% between 1990 and 2015, over 320 km^2. We highlight the applicability of earth observation data in accurately quantifying urban area for sustainable targeted planning practices.

Keywords: unsustainable development; urban expansion; remote sensing; Landsat

1. Introduction

Over the last 15 years, Perth has experienced exponential economic growth with Gross State Product (GSP) increasing 218% [1]. Originally labelled as the 'Cinderella State' due to its remote location and perceived neglect from the rest of Australia, Western Australia (WA) has experienced sustained discovery and extraction of natural resources since the beginning of the 21st century [2]. In response to a growing resource sector, the city of Perth has undergone extensive urban expansion at what Dhakal (2014) identified as an unsustainable rate [3]. To this end, the Western Australian Planning Commission (WAPC) identified that Perth's urban footprint has increased from 631 km^2 to 870 km^2 in the 10 years between 2002 and 2012 [4,5]. However, these figures should be considered with caution as data used in early estimates represent land parcel (Cadastral) valuations only (provided by the Western Australian Value General's Office), with later estimates (from 2009) based on multiple urban zoning classifications, and more recently (from 2010) spatial modelling taking into account land valuation and zoning [6,7]. The use of varied data and methods impacts confidence in the ability of the Commission's estimates to represent actual change in urban extent, especially when urban zoning information includes land identified for growth but not necessarily developed. Such inconsistencies could have potential to misinform future development decisions. Consequently, here we present a spatiotemporal assessment of change in areal urban growth based upon medium resolution remote sensing through a single classification model. This provides the first accurate depiction of urban expansion for one of the world's fastest growing cities—Perth, WA. We present our findings and discuss the implications of more accurately classified urban extents in facilitating scientifically evidence-based adaptive and targeted planning policies to help reduce environmental and socio-economic consequences of poorly planned development.

1.1. Earth Observation for Monitoring Urban Change

Mapping the spatial extent and temporal profile of urban growth from medium resolution satellite imagery facilitates a consistent, detailed characterisation of the actual urban footprint of a city [8,9]. Other conventional spatial datasets such as Cadastral data provide information on freehold and Crown land parcel boundaries including attributes such as ownership and value for a singular temporal period [10]. However, attributed data for a singular year provides an ineffective portrayal of actual parcel land cover and temporal change. Thus, the methods and results presented in this study provide foundational information for the development of planning regulations that ensure sustainable growth of our cities, particularly in the reduction of environmental risks from ever-increasing expansion along the wildland–urban interface [11]. Specifically, Earth Observation (EO) data allows spatially detailed identification of locations where (un)sustainable urban growth is occurring which enables expansion limits to be imposed through targeted policies [12]. In this theme, Schneider et al. (2005) determined the spatial distribution of development zones from 1978 to 2002 in Chengdu, Sichuan province, China in response to the Go West policy of the 1990s, aimed at economically boosting the West of the country [13]. Whilst the policy was successful in raising Gross Domestic Product (GDP) levels, urbanisation concurrently increased, generating issues of urban management, including service, infrastructure and resource deficiency. Their results indicated spatial clustering, specialisation of land use and peri urban development (not considered by the original policy) which were subsequently used to tailor policy in remediating issues, facilitating sustainable future urban development [13,14]. Similarly, Hepinstall-Cymerman et al (2013) used classified Landsat data to monitor urban growth in regards to imposed growth boundaries in the Central Puget Sound, Washington, USA [15]. Surprisingly, more new development occurred outside the growth boundaries than inside within their last time period, illustrating the ineffectiveness of the imposed policy leading to economic and ecological consequences, including a loss of avian diversity in native forest species [15,16]. These studies highlight the potential effectiveness of EO data in consistently monitoring the spatiotemporal dynamics of urban development for applied policy outcomes and ensuring sustainable future planning decisions, for which such outputs are unachievable from traditional datasets.

1.2. The Case of Perth

Perth's dramatic urban expansion can be attributed to Australia's minerals and energy boom commencing at the turn of the century. Queensland (QLD) and WA were at the forefront of the boom contributing the largest proportion of the nation's resources output, valued at 3.3% of GDP [1]. In WA, mining and petroleum extraction dominate exports, peaking at 95% of the state's export earnings between 2010 and 2011 [17]. The increase in extraction was predominantly attributable to greater demand for raw materials from China, resulting in steady growth of the WA mineral and petroleum industry from AUD 4.7 billion in 1996 to a peak of AUD 121.6 billion mid-2013. However, in 2009, a 10.3% reduction in the overall value of mineral and petroleum resources resulted from falling commodity prices and the 2007–2009 global financial crisis [17]. Again in 2012, a further 9% reduction in resource value was observed as uncertainty in global economic conditions increased [17]. The largest decline to date occurred between 2014 and 2015, with an additional 22% reduction in the value of mineral and petroleum resources as a result of surplus capacity, decreased demand, and decline in the value of the Australian dollar [17]. The temporal trend in resource value indicates a stagnation and decline since late 2013 (Figure 1).

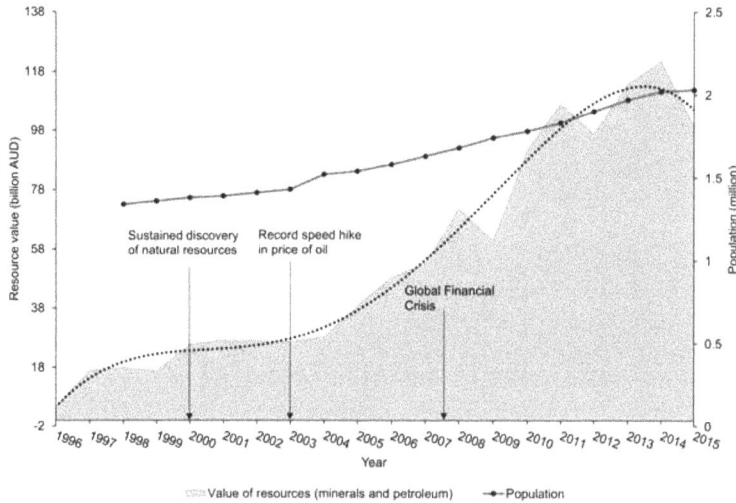

Figure 1. Timeline of natural resource value (based on Department of Mines and Petroleum annual reports) fitted with a fourth order polynomial trend line and population (based on Australian Bureau of Statistics data) also indicating key milestones.

Perth is described as one of the most isolated cities in the world (pop. > 1 million) and was Australia's fastest growing metropolis between 2007 and 2014; however, subsequent to a decline in natural resource value, a slowdown in population expansion soon followed (Figure 1) [2]. As a result, 2015 population statistics highlight the lowest population increase since records began with a 0.5% increase from the previous year [1,2]. In comparison to other Australian state capitals, based on the Australian Bureau of Statistics (ABS) 2011 population grid, Perth exhibits a relatively sparse spatial distribution of population with a maximum population density of only 3662 people per square kilometre (Melbourne 10,827; Sydney 14,747). Such low density population has generated high demand for dispersed housing, amenities and services, and has influenced changes to Perth's land use patterns in a non-strategic, "lot-by-lot fashion" based on a car-dependent lifestyle [3]. Anthropogenic modifications of the landscape from vegetation cover to human-made impervious surfaces represent a critical driving force in both local and global environmental change [18,19]. For example, abrupt, poorly planned and uncontrolled urban expansion can lead to environmental impacts which degrade ecological systems including habitat fragmentation and socio-economic issues that deteriorate efficiency of amenity provisioning, both of which can exacerbate localised climate change [11,20]. Identifying impacts of Land Use and Land Cover (LULC) change on socio-ecological systems is vital for future sustainable urban development; as reflected in the "sustainable cities and communities" 2030 sustainable development goal and the effective land use planning criteria of the City Resilience Framework (CRF) [19,21]. It is essential for Perth to adapt current practices of outward suburban expansion to achieve more sustainable urban growth and become city-smart for accommodating the predicted additional half a million new residents by 2031, which will result in an overall population exceeding 2.2 million [5].

2. Materials and Methods

2.1. Data Preprocesing

EO data have been extensively used to monitor the sustainability of urban areas [22,23]. However, accurate identification and temporal monitoring of urban land is frequently precluded due to the coarse resolution (300 m–1 km) of a number of commonly used remotely sensed datasets including night time lights (1 km) and the Moderate Resolution Imaging Spectroradiometer (MODIS) land cover product (0.083°) [22,24]. Whilst 30 m resolution data (e.g., Landsat) are more suitable to detect nuances of urban development the majority of studies and classified products which have used these finer resolution products implement large temporal windows, negating the possibility of detailed temporal urban characterisation e.g., GloeLand30 [25–29]. This research provides the first comprehensive temporal evolution analysis quantifying land cover change and associated urban expansion for the Perth Metropolitan Region (PMR) using 30 m Landsat imagery, the longest temporal record of medium spatial resolution imagery, for seven sequential time snapshots between 1990 and 2015.

Cloud free imagery was acquired in or close to the month of July for 1990, 2000, 2003, 2005, 2007, 2013 and 2015. Analysis of imagery acquired from WA winter season coincided with peak green-up which provided the greatest contrast between spectrally similar surfaces (e.g., bare earth and urban) [30–32]. Imagery date selection was founded upon the strong positive relationship between Australian soil moisture (related to rainfall) and the Normalised Difference Vegetation Index (NDVI) [33], which exhibits an approximate one month lag between peak soil moisture and peak NDVI [33].

Productive photosynthesising plants use energy in the visible red (VIS) portion of the electromagnetic spectrum whilst reflecting in the near-infrared (NIR) region. NDVI ((NIR -VIS)/ (NIR + VIS)) is a representative measure of growth allowing for the identification of green, healthy vegetation [33–35], as illustrative of Southwest WA's winter months. A total of 14 images from Landsat 5 Thematic Mapper (TM) (eight images), Landsat 7 Enhanced Thematic Mapper Plus (ETM+) (two images) and Landsat 8 Operational Land Imager (OLI) (four images) were acquired for the specified years. Seamless images were produced based on Voroni diagrams that locate the bisector between images; adjacent edges were identified as seamlines constraining effective mosaic polygons that specify inclusion pixels for the final mosaicked product, permitting less visible boundaries through blending overlapping pixels [36]. Mosaicked images were subsequently clipped to the original PMR study area boundary.

The atmospherically corrected Landsat data used in this study were obtained from the Landsat Ecosystem Disturbance Adaptive Processing System (LEDAPS) and the Landsat 8 Surface Reflectance (L8SR) algorithm [37,38]. Some inherent residual noise remained, for example, due to the differences in modelled atmospheric correction parameters [39]. To correct for this, surface reflectance values were standardised as[1]:

$$p_{i,b} = \frac{p_{x,b}}{max_b} \tag{1}$$

where $p_{i,b}$ is the standardised pixel value i, from band b based on the original surface reflectance x, standardised through division of a priori specific upper reflectance limit for each band ($max b$): 0.1 (blue; 0.48 μm), 0.11 (green; 0.56 μm), 0.12 (red; 0.66 μm), 0.225 (near-infrared; 0.84 μm), 0.205 (shortwave-infrared; 1.65 μm), 0.150 (shortwave-infrared 2; 2.22 μm) [40]. Standardised values were then normalised per pixel j through cross band sum division:

$$p_{j,b} = \frac{p_{i,b}}{\Sigma_i \, p_{i,b}} \tag{2}$$

---

[1]   Using the Interactive Data Language (IDL) version 8.3

where $\sum\limits_{i} p_{i,b}$ is the sum of each standardised pixel across all bands [40]. Normalised Landsat data obtained a statistically significant reduction of spectral variation per land cover class within (inter) and between (intra) each image (see Figure S1).

2.2. Data Classification

The normalised Landsat imagery was classified using the Import Vector Machine (IVM) which builds upon the popular Support Vector Machine (SVM) methodology[2] [41]. In order to obtain the optimum classification, the IVM algorithm explores all possible subsets of training data for optimal selection (termed import vectors) which are derived through successively adding training data samples until a given convergence criterion is met [41]. Data samples are selected according to their contribution to the classification solution. However, a pure forward system is unable to remove import vectors that become obsolete after addition of other vectors. Therefore the implemented version of IVM utilised here is a hybrid forward/backward strategy that adds import vectors whilst concurrently testing if they can be removed in each step, thus leading to a sparse and more accurate solution [41]. Furthermore, the IVM selects data points from the entire distribution resulting in a smoother decision boundary which is based on the optimal separating hyperplane in multidimensional space compared to that of SVM algorithms [42]. The benefits of the IVM algorithm have resulted in this approach being successfully applied in a number of studies (e.g., [42–45]) due to its accuracy and performance advantages over alternative methodologies including SVM and the traditional Maximum Likelihood (ML) classifiers [44,45].

Model training samples were selected using the July 2005 Landsat 5 TM image coinciding with the month post maximum rainfall of all considered Landsat 5 TM and 7 ETM+ to facilitate optimum spectral seperability[3] [33]. Land cover was defined as high albedo urban (e.g., concrete), low albedo urban (e.g., asphalt) or other. Two urban classes were initially identified in order to reduce confusion between spectrally similarly classes (e.g., urban and bare earth) being merged post-classification to represent complete urban coverage [46]. For each class, 250 pixels were randomly selected as training data, which is consistent with Foody and Mathur (2006) and Pal and Mather (2003) (see Supplementary S2). Training data parametrised the IVM algorithm, creating a classification model of spectral profiles that are compared to Landsat spectral profiles for classification. The classification model was then applied to all Landsat 5 TM and Landsat 7 ETM+ images obtaining similar spectral wavebands, considered to be equivalent [47]. However, due to Landsat 8 OLI sampling different spectral regions, a new classification model was developed using the same training areas, as these were deemed to remain representative of the land cover, but with Landsat 8 OLI spectral wavebands [47,48]. Validation was performed through an accuracy assessment based on an independent dataset (Google Earth high resolution imagery) consistent with Landsat acquisition months following previously published methods (e.g., [22,49–53]). For each land use category, 50 pixels per class per year were visually identified and classified based on the majority land cover within the coincident Landsat pixel from Google Earth imagery for the available years: 2000, 2003, 2005, 2007, 2013 and 2015 consistent with recommended land cover accuracy sample size of Congalton (2001) [54].

3. Results

The spatial footprint of PMR development has increased 45% between 1990 and 2015, over 320 km^2 (Figures 2 and 3), with a 37% increase occurring since 2000. The classification accuracy assessment indicates an average overall accuracy of 84.1% and Kappa Coefficient of 0.73 being comparable to other studies (e.g., [52,55–57]) (see Tables S1 and S2). Urban expansion mirrors population increase

---

2   Using the open source Environmental Mapping and Analysis Program version 2.1.1 (EnMAP)
3   Achieved in the ENvironment for Visualizing Images software version 5.2 (ENVI)

and as population growth has slowed, urban development has concurrently exhibited a levelling trend compared to expansion previously observed (Figure 3).

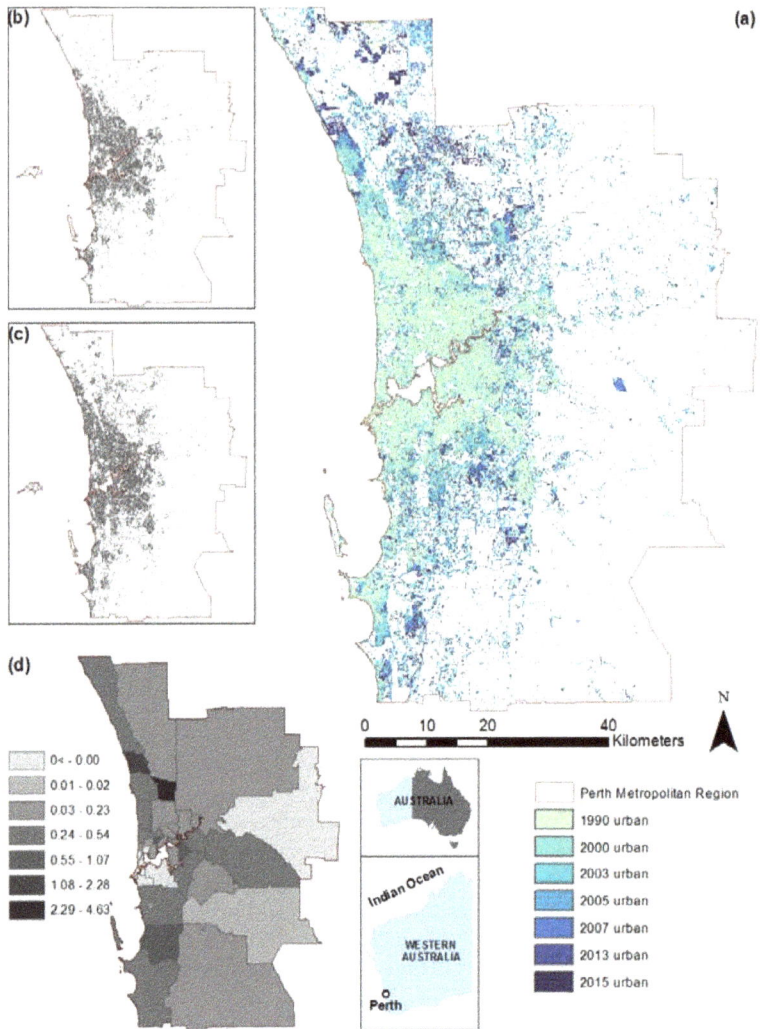

Figure 2. Urban expansion within the Perth Metropolitan Region (PMR) between 1990 and 2015. Vast urban growth has been observed in PMR with graduating colours exhibiting outward expansion (**a**); (**b**) and (**c**) exhibit static snapshots of urban extent from 2000 (**b**) and 2015 (**c**); whilst (**d**) depicts percentage of urban change per subnational administrative boundary (Local Government Area; LGA).

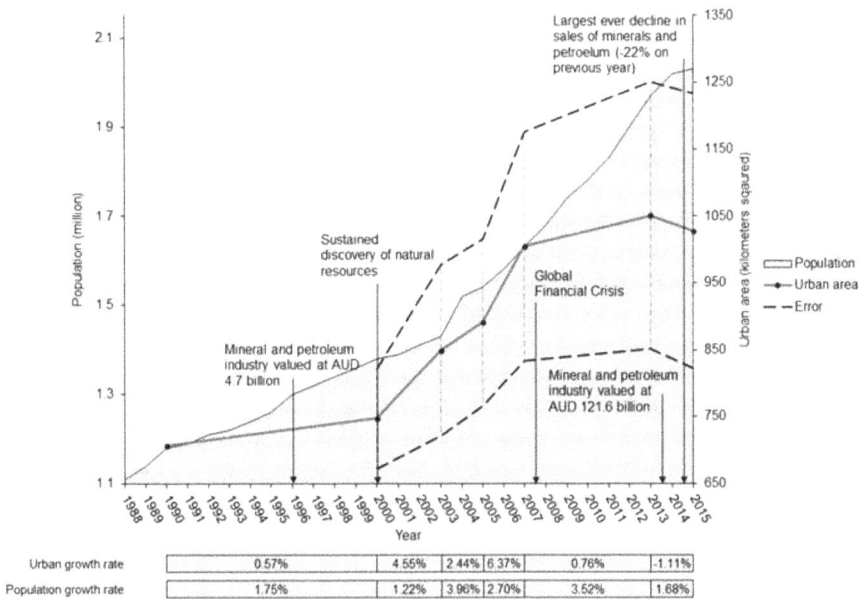

Figure 3. Time line of urban expansion in kilometers squared derived from Earth observation data with associated classification error derived from validation data (points indicating classified image years). Alongside population data in millions per year since 1988 (based on Australian Bureau of Statistics data, 2015 data is projected) with key natural resource milestones indicated, and average annual urban and average annual population growth rate indicated between classified image years.

WAPC's urban estimates of the PMR from Directions 2031 (the strategic plan for the Perth and Peel region) were provided for comparison to those produced within this study[4] [5]. WAPC's estimates note an expansion from 637 km^2 to 813 km^2 between 2001 and 2012. Our results indicate an expansion of 747 km^2 to 1050 km^2 from 2000 to 2013 illustrating an overall underestimation by WAPC figures (Figure 4). Within suburban areas surrounding the two major cities in the metropolitan region, Perth and Fremantle, WAPC's estimates underrepresent the amount of urban area derived from EO, being more pronounced in 2013 than 2000. The Local Government Area (LGA)[5] of Stirling South Eastern represented the maximum overestimation in 2013 urban area with 34% (2000: 10%) additional urban area per km^2 of LGA established on a difference of 2.89 km^2, 40.2% (2000: 0.83 km^2, 15%) between EO data and WAPC's estimates. Outer Northern and Southern LGA WAPC urban values were consistently underestimated, with the LGA of Belmont representing the maximum underestimation of percent per km^2 of LGA in 2013 with 24% (2000: 13%) due to a difference of 9.37 km^2, 40.39% (2000: 5 km^2, 26.46%). Prior to 2009, WAPC's estimates were solely based upon land parcel valuations from the Western Australian Value General's Office, consequently valuation thresholds designating land to urban may have been inappropriately applied to outer suburban LGAs, where land might be developed but less valuable than central LGAs.

For urban estimates post 2005, two urban land zones, urban and urban deferred, are used within the Perth Metropolitan Region Scheme (MRS), the division of the State Planning Policy Framework applicable to the PMR, pursuant to the Planning and Development Act (2005) that inform recent WAPC

---

4   Analysed in ArcGIS version 10.2.2
5   Outlines of LGAs are displayed in Figure S2

land parcel based estimates [58,59]. Urban land refers to locations where activities in line with urban development are permitted, but not necessarily constructed (e.g., housing and commercial use) whilst urban deferred represents land suitable for future development with remaining planning, servicing or environmental issues [59,60]. For land to be assigned urban deferred, it must obtain characteristics of the urban zone including being able to provide essential services, a logical progression of development, and able to satisfy regional requirements (e.g., roads and open spaces). The 2012 WAPC estimates were derived from stock of land zoned urban or urban deferred, cadastral land plot and value information, conditional subdivision approvals, and ongoing regional rezoning and subdivisions [61]. Similarly to 2000, valuation data may misrepresent suburban urban land cover resulting in overestimation. Inclusion of additional variables that are unrepresentative of actual land cover change (e.g., rezoning and conditional approvals) could exacerbate differences between WAPC and EO derived urban estimates (Figure 4b), leading to the potential confounding of errors in WAPC estimates.

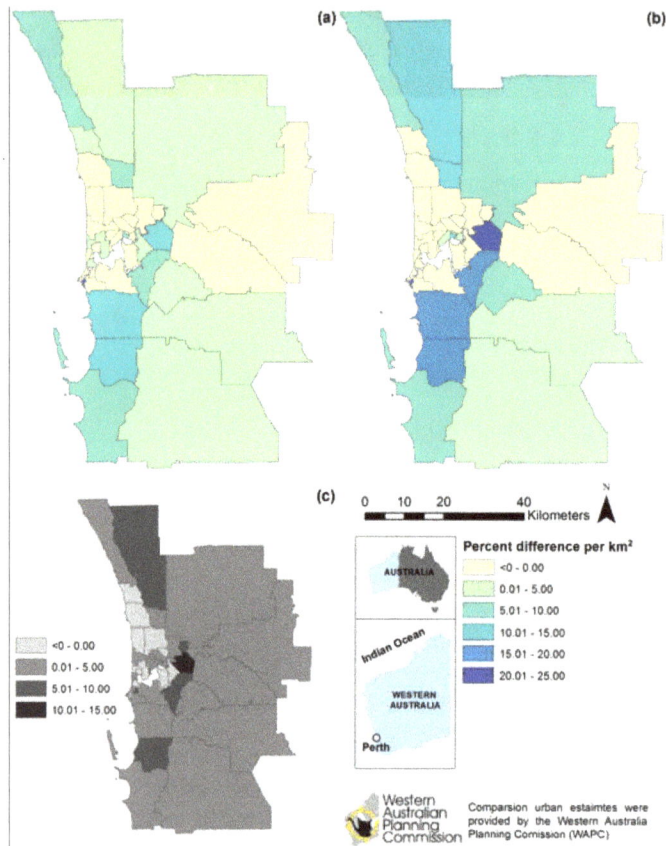

Figure 4. Percentage differences relative to local government area size, permitting a change metric standardised by Local Government Area (LGA) area between Earth Observation (EO) and the Western Australian Planning Commission's (WAPC) urban estimates for: (**a**) 2000 (EO) and 2001 (WAPC) and (**b**) 2012 (WAPC) and 2013 (EO), whilst (**c**) depicts the percentage difference in the relative urban rate of change (km^2 per LGA area) between 2000 and 2013 (EO) and 2001 and 2012 (WAPC). Positive values indicate underestimation by WAPC whilst negative values represent overestimation by WAPC.

4. Discussion

WA state government planning documentation states that the majority of new development within the PMR has occurred as low-density suburban growth, responding to consumer preferences and market forces [4]. Additionally, sustainable policy objectives suggest that new development should be managed and focused on current communities, making the most efficient use of existing urban areas [4,5]. Planning policy research has highlighted issues of outward urban expansion as being costly in economic, environmental and social terms based on dispersed service requirements, habitat fragmentation and neighbourhood segregation [20]. Thus, urban expansion in the PMR may result in further economic, social and environmental costs associated with servicing and maintaining low-density lifestyles, owing to the rapid outward urban growth estimates between 2000 and 2007 [11,20].

In contrast, the witnessed slowdown of urban growth, population and natural resource value since 2013 indicates the possibility that the 'boom' of previous years has reached a turning point. Stagnation of urban growth implies that issues associated with spatially distributed urban areas might be contained to the current urban extent. Nevertheless, it is conceivable that prosperous future economic circumstances could initiate growth at a rate previously observed, and that the economic slowdown might be a temporary hiatus responding to current economic conditions [62]. For example, in 2014–2015, WA continued to attract the largest proportion of state mineral exploration expenditure at 58%, with QLD (the second ranked stated) obtaining only 20% [17]. Furthermore, as of September 2015, WA had an estimated AUD 171 billion in mineral and petroleum projects under construction, with a further AUD 110 billion allocated for future expansion [17]. Comparatively, during the peak (mid-2013) in terms of total sales, WA only had an estimated AUD 160 billion worth of projects under construction and a further AUD 108 billion for future development [17]. Whilst 2014–2015 observed the greatest decline in total sales of resources, sustained investment and improved global economics could reinvigorate the industry and reinitiate urban expansion within the PMR.

Future development (urban and urban deferred) is guided by Directions 2031 amending the MRS and local planning schemes [5,63–66]. WAPC aims to achieve 47% of future development as infill and a 50% increase in average residential density by 2050 of 10 dwellings per urban zoned hectare and 15 per new urban zoned hectare [5]. In monitoring progress towards the infill target, zoned development land within the PMR is considered, including residential, industrial and commercial land uses [4]. Densities are defined as infill or Greenfield if above or below an undocumented residential threshold from census data [60]. Initial results from delivering directions 2031, 2014 report indicate the requirement of a significant increase in infill development if the above targets are to be met [67]. Similarly, average residential density monitoring has been achieved with land valuation data (from the Valuer General's Office) for major activity centres, being unrepresentative of actual density change and providing an incomplete metropolitan comparison [67]. Inclusion of EO data would permit quantitative evidence of urban expansion, infill and density at a higher spatial and temporal frequency than current census based estimates. This would facilitate credible, evidence-based efficient targeted action founded upon improved representative urban area, insuring infill and density attainment. In this theme, Schneider et al. (2005) and Hepinstall-Cymerman et al. (2013) used spatial metrics (e.g., urban area mean patch size) based on classified Landsat data in either pre-defined census units [15] or development corridors [13] to monitor development type (infill or expansion) over time for adapting inappropriate static urban development policy. Using EO derived land cover data in this manner aids in understanding dynamics of the urban environment through monitoring, planning and mitigating land use changes that impact natural assets and increase vulnerability of city systems [14,15,68]. Information of this sort aligns with criteria of the CRF in improving city resilience from effective land use planning, possible at lower expense and higher temporal frequencies than in situ measurements [21].

The universal methodology implemented within this research lends itself to credibly inform policy in a similar manner in other global cities through monitoring urban expansion in order to identify rapid, unsustainable development. For example, Jakarta, obtaining the world's second largest urban area with a population of 28 million, has yet to have any quantitative urban area delineation [69,70]. Identification of actual urban growth in developing cities is vital to city planners, environment managers and policy makers due to the difference between planned growth and actual growth [14]. Such information could be of critical importance for regulating urban expansion due to extreme poverty and high level of risk to environmental hazards, such as that posed from flooding [71,72]. EO data presents many opportunities for added value within urban planning policy, and additional analyses could be pursued into specific human-induced environmental issues, such as detecting thermal changes in the urban environment for planning issues associated with urban heat islands (e.g., cooling provisions) and their impact on human health (e.g., air quality).

5. Recommendations

Consistent and accurate LULC estimates are a vital aspect of sustainable urban development throughout the world, especially considering the predicted additional 2.5 billion city dwellers by 2050. LULC models that require agents that are representative of land use decisions can often fail in practicality due to the difficulty in quantifying driving forces of change and multi-level relationships. Models of this nature are also temporally independent, with each annual iteration implementing new data or data not representative of actual LULC change. EO data provides a replicable detailed representation of the complete urban extent requiring no additional data. The use and application of EO data reported within this paper highlights several improvements to WAPC policy for consistent urban area estimations with associated accuracy measures. Therefore it is recommended that planning authorities, such as WAPC, integrate EO data to achieve the following: (1) provide scientific urban estimates based on a temporally consistent model within future regional structure plans, metropolitan region and local planning schemes; (2) monitor infill development at a higher temporal frequency than census years for policy targeting to meet key goals; (3) monitor urban density through areal urban expansion compared to current metrics using land valuations; and (4) restrict development based on temporal urban analysis that degrades amenity efficiency and ecological systems whilst promoting development in locations to maximise efficiency and long-term sustainability. Additional EO datasets (e.g., finer resolution Sentinel 2 satellite imagery or aerial imagery) could be used to refine planning decisions based on areas of concern identified from Landsat. For example, finer spatial resolution datasets could facilitate enhanced feature extraction, optimising sustainable planning decisions through the identification of candidate infill sites. EO data of this nature provides an essential tool for timely planning policy that is adaptive to changes in urban landscape to mitigate socio-environmental issues associated with poorly planned urban areas for the future sustainability of our cities.

Supplementary Materials: The following are available online at www.mdpi.com/2073-445X/6/1/9/s1. Additional methodological detail, full accuracy results and Local Government Area (LGA) outlines are reported in the supplementary documentation. Figure S1 Inter year classification reflectance variation categorised by classified output for each spectral band for: pre (a) and post (b) normalisation correction. Table S1 Classification accuracy and associated Kappa Coefficient per year of classified Landsat. Table S2 Producer's and User's accuracy per year of classified Landsat imagery. Figure S2 Local Government Areas (LGAs) located in Perth Metropolitan Region (a); with (b) exhibiting LGAs South and West of the Swan River and (c) LGAs North and East of the Swan River. The classified data reported in this paper (doi:10.1594/PANGAEA.871017) are archived at https://doi.pangaea.de/10.1594/PANGAEA.871017, the pangea open access data publisher for earth and environmental science.

Acknowledgments: This work was supported by the Economic and Social Research Council [grant number ES/J500161/1]. We would like to thank the World University Network (WUN) for facilitating institutional visits, the United States Geological Survey (USGS) for providing Landsat surface reflectance data and the Western Australia Department of Planning, in particular Matt Devlin and Lisl van Aarde for urban estimates and supplementary planning information.

Author Contributions: All authors assisted in conceiving and designing the experiments lead by Andrew MacLachlan; Andrew MacLachlan performed the experiments and analyzed the data; Eloise Biggs, Gareth Roberts and Bryan Boruff contributed reagents/materials/analysis tools; Andrew MacLachlan wrote the paper with input and revisions from Eloise Biggs, Gareth Roberts and Bryan Boruff.

Conflicts of Interest: The authors declare no conflict of interest and the founding sponsors had no role in the design of the study; in the collection, analyses, or interpretation of data; in the writing of the manuscript, and in the decision to publish the results.

References

1. *ABS Australian National Accounts 1988–2015*; Australian Bureau of Statistics: Belconnen, ACT, Australia, 2015.
2. Kennewell, C.; Shaw, B.J. Perth, Western Australia. *Cities* **2008**, *25*, 243–255. [CrossRef]
3. Dhakal, S.P. Glimpses of Sustainability in Perth, Western Australia: Capturing and Communicating the Adaptive Capacity of an Activist Group. *Cons. J. Sustain. Dev.* **2014**, *11*, 167–182.
4. *Western Australian Planning Commission Perth and Peel @ 3.5 Million*; Department of Planning, Government of Western Australia: Perth, WA, Australia, 2015.
5. *Western Australian Planning Commission Directions 2031 and beyond: Metropolitan Planning beyond the Horizon*; Department of Planning, Government of Western Australia: Perth, WA, Australia, 2010.
6. *Western Australian Planning Commission Urban Growth Monitor: Perth Metropolitan, Peel and Greater Bunbury Regions 2009*; Department of Planning, Government of Western Australia: Perth, WA, Australia, 2009.
7. *Western Australian Planning Commission Urban Growth Monitor: Perth Metropolitan, Peel and Greater Bunbury Regions 2010*; Department of Planning, Government of Western Australia: Perth, WA, Australia, 2010.
8. Angiuli, E.; Trianni, G. Urban Mapping in Landsat Images Based on Normalized Difference Spectral Vector. *IEEE Geosci. Remote Sens. Lett.* **2013**, *11*, 661–665. [CrossRef]
9. Bagan, H.; Yamagata, Y. Landsat analysis of urban growth: How Tokyo became the world's largest megacity during the last 40years. *Remote Sens. Environ.* **2012**, *127*, 210–222. [CrossRef]
10. Thompson, R.J. A model for the creation and progressive improvement of a digitalcadastral data base. *Land Use Policy* **2015**, *49*, 565–576. [CrossRef]
11. Turner, B.; Lambin, E.; Reenberg, A. The emergence of land change science for global environmental change and sustainability. *Proc. Natl. Acad. Sci. USA* **2010**, *103*, 13070–13075. [CrossRef] [PubMed]
12. Bettencourt, L.; West, G. A unified theory of urban living. *Nature* **2010**, *467*, 9–10. [CrossRef] [PubMed]
13. Schneider, A.; Seto, K.C.; Webster, D.R. Urban growth in Chengdu, western China: Application of remote sensing to assess planning and policy outcomes. *Environ. Plan. B Plan. Des.* **2005**, *32*, 323–345. [CrossRef]
14. Patino, J.E.; Duque, J.C. A review of regional science applications of satellite remote sensing in urban settings. *Comput. Environ. Urban Syst.* **2013**, *37*, 1–17. [CrossRef]
15. Hepinstall-Cymerman, J.; Coe, S.; Hutyra, L.R. Urban growth patterns and growth management boundaries in the Central Puget Sound, Washington, 1986–2007. *Urban Ecosyst.* **2013**, *16*, 109–129. [CrossRef]
16. Hepinstall, J.A.; Alberti, M.; Marzluff, J.M. Predicting land cover change and avian community responses in rapidly urbanizing environments. *Landsc. Ecol.* **2008**, *23*, 1257–1276. [CrossRef]
17. *Western Australian Mineral and Petroleum Statistics Digest 1984–2015*; Department of Mines and Petroleum, Government of Western Australia: Perth, WA, Australia, 2015.
18. Kalnay, E.; Cai, M. Impact of urbanization and land-use change on climate. *Nature* **2003**, *423*, 528–531. [CrossRef] [PubMed]
19. Vitousek, P.; Mooney, H.; Lubchenco, J.; Melillo, J. Human Domination of Earth Ecosystems. *Science* **1997**, *277*, 494–498. [CrossRef]
20. Downs, A. Smart Growth: Why We Discuss It More than We Do It. *J. Am. Plan. Assoc.* **2005**, *71*, 367–378. [CrossRef]
21. *ARUP City Resilience Framework—100 Resilient Cities*; The Rockefeller Foundation: New York, NY, USA, 2015.
22. Song, X.-P.; Sexton, J.O.; Huang, C.; Channan, S.; Townshend, J.R. Characterizing the magnitude, timing and duration of urban growth from time series of Landsat-based estimates of impervious cover. *Remote Sens. Environ.* **2016**, *175*, 1–13. [CrossRef]
23. Li, X.; Gong, P.; Liang, L. A 30-year (1984–2013) record of annual urban dynamics of Beijing City derived from Landsat data. *Remote Sens. Environ.* **2015**, *166*, 78–90. [CrossRef]
24. Potere, D.; Schneider, A.; Angel, S.; Civco, D. Mapping urban areas on a global scale: Which of the eight maps now available is more accurate? *Int. J. Remote Sens.* **2009**, *30*, 6531–6558. [CrossRef]
25. Hu, Y.; Jia, G.; Hou, M.; Zhang, X.; Zheng, F.; Liu, Y. The cumulative effects of urban expansion on land surface temperatures in metropolitan Jingjintang, China Yonghong. *J. Geophys. Res. Atmos.* **2015**, *120*, 9932–9943. [CrossRef]

26. Masek, J.G.; Lindsay, F.E.; Goward, S.N. Dynamics of urban growth in the Washington DC metropolitan area, 1973–1996, from Landsat observations. *Int. J. Remote Sens.* **2000**, *21*, 3473–3486. [CrossRef]

27. Van de Voorde, T.; Jacquet, W.; Canters, F. Mapping form and function in urban areas: An approach based on urban metrics and continuous impervious surface data. *Landsc. Urban Plan.* **2011**, *102*, 143–155. [CrossRef]

28. Xian, G.; Homer, C.; Bunde, B.; Danielson, P.; Dewitz, J.; Fry, J.; Pu, R. Quantifying urban land cover change between 2001 and 2006 in the Gulf of Mexico region. *Geocarto Int.* **2012**, *27*, 479–497. [CrossRef]

29. Suarez-Rubio, M.; Lookingbill, T.R.; Elmore, A.J. Exurban development derived from Landsat from 1986 to 2009 surrounding the District of Columbia, USA. *Remote Sens. Environ.* **2012**, *124*, 360–370. [CrossRef]

30. Herold, M.; Gardner, M.; Hadley, B.; Roberts, D. The spectral dimension in urban land cover mapping from high-resolution optical remote sensing data. In Proceedings of the 3rd Symposium on remote Sensing of Urban Areas, Istanbul, Turkey, 11–13 June 2002; Volume 6, pp. 1–8.

31. Varshney, A.; Rajesh, E. A Comparative Study of Built-up Index Approaches for Automated Extraction of Built-up Regions From Remote Sensing Data. *J. Indian Soc. Remote Sens.* **2014**, *42*, 659–663. [CrossRef]

32. Lu, D.; Moran, E.; Hetrick, S. Detection of impervious surface change with multitemporal Landsat images in an urban-rural frontier. *ISPRS J. Photogramm. Remote Sens.* **2011**, *66*, 298–306. [CrossRef] [PubMed]

33. Chen, T.; de Jeu, R.A.M.; Liu, Y.Y.; van der Werf, G.R.; Dolman, A.J. Using satellite based soil moisture to quantify the water driven variability in NDVI: A case study over mainland Australia. *Remote Sens. Environ.* **2014**, *140*, 330–338. [CrossRef]

34. Myneni, R.B.; Keeling, C.D.; Tucker, C.J.; Asrar, G.; Nemani, R.R. Increased plant growth in the northern high latitudes from 1981 to 1991. *Nature* **1997**, *386*, 698–702. [CrossRef]

35. Piao, S.; Wang, X.; Ciais, P.; Zhu, B.; Wang, T.; Liu, J. Changes in satellite-derived vegetation growth trend in temperate and boreal Eurasia from 1982 to 2006. *Glob. Chang. Biol.* **2011**, *17*, 3228–3239. [CrossRef]

36. Pan, J.; Wang, M.; Li, D.; Li, J. Automatic Generation of Seamline Network Using Area Voronoi Diagrams With Overlap. *IEEE Trans. Geosci. Remote Sens.* **2009**, *47*, 1737–1744. [CrossRef]

37. Hansen, M.C.; Loveland, T.R. A review of large area monitoring of land cover change using Landsat data. *Remote Sens. Environ.* **2012**, *122*, 66–74. [CrossRef]

38. *USGS Product Guide Provisional Landsat 8 Surface Reflectance Product*; Department of the Interior U.S. Geological Survey: Sunrise Valley Drive Reston, VA, USA, 2015; pp. 1–27.

39. Ju, J.; Roy, D.P.; Vermote, E.; Masek, J.; Kovalskyy, V. Continental-scale validation of MODIS-based and LEDAPS Landsat ETM+ atmospheric correction methods. *Remote Sens. Environ.* **2012**, *122*, 175–184. [CrossRef]

40. Sexton, J.O.; Song, X.-P.; Huang, C.; Channan, S.; Baker, M.E.; Townshend, J.R. Urban growth of the Washington, D.C.–Baltimore, MD metropolitan region from 1984 to 2010 by annual, Landsat-based estimates of impervious cover. *Remote Sens. Environ.* **2013**, *129*, 42–53. [CrossRef]

41. Roscher, R.; Förstner, W.; Waske, B. I²VM: Incremental import vector machines. *Image Vis. Comput.* **2012**, *30*, 263–278. [CrossRef]

42. Braun, A.C.; Weidner, U.; Hinz, S. Classification in high-dimensional feature spaces-assessment using SVM, IVM and RVM with focus on simulated EnMAP data. *IEEE J. Sel. Top. Appl. Earth Obs. Remote Sens.* **2012**, *5*, 436–443. [CrossRef]

43. Suess, S.; van der Linden, S.; Leitao, P.J.; Okujeni, A.; Waske, B.; Hostert, P. Import Vector Machines for Quantitative Analysis of Hyperspectral Data. *IEEE Geosci. Remote Sens. Lett.* **2014**, *11*, 449–453. [CrossRef]

44. Braun, A.C.; Weidner, U.; Hinz, S. Support vector machines, import vector machines and relevance vector machines for hyperspectral classification—A comparison. In Proceedings of the IEEE 3rd Workshop on Hyperspectral Image and Signal Processing: Evolution in Remote Sensing (WHISPERS), Lisbon, Portugal, 6 June 2011; pp. 1–4.

45. Roscher, R.; Waske, B.; Forstner, W. Kernel discriminative Random fields for land cover classification. In Proceedings of the 2010 IAPR Workshop on Pattern Recognition in Remote Sensing (PRRS), Istanbul, Turkey, 22 August 2010.

46. Hu, X.; Weng, Q. Estimating impervious surfaces from medium spatial resolution imagery using the self-organizing map and multi-layer perceptron neural networks. *Remote Sens. Environ.* **2009**, *113*, 2089–2102. [CrossRef]

47. Flood, N. Continuity of Reflectance Data between Landsat-7 ETM+ and Landsat-8 OLI, for Both Top-of-Atmosphere and Surface Reflectance: A Study in the Australian Landscape. *Remote Sens.* **2014**, *6*, 7952–7970. [CrossRef]

48. Roy, D.P.; Kovalskyy, V.; Zhang, H.K.; Vermote, E.F.; Yan, L.; Kumar, S.S.; Egorov, A. Characterization of Landsat-7 to Landsat-8 reflective wavelength and normalized difference vegetation index continuity. *Remote Sens. Environ.* **2016**, *185*, 57–70. [CrossRef]

49. Dorais, A.; Cardille, J. Strategies for incorporating high-resolution google earth databases to guide and validate classifications: Understanding deforestation in Borneo. *Remote Sens.* **2011**, *3*, 1157–1176. [CrossRef]

50. Cunningham, S.; Rogan, J.; Martin, D.; DeLauer, V.; McCauley, S.; Shatz, A. Mapping land development through periods of economic bubble and bust in Massachusetts using Landsat time series data. *GIScience Remote Sens.* **2015**, *52*, 397–415. [CrossRef]

51. Sun, G.; Chen, X.; Jia, X.; Yao, Y.; Wang, Z. Combinational Build-Up Index (CBI) for Effective Impervious Surface Mapping in Urban Areas. *IEEE J. Sel. Top. Appl. Earth Obs. Remote Sens.* **2016**, *9*, 2081–2092. [CrossRef]

52. Bagan, H.; Yamagata, Y. Land-cover change analysis in 50 global cities by using a combination of Landsat data and analysis of grid cells. *Environ. Res. Lett.* **2014**, *9*, 064015. [CrossRef]

53. Zhu, Z.; Woodcock, C.E. Continuous change detection and classification of land cover using all available Landsat data. *Remote Sens. Environ.* **2014**, *144*, 152–171. [CrossRef]

54. Congalton, R.G. Accuracy assessment and validation of remotely sensed and other spatial information. *Int. J. Wildl. Fire* **2001**, *10*, 321–328. [CrossRef]

55. Gislason, P.O.; Benediktsson, J.A.; Sveinsson, J.R. Random forests for land cover classification. *Pattern Recognit. Lett.* **2006**, *27*, 294–300. [CrossRef]

56. Sundarakumar, K.; Harika, M.; Begum, S.K.A.; Yamini, S.; Balakrishna, K. Land Use And Land Cover Change Detection And Urban Sprawl Analysis Of Vijayawada City Using Multitemporal Landsat. *Int. J. Eng. Sci. Technol.* **2012**, *4*, 170–178.

57. Luo, J.; Du, P.; Alim, S.; Xie, X.; Xue, Z. Annual Landsat analysis of urban growth of Nanjing City from 1980 to 2013. In Proceedings of the 2014 3rd International Workshop on Earth Observation and Remote Sensing Applications (EORSA), Changsha, China, 11–14 June 2014; pp. 357–361.

58. *Western Australian Planning Commission Draft State Planning Policy 1, State Planning Framework (Variation No. 3)*; Department of Planning, Government of Western Australia: Perth, WA, Australia, 2016.

59. *Western Australian Planning Commission Development Control Policy 1.9*; Department of Planning, Government of Western Australia: Perth, WA, Australia, 2010; pp. 1–5.

60. *Western Australian Planning Commission Western Australian Planning Commission Urban Growth Monitor: Perth Metropolitan, Peel and Greater Bunbury Regions 2016*; Department of Planning, Government of Western Australia: Perth, WA, Australia, 2016.

61. *Western Australian Planning Commission Urban Growth Monitor: Perth Metropolitan, Peel and Greater Bunbury Regions 2012*; Department of Planning, Government of Western Australia: Perth, WA, Australia, 2012.

62. Perry, M.; Rowe, J.E. Fly-in, fly-out, drive-in, drive-out: The Australian mining boom and its impacts on the local economy. *Local Econ.* **2014**, *30*, 139–148. [CrossRef]

63. *Western Australian Planning Commission Central Sub-Regional Planning Framework*; Department of Planning, Government of Western Australia: Perth, WA, Australia, 2015.

64. *Western Australian Planning Commission North-West Sub-Regional Planning Framework*; Department of Planning, Government of Western Australia: Perth, WA, Australia, 2015.

65. *Western Australian Planning Commission North-East Sub-Regional Planning Framework*; Department of Planning, Government of Western Australia: Perth, WA, Australia, 2015.

66. *Western Australian Planning Commission South Metropolitan Peel Planning Framework*; Department of Planning, Government of Western Australia: Perth, WA, Australia, 2015.

67. *Western Australian Planning Commission Delivering Directions 2031 Report Card 2014*; Department of Planning, Government of Western Australia: Perth, WA, Australia, 2014.

68. Miller, R.B.; Small, C. Cities from space: Potential applications of remote sensing in urban environmental research and policy. *Environ. Sci. Policy* **2003**, *6*, 129–137. [CrossRef]

69. Pravitasari, A.E.; Saizen, I.; Tsutsumida, N.; Rustiadi, E.; Pribadi, D.O. Local Spatially Dependent Driving Forces of Urban Expansion in an Emerging Asian Megacity: The Case of Greater Jakarta (Jabodetabek). *J. Sustain. Dev.* **2015**, *8*, 108–120. [CrossRef]

70. Seto, K.; Fragkias, M.; Guneralp, B.; Reilly, M. A next-generation approach to the characterization of a non-model plant transcriptome. *Curr. Sci.* **2011**, *101*, 1435–1439.

71. Marfai, M.A.; Sekaranom, A.B.; Ward, P. Community responses and adaptation strategies toward flood hazard in Jakarta, Indonesia. *Nat. Hazards* **2014**, *75*, 1127–1144. [CrossRef]

72. Suryahadi, A.; Sumarto, S. Poverty and Vulnerability in Indonesia before and after the Economic Crisis. *Asian Econ. J.* **2003**, *17*, 45–64. [CrossRef]

land

MDPI

Article

Agricultural Land Fragmentation at Urban Fringes: An Application of Urban-To-Rural Gradient Analysis in Adelaide

Suranga Wadduwage *, Andrew Millington, Neville D. Crossman and Harpinder Sandhu

School of the Environment, Flinders University, GPO Box 2100, Adelaide, SA 5001, Australia; andrew.millington@flinders.edu.au (A.M.); neville.crossman@gmail.com (N.D.C.); harpinder.sandhu@flinders.edu.au (H.S.)
* Correspondence: suranga.wadduwage@flinders.edu.au

Academic Editors: Harini Nagendra and Monika Kopecka
Received: 23 January 2017; Accepted: 11 April 2017; Published: 16 April 2017

Abstract: One of the major consequences of expansive urban growth is the degradation and loss of productive agricultural land and agroecosystem functions. Four landscape metrics—Percentage of Land (PLAND), Mean Parcel Size (MPS), Parcel Density (PD), and Modified Simpson's Diversity Index (MSDI)—were calculated for 1 km × 1 km cells along three 50 km-long transects that extend out from the Adelaide CBD, in order to analyze variations in landscape structures. Each transect has different land uses beyond the built-up area, and they differ in topography, soils, and rates of urban expansion. Our new findings are that zones of agricultural land fragmentation can be identified by the relationships between MPS and PD, that these occur in areas where PD ranges from 7 and 35, and that these occur regardless of distance along the transect, land use, topography, soils, or rates of urban growth. This suggests a geometry of fragmentation that may be consistent, and indicates that quantification of both land use and land-use change in zones of fragmentation is potentially important in planning.

Keywords: urban-to-rural gradients; agricultural land-use; land fragmentation; urban fringe; Mean Parcel Size; Parcel Density

1. Introduction

Projections suggest that over two-thirds of the world's population will live in urban centres by 2050 [1], and that a major part to this growth will be due to people migrating from the countryside [2–4]. Over the last 30 years, the global rate of urban land occupation [5,6] has been double the rate of urban population growth [7]. Agricultural land loss due to urbanization has been highlighted by a number of researchers [8–14], and has raised a number of environmental concerns; e.g., declining quality of soil and water assets, loss of natural habitat, decreased plant and animal diversity, and compromised ecological functions [15,16]. The urban sprawl that can be anticipated (given urban population projections) will increase demands for land for housing, industry and infrastructure; thereby consuming more agricultural land at the edges of cities [2,17,18]. This will lead to irreversible and unsustainable land–use transitions at the cost of productive agricultural land in peri-urban areas [19–21], where open spaces and scarce remnant ecosystems with high ecological and conservation values are already threatened [22].

Urban fringes—the transitional zones between urban and rural areas [23]—are characterized by highly dynamic, spatially heterogeneous land-use and land-cover changes [24,25]. This takes place because of the relatively lower land prices in these zones and the high frequency of land tenure change [26,27]. Compared to urban environments, the faster rates of housing and infrastructure growth and the higher proportion of remnant 'green' spaces lead to different landscape structures at the fringe. Research has also demonstrated that urban growth leads to increased land fragmentation [28] and landscape diversity [29] in these areas. The diverse arrays of land uses that result from these processes create spatially heterogeneous, complex land-use configurations [30–34]. However, a concern for planners and people implementing land management policies in urban fringe environments is that the quantitative land-use data they require is often accompanied by relatively low levels of accuracy [35,36].

A recent development in understanding the influence of urbanization on land use has been the use of urban-to-rural gradient analysis [34,37,38]. This concept originated as a combination of elements drawn from landscape ecology and urban ecology [39,40], and has been used to synthesize complex anthropogenic land transitions worldwide [31,34,41–47]. The continuous representation of land-use intensity and the spatial arrangement of land use along gradients is more effective in land-use planning than conventional, discrete spatial measurements [48]. Urban-to-rural gradient analysis is also useful for examining gradual landscape change at urban fringes. The approach has other advantages, e.g., in environmental modeling it is used to minimize subjectivity in categorizing variability, and in describing ecological processes at urban fringes [49]. It is also used to represent land-use as a gradient and for measuring the spatial attributes of land parcels along gradients, both of which improve our ability to interpret landscapes [31,50]. Geographically-referenced points along gradients enable spatial and non-spatial data to be aggregated for systematic landscape comparisons [51–53]. Finally, these continuous information gradients can be utilized to understand landscape structures and potential land-use variations in complex land systems.

Landscape metrics calculated along these gradients have been used to identify land structure elements, and their changing patterns, to describe the effects of urban development at the margins of several cities [31,34,42]. Vizzari and Sigura [48] claim that gradient analyses enable interactions between land-use types to be identified precisely when exploring land transitions. In this research, landscape structure is defined as the spatial configuration of land parcels (i.e., their size and spatial arrangement) and their composition (land-use presence and amount of each land parcel in the landscape) [54].

This paper reports the application of urban-to-rural gradient analysis to understand agricultural land fragmentation at the urban fringes of Adelaide. In previous research, landscape metrics have been plotted along transects, but the relationships between them have not been integrated into gradient analyses. Four landscape metrics—Parcel Density (PD), Mean Parcel Size (MPS), Percentage of Land (PLAND) and Modified Simpson's Diversity Index (MSDI)—were used to quantify and characterize land fragmentation along transects extending from the Adelaide CBD into surrounding rural areas. A novel element of the research is the quantitative analysis of agricultural land-use presence in zones of active land fragmentation at the urban fringe. In this context, urban-to-rural transects were used as georeferenced land-use information gradients that integrate measurements of land-use, while simultaneously examining landscape structure and land-use changes.

2. Materials and Methods

2.1. Study Area

Adelaide—the capital of South Australia—is a coastal city surrounded by sprawling residential and modern industrial suburbs to the north and south. In addition, satellite towns to the east and north, Mount Barker and Gawler (Figure 1), are being incorporated into the urban fabric of the metropolitan area. Adelaide's fringes are urban frontiers that impinge on intensive horticulture and dryland agriculture in the northern plains; a conservation green belt with mixed agricultural land use in the Adelaide Hills to the east; and traditional agricultural areas focused around high value, globally-recognized wine regions to the south (McLaren Vale) and north-east (Barossa Valley). Population growth and economic diversification are increasing the demand for land for housing, transport and industrial infrastructure. In turn, this has led to significant pressure on adjacent productive agricultural land.

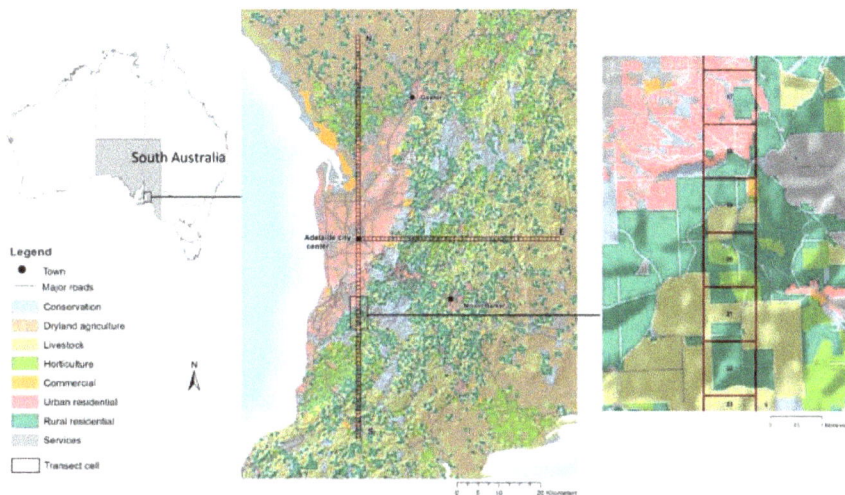

Figure 1. Land-use distribution in Adelaide and its surrounding areas (Source: DPTI 2014). The urban-to-rural transects are overlain in red. The inset map to the right shows an enlargement of the urban-to-rural transect south of the city.

The variations in rural land use at the northern, eastern and southern margins of Adelaide provide a heterogeneous setting in which to test urban-to-rural gradient analysis. Transects were used to sample land-use gradients 50 km outwards from the Adelaide CBD in northerly, easterly and southerly directions (Figure 1).

Previous researchers using gradient analysis [31,55] have mapped urban-to-rural gradients along transport corridors. It is probable that this leads to a bias toward the investigation of urban land use. However, as this paper's research focus is on the incorporation of different types of agricultural land into an expanding urban area, it was decided to maximize the agricultural land use considered in the gradient analysis. Therefore, they were not oriented along main routes out of Adelaide, but in three cardinal directions. In fact, there are many routes out of Adelaide, which are orientated in a variety of directions. Therefore, each of these transects has some transport corridor influence. The transects were sampled over 50 km so that they are comparable and of sufficient length to cover all the types of parcels where agricultural land is being incorporated into the urban fabric of the city.

This study uses a single statewide cadastral dataset produced by the South Australian Government's Department of Planning, Transport and Infrastructure (DPTI) in 2014, which is publically available online (http://data.sa.gov.au). The primary purpose of this dataset is to assess council rates and levies based on land parcel valuations. The attributes of the dataset that are pertinent to this research are: land parcel identity codes; land-use categories; and the land-use classes occurring in each of the land parcels. It contains nineteen land-use categories (Table 1), which were regrouped into eight land-use classes for the purposes of this research. Sixteen categories were regrouped into five land-use classes—Conservation, Urban residential, Rural residential, Commercial and Services. Three categories—Dryland agriculture, Livestock land and Horticulture land—were not changed.

Table 1. Scheme used to reclassify land-use categories in the cadastral dataset (2014) to land-use classes for this research.

Original Land-Use Categories *	Reclassified Land-Use Classes (the numbers in parentheses are used in subsequent graphs)
Reserve, Forestry, Vacant	Conservation (1)
Agriculture	Dryland agriculture (2)
Livestock	Livestock (3)
Horticulture	Horticulture (4)
Commercial, Food Industry, Mine and Quarry, Public Institution,	Commercial (5)
Residential, Non private residential, Vacant residential	Urban residential (6)
Rural residential	Rural residential (7)
Education, Golf, Recreation, Utility Industry	Services (8)

* Land categories defined in the South Australian government cadastral data set in 2014.

2.2. Urban-To-Rural Gradients at Urban Fringes

Urban-to-rural gradients [34] were used to visualize and analyze land use along three 50 km long transects, each of which comprise 50 1 km × 1 km cells. ArcGIS© 10.2.1 (ESRI: Redlands, CA, USA) was used for all spatial data analyses. The 1 km^2 cell-based transects were produced using the Fishnet tool by defining the spatial areas for cell references. They were overlain on the cadastral dataset and land-use information extracted for each cell. These data were then compiled using the tabulation tool in ArcGIS spatial analyst extension. Each cell in the resulting dataset includes a unique identifier and the areas of each of land-use class (Table 1) within each cell.

Landscape metrics have been used extensively in conservation biology, but their application in land-use research to measure, characterize, analyze, and visualize landscape structure is far less common, particularly in urban areas [41,56–58]. Four landscape metrics were calculated from the attributes for each cell in the three transects (Table 2). The percentage of each land-use class in each cell (PLAND) provides data on compositional changes in land use along the gradients. MPS and the PD are measurements of key spatial features along the transects. Finally, MSDI is a measure of the proportional abundance of the land-use classes in each cell, and is an indicator of land-use diversity. Plots of each of the metrics for each gradient enabled landscape structures to be visualized and analyzed.

Table 2. Landscape metrics used for spatial feature characterization.

Metric	Description	Range	Equation
Percent of land-use coverage (PLAND) [%]	The proportion of the total area occupied by a particular land-use class.	0< PLAND ≤ 100	$P_i = \frac{\sum_{j=1}^{n} a_{ij}}{A}(100)$
Modified Simpson's Diversity Index (MSDI)	A measurement of land-use diversity in a cell determined by the distribution of the proportional abundance of different land-use types (parcel richness) extensively.	MSDI ≥ 0	$MSDI = -ln \sum_{i=1}^{n} P_i^2$
Mean Parcel Size (MPS) [ha]	The average area of all land parcels in the landscape.	MPS > 0	$MPS = \frac{\sum_{j=1}^{N} a_j}{N}\frac{1}{10,000}$
Parcel Density (PD) [N/km^2]	The number of land parcels per 100ha.	PD > 1	$PD = \frac{N}{A}(10,000)(100)$

P_i = proportion of the landscape occupied by parcel land-use type i, a_{ij} = area (m^2) of parcel ij, a_j = area (m^2) of parcel j, A = total area of the landscape (m^2)—cell, i = land-use class (1–8), j = number of parcels, $n = n_i$ = number of parcels in the landscape (cell) of parcel land-use type I, N = number of parcels in the landscape. (McGarigal and Marks, 1995).

2.3. Landscape Matrix Analysis

The relationships between MPS and PD were investigated to examine the extent of land fragmentation with distance along each transect. The associations between MPS and PD demonstrate probable land structure variations in the landscape, and trend lines were used to visualize the nature of the relationships between MPS and PD.

The study area contains the following median land parcel areas: LL (Livestock land) (59 ha), DL (Dryland cultivation) (50 ha), and HL (Horticultural land) (12 ha). HL has a minimum size of 2.5 ha, which probably represents intensive irrigated vegetable cultivation or small vineyards. The median (12 ha) to minimum (2.5 ha) size of HL land parcels allows the range in the number of agriculture-based land parcels which are likely to occur in a 1 km^2 (100 ha) cell to be estimated. Horticultural land (HL) was used to define the PD range between 7 and 35 N/km^2, because it is the agricultural land-use type with the smallest median parcel size. Therefore, it is the land-use class that will provide the maximum number of land parcels in a 1 km^2 (100 ha) cell. It is believed that this range of values indicates a high potential for transforming agricultural to urban land-uses at urban fringes. This is due to high property values, proximity to built-up areas, and that they frequently experience government-promoted land subdivision and land re-zoning for urban development. Rauws and De Roo [26] have identified these land-use change drivers as "pull factors" which are influenced by urban economies converting non-urban land uses to urban form at the peri-urban areas. Therefore, in the scatter diagrams, a common range of PD from 7 to 35 N/km^2 is used; where a 1 km^2 cell can have 7 to 35 land parcels/km^2 that are highly vulnerable to change in land use. The agriculture-based land parcel information associated with the cells from the land cadastral dataset was extracted within this range of patch densities.

3. Results

Landscape metric values were plotted along the three urban-to-rural gradients; north (N), east (E) and south (S); PLAND in Figure 2, MPS and PD in Figure 3, and MSDI in Figure 4. PLAND values for the eight land-use types (Figure 2) illustrate the variations in land-use composition along the transects, thereby demonstrating the urban, peri-urban and rural characteristics of these transects. The PLAND values along these three transects show high percentages of urban land uses near the city centre, a gradual change to higher percentages of agricultural land uses at the end of the transects, and a heterogeneous mix of land-use types in the peri-urban areas. MPS and PD have a negative relationship (Figure 3), with greater MPS values being associated with lower PD values. Figure 5a illustrates the association between MPS and PD of the land parcels for each transect. Figure 5b shows the relationship between PD and MPS in the ranges 0–40 N/km^2 and 0–80 ha, respectively, for each transect. MSDI is somewhat similar between transects (Figure 4), and shows that diversity generally declines with distance from the CBD. However, it is noteworthy that the southern transect has relatively lower landscape diversity than the other two.

3.1. Agricultural Land-Use Presence

The PLAND values for Dryland agriculture (DL), Livestock land (LL) and Horticulture (HL) along the three transects are shown in Figure 6. The northern transect shows three distinctly different zones of land use. The built-up area, between 0–15 km, has low agricultural PLAND for the three agricultural land uses, and high PD and low MPS. Between 15 and 37 km the agricultural land-use percentages are HL (61.4%), DL (31.6%), and LL (6.8%). These represent mainly intensive vegetable production, rain-fed cereal cultivation, and sheep and horse grazing, respectively. This 22-km long zone presents a typical urban fringe landscape structure, with increasing MPS and decreasing PD. The landscape beyond the fringe (>37 km) is dominated by Dryland agriculture, and mainly comprises rain-fed wheat, barley and olive groves, which occupy large land parcels in a rural landscape. Land-use presence in the zones of high fragmentation is provided in Figure 7.

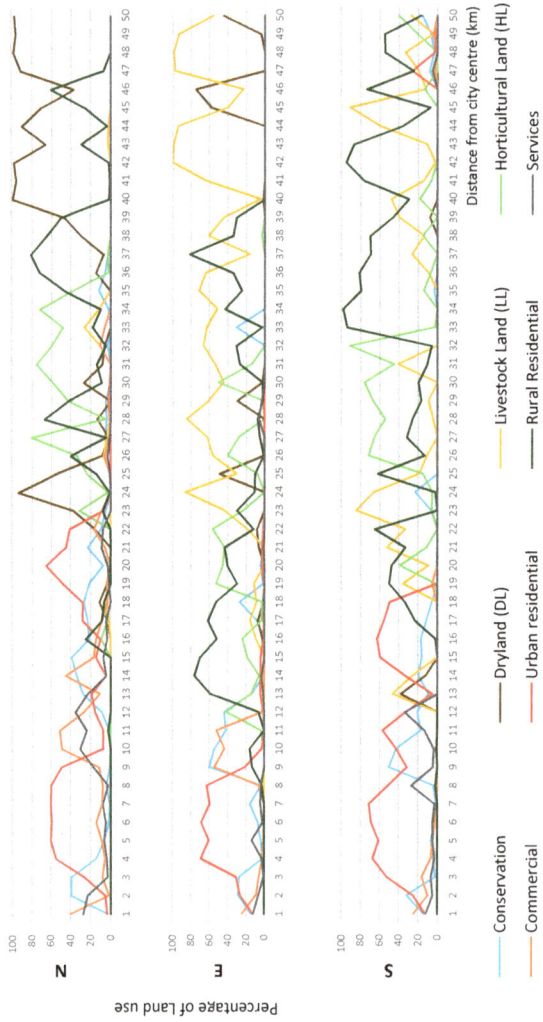

Figure 2. PLAND: north (N), east (E) and south (S) transects.

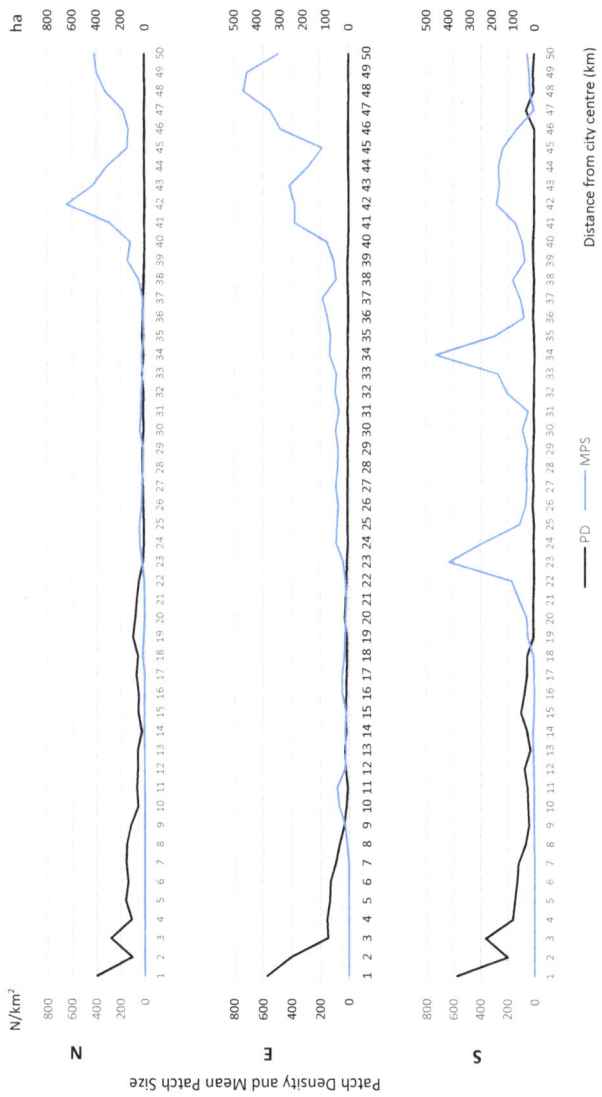

Figure 3. Patch Density (PD) and Mean Patch Size (MPS): north (N), east (E) and south (S) transects.

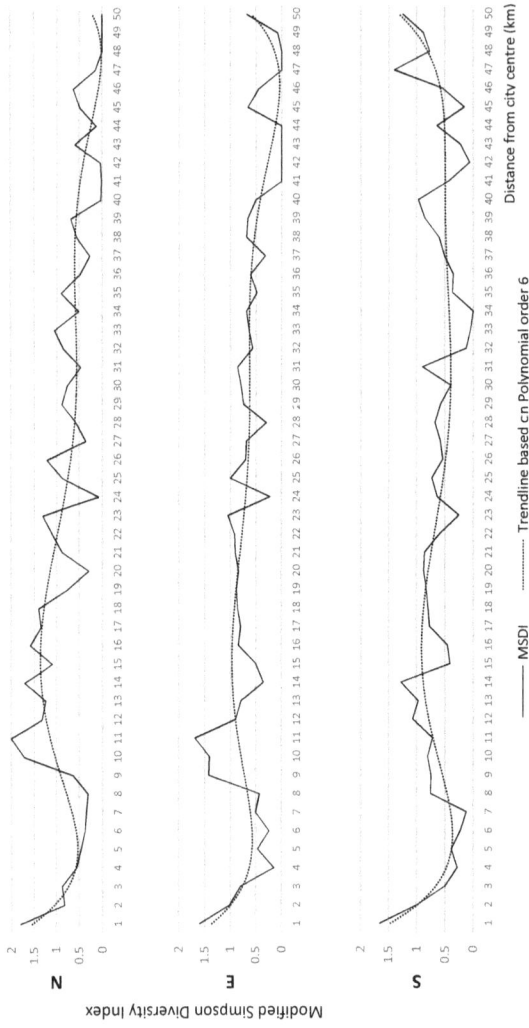

Figure 4. Modified Simpson Diversity Index: north (N), east (E) and south (S) transects.

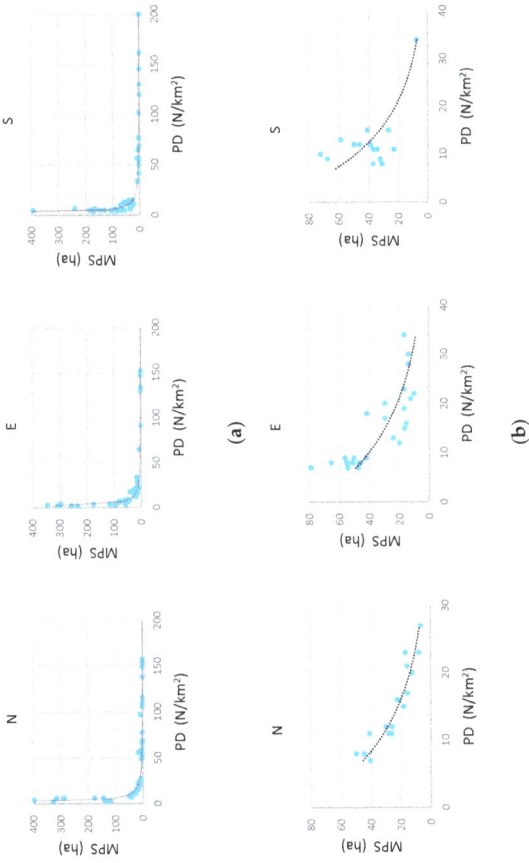

Figure 5. (a) MPS-PD plots: north (N), east (E) and south (S) transects. (b) Enlargements of MPS_PD plots in the 7 < PD < 35 range.

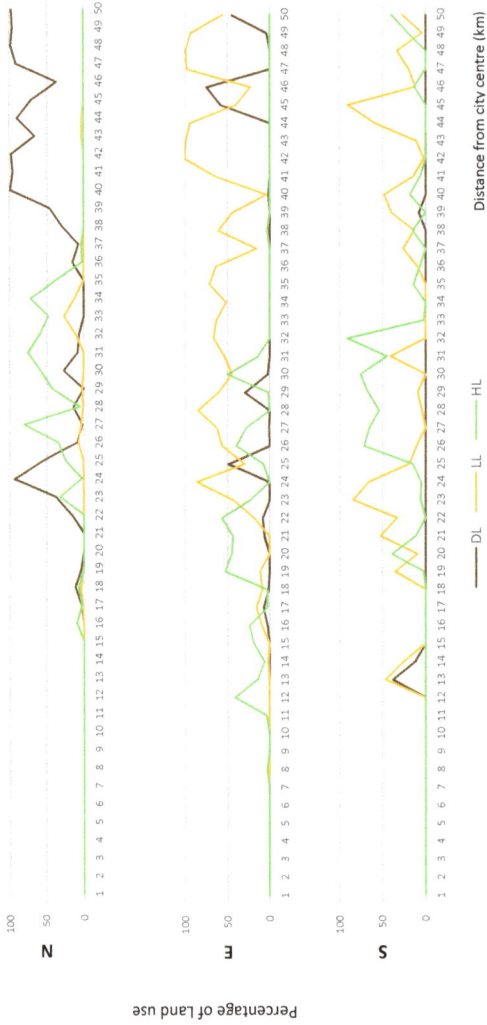

Figure 6. Percentage of the total land area occupied by each agricultural land-use type: north (N), east (E) and south (S) transects.

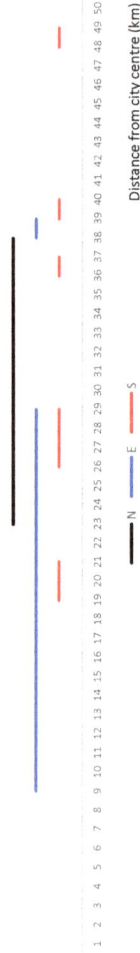

Figure 7. Areas prone to land fragmentation between PD 7 and 35 for each transect.

The first 10 km of the eastern transect represents the built-up areas of eastern Adelaide. The three agricultural land-uses of the Adelaide Hills—sheep and cattle rearing (LL, 52.7%); vegetable cultivation, fruit orchards and wineries (HL, 38.5%); and rain-fed crops (DL, 8.7%) characterize the transect from 11 to 32 km. The MPS of the land parcels in this hilly terrain are relatively small. Livestock land and Dryland cultivation dominate the transect beyond 33 km.

The southern transect is significantly different from the northern or eastern transects in terms of agricultural land use. Beyond the built-up area, which covers the first 18 km of the transect, LL and HL have much higher shares of the overall land use than DL. The landscape from 18 to 33 km has an agricultural land use split of HL (60.1%), LL (39.2%), and DL (0.7%). This combination characterizes the complex land use of McLaren Vale, which has transitioned from a mixed grazing and horticultural region, to one of vineyards and olive groves, with some grazing being retained at the margins. The amount of LL increases in the landscape beyond 33 km. However, in these final 17 km, PLAND values of Rural residential land and Urban residential land increase, leading to correspondingly higher MSDI values. The changes in these metrics demonstrate the influence of the town of Victor Harbor, which is located beyond the end of the transect. Table 3 summarizes agricultural land presence in the three transects:

Table 3. Summary of the agricultural land along the three gradients.

Transect	Built-Up Area	Urban Fringe Areas	Rural Areas
North	0–15 km. Low PLAND, high PD and low MPS for DL, LL and HL	15–37 km. HL (61.4%), DL (31.6%) and LL (6.8%) representing mainly intensive vegetable production, rain-fed cereal cultivation, and sheep and horse grazing respectively.	>37 km. Dominated by DL (rain-fed wheat, barley and olives) which occupies large land parcels.
East	0–10 km. Low PLAND, high PD and low MPS for DL, LL, HL.	11–32 km. LL (52.7%), HL (38.5%) and DL (8.7%) representing sheep and cattle rearing; vegetable cultivation, orchards and wineries; and rain-fed crops respectively. Relatively small MPS compared to other rural areas due to hilly terrain.	>32 km. Dominated by LL and DL.
South	0–18 km. Low PLAND, high PD and low MPS for DL, LL, HL.	18–33 km. HL (60.1%), LL (39.2%) and DL (0.7%) representing the complex land use of McLaren Vale which has transitioned from a mixed grazing and horticulture region to a vineyards and olive groves with some grazing retained at the margins.	>33 km. High proportions of land in LL (cattle grazing). Increase in PLAND for residential uses, and higher MSDI values at the end of the transect due to the influence of the town of Victor Harbor.

The total amount of agricultural land in each transect is summarized in Figure 8. The eastern transect has the highest amount of agricultural land (2558 ha, 51.2%), comprised of 11% DL, 70% LL and 19% HL. The southern transect has the lowest amount of agricultural land (1583 ha, 31.6%: 4% DL, 53% LL, 3% HL). The northern transect has 1979 ha (39.6%) under the three types of agricultural land-use, and is dominated by Dryland cultivation, accounting for 66% of all agricultural land.

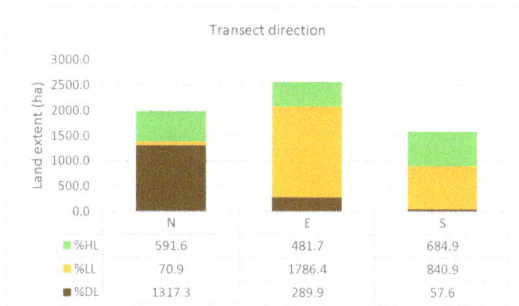

Figure 8. Total agricultural land extent and the land use type percentages in the north (N), east (E) and south (S) transects.

3.2. Agricultural Land Fragmentation

MPS and PD were used to characterize agricultural land fragmentation along each transect. In considering the zone where PD ranges from 7 to 35 N/km^2, the critical zones for land fragmentation in the northern and eastern gradients extend for 15 km and 20 km, respectively (Figure 7). This zone is disjunctive in the southern transect, and extends from 19 km to the end of the transect. Figure 9 shows the amount of land occupied by the agricultural land uses in the zones of land fragmentation for each transect, while Figure 9 shows the corresponding percentage data. The total amounts of agricultural land of all types in the zones of high fragmentation are: 935.1 ha, 1311.9 ha and 825.7 ha for the northern, eastern and southern transects, respectively. Figure 10 displays the amount of each class of agricultural land in the zones of fragmentation. Horticultural land comprises a large component in each transect, and dominates the northern transect. Livestock grazing accounts for the highest proportions of agricultural land in the zones of high fragmentation in the eastern and southern transects, but is a minor element in the northern transect. Dryland agriculture has a low presence in the zones of fragmentation in all three transects. This is only encountered with any frequency in the northern transect, where there is significant contemporary urban fringe formation on land formerly used for rain-fed cereal cultivation on the Northern Adelaide Plains.

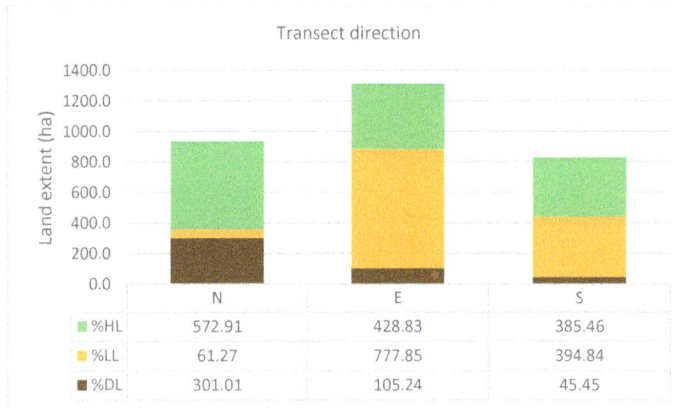

Figure 9. The agricultural land extent and the land use type percentages in the zones of high land fragmentation.

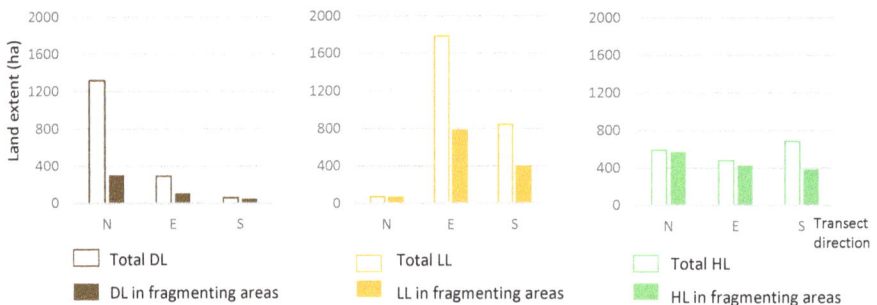

Figure 10. The extent of DL, LL, HL in zones of high land fragmentation: north (N), east (E) and south (S) transects.

4. Discussion

Weng [42], in applying gradient analysis, found that landscape fragmentation is positively correlated with the degree of urbanization, and results in agricultural land loss at urban fringes. Therefore, as agricultural land is generally the major land-use category beyond the fringe, it is the major land reserve for meeting the land demands of urban development in sprawling cities such as Adelaide. Moreover, fragmentation is the key spatial manifestation of the process of incorporating agricultural land into transitional, urban fringe landscapes.

This research confirms the presence of agricultural land along all three gradients, and that fragmentation can be easily visualized and quantified using a combination of gradient analysis and landscape metrics. It is the first application of these techniques in the Australian context. More importantly, this research provides an advance on previous analyses of the incorporation of agricultural land into the urban fabric of cities, by comparing the conversion processes acting on three types of agricultural land (Dryland agriculture, Livestock grazing and Horticulture).

4.1. Land Structure Analysis along Gradients

This research presents a novel method for investigating agricultural land fragmentation at the urban fringe, by analyzing the associations between mean patch size and patch density. Notwithstanding the differences in the land-use geographies along the transects, scatter plots of MPS and PD for the three transects showed similar patterns of cell organization with respect to patch density and mean patch size. Cells associated with the horizontal parts of the trend lines (Figure 5) indicate low levels of association between MPS and PD; e.g., a decline in MPS from 400 ha to approximately 100 ha leads to very little increase in PD, which remains at <7 N/km^2. When PD reaches approximately 35 N/km^2, further increases are not accompanied by significant changes in MPS, i.e., the vertical parts of the trend lines in Figure 5. This means that the zone bounded by PD values of 7 to 35 N/km^2 is a critical zone of land fragmentation in which the relationship between MPS and PD is very sensitive. For example, a decrease of one hectare in MPS leads to an increase in PD of 0.52 N/km^2 (in the northern transect), 0.54 N/km^2 (eastern transect), and 0.33 N/km^2 (southern transect) in this study.

The cell values that correspond to the zone of high fragmentation are well distributed in the northern and eastern transects (Figure 5b). This indicates that large land parcels are being fragmented in a regular and incremental manner to create progressively smaller parcels, and that the resulting increases in PD are responses to rapid urban development to the north and east of Adelaide. More clustered cell values in the southern transect indicate a differently organized landscape structure. The corresponding cell values in the southern transect cluster between a MPS of 22–70 and PD of 7–5. It is believed that this pattern derives from an urban fringe that is characterized by larger land parcels that can be attributed to the size of vineyards and planning restrictions on the post-sale use of vineyards due to the implementation of the *Character Preservation (McLaren Vale) Act 2012* [59]. That is, fragmentation is not occurring at the same rate or in the same way as it is on the eastern and northern fringes of the city.

Overall, the results demonstrate that it is the contemporary land-use transformation processes that explain the landscapes metrics measured. This validates the use of landscape metrics derived for cells along transects to characterize landscape structures. For example, the urban fringe to the south has a lower MPS for agricultural land than in the north and east; and that the difference is due to the high frequency of vineyards in the south compared to the dominance of dryland cereal fields in the north, and extensive grazing areas and fruit orchards in the east. Furthermore, this method can be used to understand the influences of regional towns on land-use transitions—a point that is rarely considered in peri-urban studies [48]. For example, in the southern transect, the influence of the town of Victor Harbor on land fragmentation and land-use changes near the end of the transect is clear in comparison to the other two transects.

If landscape metrics are to be used effectively in assessing land fragmentation at urban fringes, it is imperative that they are calculated for all cells and plotted along the entire transect, rather than simply focusing on the peri-urban areas. This allows emerging and existing areas of fragmentation to be identified objectively through the behavior of metrics.

4.2. Agricultural Land in the Areas of Fragmentation

The agricultural land types in the zones of high land fragmentation are, in order of decreasing area, Horticulture, Livestock grazing and Dryland cultivation. This differs from the total distribution of agricultural land along the three transects, which in order of decreasing area, are Livestock grazing, Horticulture and Dryland cultivation. This change in order highlights the importance of quantifying the agricultural land in high fragmentation zones, rather than analyzing agricultural land along an entire transect—particularly if the results are being used to make strategic land-use decisions regarding urban fringes.

It can be argued that quantifying agricultural land in fragmenting areas, instead of the total land presence, will improve planners' understanding of the vulnerability of agricultural land in these transitional landscapes. For example, land under Dryland cultivation has the highest land-use presence in the northern transect, but only 20% of that land-use class in the transect is prone to fragmentation. The fact that agricultural land fragmentation occurring at the fringes of Adelaide can be identified and characterized using gradient analysis and landscape metrics (regardless of the different characteristics of the northern, eastern and southern transects) is testament to the robustness of the method. Moreover, different spatial configurations of land parcel arrangements can be identified. Figure 10 provides data on the proportions of different land-use classes in the three transects. These data reveal the importance of quantifying the individual land-use class measurements to identify the detailed land structure elements in these complex landscapes. Vizzari and Sigura [48] argue that whole gradient analysis is required in rural-to-urban analyses. Urban expansion in Australian cities occurs in less complex landscapes than those in world regions chararcterised by high levels of urbanization or rapid urbanization and unprecedented levels of development in tangled webs of complex rural-urban transitions [10], e.g., Japan, eastern China, south-east Asia, western Europe, and parts of North America. Nevertheless, the usefulness of whole gradient analysis is again emphasised in this research.

Land-use responses to urbanization stimuli are dependent on geographical location, and land ownership and land-use policies as integral parts of complex land systems [60]. Though this research illustrates a higher probability of land fragmentation in some types of horticultural land, other areas are much less affected, e.g., protected heritage wine making regions with large capital investments. This indicates that other attributes of land-use classes are important in determining the extent of fragmentation. In the northern transect, many intensively-cultivated vegetable farms are proximate to built-up areas, have relatively small investments currently, are operated by ageing land owners who are contemplating selling their farms, and are located in areas where local councils are actively re-zoning land. Therefore, it is land attributes that lead to fragmentation, rather than simply the spatial characteristics. This demonstrates the importance of integrating local knowledge and current urban development policies into future urban-to-rural gradient analyses to improve outcomes.

The method outlined in this paper can be applied to different geographies, where a land dataset (or land-use maps derived from remotely sensed data) with land-use attributes exists to provide justifiable evidence for probable agricultural land transitions. An analytical approach such as this, which uses a single dataset, could overcome issues that exist with analyses based on multiple data sets [61], e.g., data incompatibility, error generation and variations in data definitions associated with previous landscape studies. Though some limitations would still exist, such as human error in data collection and spatial analysis.

Integration of the gradient method with the analysis of landscape metrics leads to two main advances. First, it improves the interpretability of transitional processes on agricultural land at city fringes by focusing measurements on particular areas (e.g., agricultural land within zones of fragmentation), while still analyzing the landscape structure in an urban-to-rural continuum. Secondly, it enhances information richness for improved peri-urban land-use planning strategies within planning and policy-making groups at different levels of land governance (e.g., local government to state level), as well as for other stakeholder groups who share common interests in effective management of peri-urban land. These include primary industries, biological conservation, natural resource management and recreational opportunities.

5. Conclusions

This research integrated landscape metrics into urban-to-rural gradient analysis to deepen our understanding of the geographies of agricultural land-use change at the urban fringes of Adelaide. The study reveals that less well-regulated horticultural land uses are the most vulnerable to urban expansion, though well-protected horticultural land experiences much lower levels of conversion and fragmentation. Land uses related to livestock grazing and rearing have a larger presence than horticulture, but are less likely to change. Dryland agriculture is the least vulnerable for urban sprawl.

The research findings confirm that integrating landscape metrics and urban-to-rural gradient analysis provides a robust method that works equally well under different natural environments, rates of urban growth, and types of land use. A new finding is that MPS and PD can be used to identify zones with high rates agricultural land fragmentation. These occur where PD ranges from 7 to 35 N/km^2, regardless of distance from the city centre, land use, topography, soils and rates of urban growth and suggest a geometry of fragmentation that may be consistent.

Integrating landscape metrics into gradient analysis has the potential to provide a wide range of stakeholders, ranging from planners to conservation and primary production groups, with a rich source of information on agricultural land-use configurations, and their interdependencies. Further it can provide them with the ability to systematically compare spatially quantifiable land-use metrics along urban-to-rural gradients. Nonetheless, we suggest there are further opportunities to test the robustness of this method in urban fringe landscapes in different types of cities around the world.

Acknowledgments: This research was funded by the doctoral study grant—Australian Postgraduate Award (APA)—from the Commonwealth Government of Australia to Suranga Wadduwage. We acknowledge the Department of Planning, Transport and Infrastructure (DPTI) in South Australia for providing raw data for this study and the constructive comments made by the international land use researchers at the Global Land Project meetings in Taiwan and at the 'Land Lab' meetings at Flinders University. We appreciate the constructive comments of the three anonymous reviewers, and the grammatical insights of Cécile Cutler.

Author Contributions: Suranga Wadduwage designed the study, analyzed the data and drafted the initial manuscript. Andrew Millington made critical suggestions on the methods, results and the discussion, and redrafted some sections of the manuscript. Neville Crossman and Harpinder Sandhu contributed to drafting the manuscript in different stages to ensure compliance to contemporary land use research themes. All four authors contributed to writing the paper.

Conflicts of Interest: The authors declare no conflict of interest. The sponsor had no role in the design of the study; in the collection, analyses or interpretation of data; in the writing of the manuscript; nor in the decision to publish the results.

References

1. United Nations. *The State of Asian and Pacific Cities 2015 Urban Transformations Shifting from Quantity to Quality*; United Nations: New York, NY, USA, 2015; p. 204.
2. Seto, K.C.; Fragkias, M.; Güneralp, B.; Reilly, M.K. A meta-analysis of global urban land expansion. *PLoS ONE* **2011**, *6*, e23777. [CrossRef] [PubMed]
3. Schneider, A.; Mertes, C.; Tatem, A.; Tan, B.; Sulla-Menashe, D.; Graves, S.J.; Patel, N.N.; Horton, J.A.; Gaughan, A.E.; Rollo, J.T.; et al. A new urban landscape in East–Southeast Asia, 2000–2010. *Environ. Res. Lett.* **2015**, *10*, 034002. [CrossRef]
4. Fragkias, M.; Langanke, T.; Boone, C.; Haase, D.; Marcotullio, P.J.; Munroe, D.; Olah, B.; Reenberg, A.; Seto, K.; Simon, D. *Land Teleconnections in an Urbanizing World*; GLP Workshop Report, GLP Report No. 5; Global Land Programme: Bern, Switzerland, 2012; p. 22.

5. Angel, S.; Parent, J.; Civco, D.L.; Blei, A.; Potere, D. The dimensions of global urban expansion: Estimates and projections for all countries, 2000–2050. *Prog. Plan.* **2011**, *75*, 53–107. [CrossRef]

6. Stefan, B.; Helmut, S.; Walter, P.; O'Brien, M.; Garcia, F.; Sims, R.; Howarth, R.W.; Kauppi, L.; Swilling, M.; Herrick, J. *ASSESSING GLOBAL LAND USE: Balancing Consumption With Sustainable Supply*; UNEP Publication: Nairobi, Kenya, 2014; p. 131.

7. Seto, K.C.; Güneralp, B.; Hutyra, L.R. Global forecasts of urban expansion to 2030 and direct impacts on biodiversity and carbon pools. *Proc. Natl. Acad. Sci. USA* **2012**, *109*, 16083–16088. [CrossRef] [PubMed]

8. Shi, Y.; Sun, X.; Zhu, X.; Li, Y.; Mei, L. Characterizing growth types and analyzing growth density distribution in response to urban growth patterns in peri-urban areas of Lianyungang City. *Landsc. Urban Plan.* **2012**, *105*, 425–433. [CrossRef]

9. Debolini, M.; Valette, E.; François, M.; Chéry, J.-P. Mapping land use competition in the rural–urban fringe and future perspectives on land policies: A case study of Meknès (Morocco). *Land Use Policy* **2015**, *47*, 373–381. [CrossRef]

10. Handayani, W. Rural-urban transition in Central Java: Population and economic structural changes based on cluster analysis. *Land* **2013**, *2*, 419–436. [CrossRef]

11. Appiah, D.O.; Bugri, J.T.; Forkuor, E.K.; Boateng, P.K. Determinants of peri-urbanization and land use change patterns in peri-urban Ghana. *J. Sustain. Dev.* **2014**, *7*, 95–109. [CrossRef]

12. Malaque, I.R.; Yokohari, M. Urbanization process and the changing agricultural landscape pattern in the urban fringe of Metro Manila, Philippines. *Environ. Urban.* **2007**, *19*, 191–206. [CrossRef]

13. Piorr, A. *Peri-Urbanisation in Europe: Towards European Policies to Sustain Urban-Rural Futures*; Synthesis Report; PLUREL [Sixth Framework Programme]; Forest & Landscape, University of Copenhagen: Copenhagen, Denmark, 2011.

14. Piorr, A.; Ravetz, J.; Tosics, I.; PLUREL. *Peri-urbanisation in Europe: Towards European Policies to Sustain Urban-Rural Futures: Synthesis Report*; Forest & Landscape: Copenhagen, Denmark, 2011.

15. Matson, P.A.; Parton, W.J.; Power, A.; Swift, M. Agricultural intensification and ecosystem properties. *Science* **1997**, *277*, 504–509. [CrossRef] [PubMed]

16. Flynn, D.F.; Gogol-Prokurat, M.; Nogeire, T.; Molinari, N.; Richers, B.T.; Lin, B.B.; Simpson, N.; Mayfield, M.M.; DeClerck, F. Loss of functional diversity under land use intensification across multiple taxa. *Ecol. Lett.* **2009**, *12*, 22–33. [CrossRef] [PubMed]

17. Xiao, J.; Shen, Y.; Ge, J.; Tateishi, R.; Tang, C.; Liang, Y.; Huang, Z. Evaluating urban expansion and land use change in Shijiazhuang, China, by using GIS and remote sensing. *Landsc. Urban Plan.* **2006**, *75*, 69–80. [CrossRef]

18. Liu, J.; Hull, V.; Batistella, M.; DeFries, R.; Dietz, T.; Fu, F.; Hertel, T.W.; Izaurralde, R.C.; Lambin, E.F.; Li, S.; et al. Framing sustainability in a telecoupled world. *Ecol. Soc.* **2013**, 18. [CrossRef]

19. Jiang, Y.; Swallow, S.K. Providing an ecologically sound community landscape at the urban–rural fringe: A conceptual, integrated model. *J. Land Use Sci.* **2015**, *10*, 323–341. [CrossRef]

20. Pham, V.C.; Pham, T.-T.-H.; Tong, T.H.A.; Nguyen, T.T.H.; Pham, N.H. The conversion of agricultural land in the peri-urban areas of Hanoi:(Vietnam): Patterns in space and time. *J. Land Use Sci.* **2014**. [CrossRef]

21. D'Amour, C.B.; Reitsma, F.; Baiocchi, G.; Barthel, S.; Güneralp, B.; Erb, K.-H.; Haberl, H.; Creutzig, F.; Seto, K.C. Future urban land expansion and implications for global croplands. *Proc. Natl. Acad. Sci. USA* **2016**. [CrossRef]

22. Crossman, N.D.; Bryan, B.A.; Ostendorf, B.; Collins, S. Systematic landscape restoration in the rural–urban fringe: Meeting conservation planning and policy goals. *Biodivers. Conserv.* **2007**, *16*, 3781–3802. [CrossRef]

23. Thapa, R.B.; Murayama, Y. Land evaluation for peri-urban agriculture using analytical hierarchical process and geographic information system techniques: A case study of Hanoi. *Land Use Policy* **2008**, *25*, 225–239. [CrossRef]

24. Seto, K.C.; Sánchez-Rodríguez, R.; Fragkias, M. The new geography of contemporary urbanization and the environment. *Annu. Rev. Environ. Resour.* **2010**, *35*, 167–194. [CrossRef]

25. Nagendra, H.; Unnikrishnan, H.; Sen, S. Villages in the city: Spatial and temporal heterogeneity in rurality and urbanity in Bangalore, India. *Land* **2013**, *3*, 1–18. [CrossRef]

26. Rauws, W.; de Roo, G. Exploring transitions in the peri-urban area. *Plan. Theory Pract.* **2011**, *12*, 269–284. [CrossRef]

27. Liu, Y.; Feng, Y.; Pontius, R.G. Spatially-explicit simulation of urban growth through self-adaptive genetic algorithm and cellular automata modelling. *Land* **2014**, *3*, 719–738. [CrossRef]
28. Irwin, E.G.; Bockstael, N.E. The evolution of urban sprawl: Evidence of spatial heterogeneity and increasing land fragmentation. *Proc. Natl. Acad. Sci. USA* **2007**, *104*, 20672–20677. [CrossRef] [PubMed]
29. Lambin, E.F.; Turner, B.L.; Geist, H.J.; Agbola, S.B.; Angelsen, A.; Bruce, J.W.; Coomes, O.T.; Rodolfo Dirzo, R.; Fischer, G.; Carl Folke, C.; et al. The causes of land-use and land-cover change: Moving beyond the myths. *Glob. Environ. Chang.* **2001**, *11*, 261–269. [CrossRef]
30. Allan, J.D. Landscapes and riverscapes: The influence of land use on stream ecosystems. *Annu. Rev. Ecol. Evol. Syst.* **2004**, 257–284. [CrossRef]
31. Luck, M.; Wu, J. A gradient analysis of urban landscape pattern: A case study from the Phoenix metropolitan region, Arizona, USA. *Landsc. Ecol.* **2002**, *17*, 327–339. [CrossRef]
32. Kuang, W.; Liu, J.; Dong, J.; Chi, W.; Zhang, C. The rapid and massive urban and industrial land expansions in China between 1990 and 2010: A CLUD-based analysis of their trajectories, patterns, and drivers. *Landsc. Urban Plan.* **2016**, *145*, 21–33. [CrossRef]
33. Lambin, E.F.; Meyfroidt, P. Land use transitions: Socio-ecological feedback versus socio-economic change. *Land Use Policy* **2010**, *27*, 108–118. [CrossRef]
34. Haase, D.; Nuissl, H. The urban-to-rural gradient of land use change and impervious cover: A long-term trajectory for the city of Leipzig. *J. Land Use Sci.* **2010**, *5*, 123–141. [CrossRef]
35. Bunker, R.; Houston, P. Prospects for the Rural-Urban Fringe in Australia: Observations from a Brief History of the Landscapes around Sydney and Adelaide. *Aust. Geogr. Stud.* **2003**, *41*, 303–323. [CrossRef]
36. Houston, P. Re-valuing the fringe: Some findings on the value of agricultural production in Australia's peri-urban regions. *Geogr. Res.* **2005**, *43*, 209–223. [CrossRef]
37. Kroll, F.; Müller, F.; Haase, D.; Fohrer, N. Rural–urban gradient analysis of ecosystem services supply and demand dynamics. *Land Use Policy* **2012**, *29*, 521–535. [CrossRef]
38. Andersson, E.; Ahrné, K.; Pyykönen, M.; Elmqvist, T. Patterns and scale relations among urbanization measures in Stockholm, Sweden. *Landsc. Ecol.* **2009**, *24*, 1331–1339. [CrossRef]
39. McGarigal, K.; Cushman, S. *The Gradient Concept of Landscape Structure*; Cambridge University Press: Cambridge, UK, 2005; Chapter 12.
40. Godron, M.; Forman, R. *Landscape Modification and Changing Ecological Characteristics, in Disturbance and Ecosystems*; Springer: Berlin/Heidelberg, Germany, 1983; pp. 12–28.
41. McDonnell, M.J.; Pickett, S.T.; Groffman, P.; Bohlen, P.; Pouyat, R.V.; Zipperer, W.C.; Parmelee, R.W.; Carreiro, M.M.; Medley, K. Ecosystem processes along an urban-to-rural gradient. *Urban Ecosyst.* **1997**, *1*, 21–36. [CrossRef]
42. Weng, Y. Spatiotemporal changes of landscape pattern in response to urbanization. *Landsc. Urban Plan.* **2007**, *81*, 341–353. [CrossRef]
43. McDonnell, M.J.; Hahs, A.K. The use of gradient analysis studies in advancing our understanding of the ecology of urbanizing landscapes: Current status and future directions. *Landsc. Ecol.* **2008**, *23*, 1143–1155. [CrossRef]
44. Larondelle, N.; Haase, D. Urban ecosystem services assessment along a rural–urban gradient: A cross-analysis of European cities. *Ecol. Indic.* **2013**, *29*, 179–190. [CrossRef]
45. Vizzari, M.; Sigura, M.; Antognelli, S.; Kovačev, I. Ecosystem services demand, supply and budget along the urban-rural-natural gradient. In Proceedings of the 43rd International Symposium on Agricultural Engineering, Actual Tasks on Agricultural Engineering, Opatija, Croatia, 24–27 February 2015.
46. Forman, R.T.; Godron, M. *Landscape Ecology*; John Wiley & Sons: New York, NY, USA, 1986; p. 619.
47. McDonnell, M.J.; Pickett, S.T.; Pouyat, R.V. *The Application of the Ecological Gradient Paradigm to the Study of Urban Effects, in Humans as Components of Ecosystems*; Springer: Berlin/Heidelberg, Germany, 1993; pp. 175–189.
48. Vizzari, M.; Sigura, M. Landscape sequences along the urban–rural–natural gradient: A novel geospatial approach for identification and analysis. *Landsc. Urban Plan.* **2015**, *140*, 42–55. [CrossRef]
49. Bridges, L.M.; Crompton, A.E.; Schaefer, J.A. Landscapes as gradients: The spatial structure of terrestrial ecosystem components in southern Ontario, Canada. *Ecol. Complex.* **2007**, *4*, 34–41. [CrossRef]

50. Warren, P.S.; Ryan, R.L.; Lerman, S.B.; Tooke, K.A. Social and institutional factors associated with land use and forest conservation along two urban gradients in Massachusetts. *Landsc. Urban Plan.* **2011**, *102*, 82–92. [CrossRef]

51. Shkaruba, A.; Kireyeu, V.; Likhacheva, O. Rural–urban peripheries under socioeconomic transitions: Changing planning contexts, lasting legacies, and growing pressure. *Landsc. Urban Plan.* **2016**. [CrossRef]

52. Joo, W.; Gage, S.H.; Kasten, E.P. Analysis and interpretation of variability in soundscapes along an urban–rural gradient. *Landsc. Urban Plan.* **2011**, *103*, 259–276. [CrossRef]

53. Díaz-Varela, E.; Roces-Díaz, J.V.; Álvarez-Álvarez, P. Detection of landscape heterogeneity at multiple scales: Use of the Quadratic Entropy Index. *Landsc. Urban Plan.* **2016**, *153*, 149–159. [CrossRef]

54. McGarigal, K.; Marks, B.J. *Spatial Pattern Analysis Program for Quantifying Landscape Structure*; Gen. Tech. Rep. PNW-GTR-351; US Department of Agriculture, Forest Service, Pacific Northwest Research Station: Corvallis, OR, USA, 1995.

55. Zhang, Z.; Tu, Y.; Li, X. Quantifying the spatiotemporal patterns of urbanization along urban-rural gradient with a roadscape transect approach: A case study in Shanghai, China. *Sustainability* **2016**, *8*, 862. [CrossRef]

56. Liu, Z.; He, C.; Wu, J. The relationship between habitat loss and fragmentation during urbanization: An empirical evaluation from 16 World Cities. *PLoS ONE* **2016**, *11*, e0154613. [CrossRef] [PubMed]

57. Wrbka, T.; Erb, K.-H.; Schulz, N.B.; Peterseil, J.; Hahn, C.; Haberl, H. Linking pattern and process in cultural landscapes. An empirical study based on spatially explicit indicators. *Land Use Policy* **2004**, *21*, 289–306. [CrossRef]

58. Millington, A.C.; Velez-Liendo, X.M.; Bradley, A.V. Scale dependence in multitemporal mapping of forest fragmentation in Bolivia: Implications for explaining temporal trends in landscape ecology and applications to biodiversity conservation. *ISPRS J. Photogramm. Remote Sens.* **2003**, *57*, 289–299. [CrossRef]

59. South Australian Government (Ed.) *Character Preservation (McLaren Vale) Act 2012*; Government of South Australia: Adelaide, SA, Australia, 2013.

60. Ornetsmüller, C.; Verburg, P.H.; Heinimann, A. Scenarios of land system change in the Lao PDR: Transitions in response to alternative demands on goods and services provided by the land. *Appl. Geogr.* **2016**, *75*, 1–11. [CrossRef]

61. Walcott, J.J.; Zuo, H.; Loch, A.D.; Smart, R.V. Patterns and trends in Australian agriculture: A consistent set of agricultural statistics at small areas for analysing regional changes. *J. Land Use Sci.* **2013**, *9*, 453–473. [CrossRef]

land

MDPI

Article

Understanding Land Use and Land Cover Dynamics from 1976 to 2014 in Yellow River Delta

Baolei Zhang [1], Qiaoyun Zhang [1], Chaoyang Feng [2], Qingyu Feng [3] and Shumin Zhang [4],*

[1] School of Geography and Environment, Shandong Normal University, 88 East Wenhua Rd, Jinan 250014, China; blzhangsd@gmail.com (B.Z.); zqylx127@163.com (Q.Z.)
[2] Chinese Research Academy of Environmental Sciences, Beijing 100012, China; fengchy@craes.org.cn
[3] Agricultural and Biological Engineering, Purdue University, 225 S. University St, West Lafayette, IN 47907-2093, USA; feng37@purdue.edu
[4] Institute of Regional Economic Research, Shandong University of Finance and Economics, 7366 East Erhuan Rd, Jinan 250014, China
* Correspondence: zhangsmsd13@gmail.com

Academic Editors: Andrew Millington, Harini Nagendra and Monika Kopecka
Received: 10 February 2017; Accepted: 7 March 2017; Published: 13 March 2017

Abstract: Long-term intensive land use/cover changes (LUCCs) of the Yellow River Delta (YRD) have been happening since the 1960s. The land use patterns of the LUCCs are crucial for bio-diversity conservation and/or sustainable development. This study quantified patterns of the LUCCs, explored the systematic transitions, and identified wetland change trajectory for the period 1976–2014 in the YRD. Landsat imageries of 1976, 1984, 1995, 2006, and 2014 were used to derive nine land use classes. Post classification change detection analysis based on enhanced transition matrix was applied to identify land use dynamics and trajectory of wetland change. The five cartographic outputs for changes in land use underlined major decreases in natural wetland areas and increases in artificial wetland and non-wetland, especially aquafarms, salt pans and construction lands. The systematic transitions in the YRD were wetland degradation, wetland artificialization, and urbanization. Wetland change trajectory results demonstrated that the main wetland changes were wetland degradation and wetland artificialization. Coastline change is the subordinate reason for natural wetland degradation in comparison with human activities. The results of this study allowed for an improvement in the understanding of the LUCC processes and enabled researchers and planners to focus on the most important signals of systematic landscape transitions while also allowing for a better understanding of the proximate causes of changes.

Keywords: land use dynamic; systematic transition; wetland change trajectory; imagery analysis; enhanced transition matrix; Yellow River Delta

1. Introduction

Land use/cover change (LUCC) is considered to be one of the most important components and driving factors of global environmental change [1–4], and it is one of the most important indicators in understanding the interactions between human activities and the environment [5,6]. Understanding the reorganization of land in order to adapt its use and spatial structure to social demands has become crucial to management and represents a major challenge to land use planning and public policies [7–9].

LUCC is defined as the transformation of the physical or biotic nature of a site, whereas land use change involves a modification in the way in which land is being used by humans [10]. These transitions can be random or systematic [11,12], with random transitions representing those characterized by abrupt changes or episodic processes of change and systematic transitions those marked by consistency and stable processes [13]. Land use transitions can be detected by statistical evaluation by comparing different temporal pattern maps. A common method employs the use of a land use/cover transition matrix, which provides a cross-tabulation matrix including change quantities and directions, and allows identification of differences between random and systematic land use transitions [14–17]. However, matrix-based land use studies mainly focus on overall gains and losses, and tend to ignore the spatial locations and swap changes of land use transitions [10,18]. Furthermore, the maps or databases of these studies are sampled or classed at discrete intervals, and the analysis tends to focus on the adjacent periods, but ignore successive process of land use transitions [19,20]. Therefore, establishing a better understanding of the fundamental processes of land use transitions requires the detection of dominant systematic land cover transitions and an illustration of the trajectory of the interest objects (land use types).

The Yellow River Delta (YRD)—located in the estuary of the Yellow River, with resource–rich territory of coastal wetlands—is the only habitat for several species of rare migratory birds and preserves natural vegetation near several big cities [21]. As the key economic development area of Shandong province and one of the most important regions of petroleum production in China, the YRD has been subject to increasing human disturbance (e.g., petroleum exploitation and production, agricultural development, and urbanization) since the early 1960s [22,23]. Moreover, the runoff and sediment discharge from the Yellow River has decreased considerably since the 1950s, resulting in frequent and prolonged channel drying in the downstream area since the 1970s [24–26]. These two stressors led to dramatic land use changes, so the detection of LUCCs and the identification of the trajectory of wetland change are fundamental for bio-diversity conservation and/or sustainable development of the YRD. In recent years, LUCCs in the YRD have received considerable attention in China, and researchers at home and abroad have conducted numerous studies surrounding the aspects of land use change [27–29], landscape dynamics [30], wetland evolution [31–33], and impacts of anthropogenic activities [34] based on qualitative, quantitative, and modeling methods. However, these studies mainly focused on the concentration of land use status before 2009 and covered much less the land use situation after 2010. More importantly, these studies paid more attention to the area changes and driving forces, but ignored the systematic transitions of LUCCs and the trajectory of wetland change.

This study aimed at the detection of LUCCs and the identification of the trajectory of wetland changes due to their importance for bio-diversity conservation and/or sustainable development of the YRD. Therefore, three specific objectives of this article were to (1) analyze spatial and temporal dynamics of land use patterns from 1976 to 2014; (2) explore the systematic transitions of land use of the YRD; and (3) illustrate the trajectory of wetland change and the driving factors. The following sections of the paper are organized in the following ways: the study area and methods section provides details on case study area, data sources, and methods to quantify land use change, and trajectories of wetland change; the results section presents the accuracy of our analysis, land use pattern detected, and the wetland changes; the discussion section provides insights regarding the comparison of our study and existing studies, the implication of our results, and ultimately this is followed by the conclusion of this study.

2. Study Area and Methods

2.1. Study Area

The YRD is the newly-formed fan-shaped delta of the Yellow River estuary area after the Yellow River was diverted into the Bohai Sea in 1855. The delta (located in 118°33′ E to 119°18′ E and from 37°26′ N to 38°09′ N) takes Ninghai as the vertex, starts from the Taoer Estuary in the north, and reaches the ZhimaiGou river in the south and Tuhai River in the west (Figure 1). It has a warm temperate continental monsoon climate with distinctive seasonality. The annual temperatures and precipitation is 11.7–12.6 °C and 530–630 mm, respectively [35]. However, average annual evaporation is almost 3.5 times the average yearly precipitation. Approximately 10.5 million tons of sand and soil discharged by the river is deposited in the delta annually, forming a vast floodplain and special wetland landscape [36]. The soil of the YRD is mainly composed of fine sand and is characteristically young, with a high groundwater table, low fertility, and a tendency towards secondary salinization and desertification [37]. The average groundwater table is generally 2–3 m, and only 0.5–1.5 m along the coastline. The natural vegetation is composed of broadleaf deciduous forest (mainly Hankow willow and weeping willow), shrubbery (mainly Chinese tamarisk), and shore coppice [38]. The YRD is one of six of the most beautiful wetlands in China and an important energy base with more than 5×10^9 t petroleum and 2.3×10^{11} m natural gas [23,39].

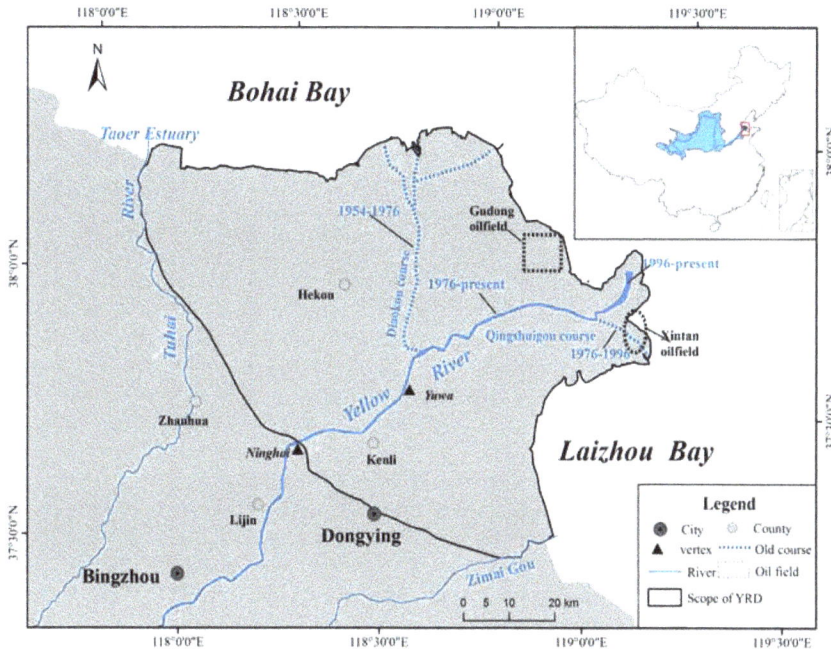

Figure 1. Location of the modern Yellow River Delta (YRD).

2.2. Data Preparation and Acquisition

2.2.1. Satellite Image Selection and Pre-Processing

Generally, the selection of Landsat images was mainly based on availability, cloud cover percentage, and correspondence [40]. However, the image features of land use types in the YRD are more likely to be affected by seasonal aspects and tidal conditions. The duration of seasonal tidal flows in YRD was always from January to April [41,42]. Therefore, the remotely sensed imageries selected in this paper (Table 1) were not only cloud–free but also during the appropriate period. All the imageries were acquired from Earth Observation and Digital Earth Science Center of Chinese Academy of Sciences. The satellite images were corrected in order to remove atmospheric effects by subtracting the radiance of a "dark pixel" within each band image [43], and then the images were geo-referenced using between 15 and 30 ground control points distributed across each image. After geo-reference, the images had a Gaussian–Krueger projection and a Root Mean Squared Error (RMSE) of less than one pixel.

Table 1. List of satellite images used in this study.

Platform	Sensor	Path/Row	Resolution(m)	AcquisitionDate
Landsat 2	Multispectral Scanner	130–134	90	2 June 1976
Landsat 5	Thematic Mapper	121–134	30	5 November 1984
Landsat 5	Thematic Mapper	121–134	30	18 September 1995
Landsat 5	Enhanced Thematic Mapper	121–134	15	2 November 2006
Landsat 8	Operational Land Imager	121–134	30	11 May 2014

2.2.2. Region Definition, Land Use Classification, and Accuracy Assessment

The YRD experiences erosion and deposition, so its scope and area is constantly changing. In this paper, the coastlines of each period were extracted to defined the YRD's scope, and an interactive interpretation technique combining an automatic boundary detection algorithm with human supervision was used to detect the land–ocean shoreline boundaries in satellite images. The coastline types of the YRD are muddy coastlines and artificial coastlines, and artificial coastlines were acquired by interpreting the construction edges manually. The muddy coastlines were established using an automatic boundary detection algorithm following the process of Tasseled Cap Transformation [44], binary converting, edge enhancement, and edge detection [26].

The detection of time intervals in land use changes required a pre-classification image analysis process (image to image comparison) of land use. Visual interpretation of land use types based on elements such as color, tone, texture, form, size, presence of shadows, and the location of infrastructures [45,46] has been the main approach for identifying land use changes because it can provide more accurate land use maps compared with automatic classification [47,48]. In this study, the land use data series were acquired by interpreting the basic land use map in 2010 and detecting the changing parts between adjacent time periods of Landsat images manually. The whole process is supported by six main stages: field investigation, establishment of land use classification, interpretation of basic map, change detection, field test and corrections, and accuracy assessment. Field investigation was conducted in May 2013 covering the entire study area to get a priori knowledge of the study area as a whole, including landform, soil, vegetation, ponds, rivers, salt fields, agriculture fields, and built-up areas. A classification system including three land use types and nine classes (Table 2) was established based on the national land classification system and the regional characteristics of the YRD while referring to wetland classification principles of previous studies [49–51]. In the process of interpretation, interpreters used ArcGIS software to identify land use types based on their understanding about the object's spectral reflectance, structure, and other ancillary information with the smallest patch of land use bigger than 25 pixels (2.25 ha) and the shortest edge longer than 3 pixels (90 m). A second round of field surveys/tests was conducted on August–September 2014 after finish detecting land use changes and from the land use maps from the five periods. Subsequently, the corrections were implemented on the land use maps from the five periods based on samples of the two field surveys, land use maps from local governments, and high resolution aerial photographs. Discrete multivariate analytical techniques were used to statistically evaluate the accuracy of the classified maps [52] and a variety of indices such as overall accuracy, producer's accuracy, user's accuracy, and kappa analysis were calculated [53].

Table 2. The classification key used in the present study.

Land Use Type	Land Use Class	Description	Code
Natural wetland	Beach	Mucky, sandy, and gravel beach located between the estuary and tidal zone	BC
	Grassland	Reeds, cattails (*Typha orientalis*), and other water-loving plant community members located in rivers and estuaries, reservoirs, and lakes of flood land	GL
	Bushland	Mainly Tamarix bush combined with the alkaline meadows such as *Suaeda heteroptera, Salicornia*, and *Suaeda sals*	BL
Artificial wetland	River	Permanent and seasonal rivers including their floodplains	RV
	Ditch and ponds	A natural or artificial pond or lake used for the storage and regulation of water, including lake, reservoirs, and ponds	DP
	Aquafarm and salt pan	Artificial built around shrimp, crabs and other aquatic ponds, etc.; Salt field in coastal areas and near estuaries	AS
Non-wetland	Woodland	Woodland composed of Populus, Salix, Black locust (*Robinia pseudoacacia*) and Salix (*Salix integra*)	WL
	Cultivated land	Arable land that is worked by plowing and sowing and raising crops	CL
	Construction land	Man-made impervious surface such as roads, urban, and rural residential land, industrial land, oil field infrastructure, etc.	AL

2.3. Quantification of LUCC Based on Transition Matrix

In order to quantify the land use/cover dynamics, post classifications (map to map comparisons) were generated involving the successive sets of images cross-referenced to define land use transition matrixes and a series of evaluation indexes. In the process of generating land use transition matrixes, the union scope of five periods with an area of 6398 km^2 was taken as the analysis scope and a new land use type of sea surface (SF) was added to represent the land that disappeared as a result of coastline erosion.

2.3.1. Land Use Transition Matrix

The land use transition matrix comes from system analysis aiming at quantitative description of the system state and state transition, and it is the most common approach used to compare maps of different sources, as it provides detailed "from-to" change class information [40]. The traditional area cross-tabulation matrix (transition matrix) was computed using overlay functions in ArcGIS 9.3 software. The computed transition matrix consists of rows that display categories at time T1 and columns that display categories at time T2 (Table 3). The notation A_{ij} is the area of the land that experiences transition from category i to category j. The diagonal elements (i.e., A_{ii}) indicate the area of the landscape that shows persistence of category i. Entries off the diagonal indicate a transition from category i to a different category j. The area of the landscape in category i in time T1 (A_{i+}) is the sum of A_{ij} over all j. Similarly, the area of the landscape in category j in time T2 (A_{+j}) is the sum of A_{ij} over all i. The losses ($A_{i+} - A_{ii}$) were calculated as the differences between row totals and persistence. The gains ($A_{+i} - A_{ii}$) were calculated as the differences between the column totals and persistence.

Table 3. A sample of land use transition matrix.

		T$_2$				A$_{i+}$	Loss
		L$_1$	L$_2$	$\ldots$	L$_n$		
T$_1$	L$_1$	A$_{11}$	A$_{12}$	$\ldots$	A$_{1n}$	A$_{1+}$	A$_{1+}$ − A$_{11}$
	L$_2$	A$_{21}$	A$_{22}$	$\ldots$	A$_{2n}$	A$_{2+}$	A$_{2+}$ − A$_{22}$
	.	.	.	.	.	.	.
	L$_n$	A$_{n1}$	A$_{n2}$	$\ldots$	A$_{nn}$	A$_{n+}$	A$_{n+}$ − A$_{nn}$
	A$_{+i}$	A$_{+1}$	A$_{+2}$	$\ldots$	A$_{+n}$		
	Gain	A$_{+1}$ − A$_{11}$	A$_{+2}$ − A$_{22}$	$\ldots$	A$_{+n}$ − A$_{nn}$		

2.3.2. Annual Rate of Change

The annual rate of change (RC_i) for each land cover category i was calculated as [5,54]:

$$RC_i = ((A_{+i}/A_{i+})^{\frac{1}{T_2-T_1}} - 1) \times 100\% \tag{1}$$

where A_{i+} and A_{+i} are the areas (in ha) of a cover class at years T_1 (initial time) and T_2 (next time step), respectively.

2.3.3. Stability Grade

To expresses the proportion of the landscape category i that had not experienced a transition to any different category of land use, the indicator stability grade (SG_i) was defined as Equation (2) and the total stability grade of the region (SG) was defined as Equation (3) [11]:

$$SG_i = \frac{2 \times A_{ii}}{A_{+i} + A_{i+}} \times 100\% \tag{2}$$

$$SG = \sum_{i=1}^{n} A_{ii} / \sum_{i=1}^{n} A_{+i} \times 100\% \tag{3}$$

2.3.4. Swap Change (SW) Percentage

Swap was a component of change which implied that a given area of a category was lost at one location, while the same area was gained at a different location. The amount of swap was calculated as two times the minimum of the gain and loss [53]. The total change for each land class was calculated as either the sum of the net change and the swap or the sum of the gains and losses. The percentage of swap change (R_{sw}) was calculated as follow [11,40]:

$$R_{sw} = \frac{2 \times Min(A_{+j} - A_{jj}, A_{j+} - A_{jj})}{A_{+j} + A_{j+} - 2A_{jj}} \times 100 \tag{4}$$

2.3.5. Selection of Main Transition

Main transitions were identified as dominant conversions with bigger proportions of the total change. The proportion of the land (P_{ij}) that experiences transition from category i to category j was calculated, and the transitions with the proportion values larger than the average values were selected as the main transition. The proportion of the transition and the comparison with the average proportion were defined as follows in Equation (5):

$$\begin{cases} P_{ij} = A_{ij} / \sum_{i=1}^{n} A_{ij} \times 100\% \ i \neq j \\ P_{ij} > \frac{100\%}{n \times (n-1)} \end{cases} \tag{5}$$

in which n represents the number of land types.

2.4. Trajectories of Wetland Change

Swetnam [55] presented a method to explore land use change characteristics or trajectory using the combinations of the three spatial indices (similarity, turnover, and diversity) to classify the land use change into six groups: stepped, cyclical, dynamic, no constant trend (NCT), and (stable) [56]. In this study, trajectory analysis was made for natural wetland, artificial wetland, and non-wetland classes because of their ecological importance. Additionally, the original six groups were clustered and reclassified in the aspect of wetland landscape change. With three land cover classes (natural wetland, artificial wetland, and non-wetland classes) and five temporal image dates (1976, 1984, 1995, 2006, and 2014), 61 out of 243 possible wetland land cover change trajectories were found. Finally, similar trajectories were clustered, resulting in six classes (Figure 2 and Table 4).

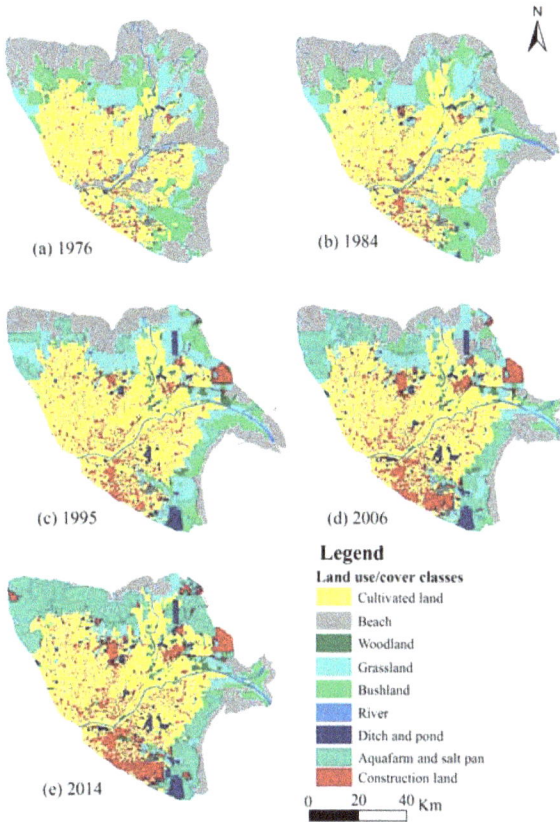

Figure 2. Land use maps based on image classifications by year: (**a**) 1976; (**b**) 1984; (**c**) 1995; (**d**) 2006; (**e**) 2014.

Table 4. Wetland change trajectories between 1976 and 2014.

No.	Description	Trajectories *
1	Stable wetland	WWWWW, RRRRR
2	wetland formation/restoration	NNNNR, NNNNW, NNNRR, NNNWW, NNRRR, NNWWW, NRRRR, NWWWW, NWNNW, NWNWW, NWWNW, NWWWW, RRNWW, WNNWW, WNWNW, WNWWW, WWNNW, WWNRR, WWRNR
3	Wetland artificialization	NNNWR, NNWRR, NWNWR, NWRRR, NWWRR, NWWWR, WNNWR, WNRRR, WRRRR, WWNWR, WWRRR, WWWNR, WWWRR, WWWWR
4	Old degradation	NRNNN, NWNNN, RRNNN, WNNNN, WRNNN, WWNNN
5	Recent degradation	NNWNN, NNWRN, NNWWN, NRRNN, NRWNN, NRWWN, NWNWN, NWWNN, NWWWN, RRRNN, RRWWN, WNNWN, WNWNN, WNWWN, WWNWN, WWRNN, WWRRN, WWWNN, WWWRN, WWWWN
6	Non-wetland	NNNNN

* The sequence represents the time periods 1976, 1984, 1995, 2006, and 2014. "W" stands for natural wetland class, "R" stands for artificial wetland, and "N" stands for non-wetland class.

3. Results

3.1. Accuracy Assessment

Figure 2 depicts the classified maps for 1976, 1984, 1995, 2006, and 2014. According to the confusion matrix report (Table 5), 90.62% overall accuracy and a Kappa Coefficient (*KC*) value of 0.89 were attained for the 2014 classified map. Similarly, overall classification accuracy levels achieved were 91.63% (with a *KC* of 0.90) for the 2006, 94.74% (with *KC* of 0.94) for the 1995, 91.43% (with *KC* of 0.89) for the 1984, and 93.16% (with *KC* of 0.91) for the 1976 image classifications. In general, the maps met the minimum accuracy requirements to be used for the subsequent post-classification operations.

Table 5. Confusion matrix (error matrix) for the 2014 classification map.

Classified Data	BC	GL	BL	RV	DP	AS	CL	WL	AL	Row Total	User's Accuracy
BC	40	2	2						2	46	87%
GL		28	2						2	32	88%
BL		2	27				1			30	90%
RV				20	2					22	91%
DP				1	18					19	95%
AS	1	2	1			56				60	93%
CL	1	3	2			1	87			94	93%
WL			2				1	23		26	88%
AL	2				2		1		39	44	89%
Column total	44	37	36	21	22	57	90	23	43	373	
Producer's accuracy	87%	88%	90%	91%	95%	93%	93%	88%	89%	91%	

Overall accuracy = 90.62%, *KC* =0.89.

3.2. Temporal Patterns for Changes in Land Use

Figure 2 presents the land use classification for the five years/moments of image analysis (1976, 1984, 1995, 2006, and 2014). The cartographic outputs show a large change in different sectors of the YRD, with diverse trajectories. The analysis indicates some systematic transitions involving great changes of coastline shapes, as well as an increase in artificial wetland and construction land with urbanization and wetland artificialization characteristics. A more detailed observation emphasizes that the landscape change is relevant, involving a significant decrease in the natural wetland area, in particular the beach, grassland, and bushland, an increase in construction areas, and a large transformation from natural wetland to artificial wetland.

Table 6 shows the evolution of land use and occupation in the YRD, as represented by the landscape patterns for the period analyzed. In analyzing areas of land use, certain systematic transitions were observed, namely: an increase in artificial wetland with ditch, pond, aquafarm, and salt pan units; an increase in artificial land involving artificial wetland and non-wetland; a decrease of natural wetland, in particular the areas and percentages of beach, grassland, and bushland. For natural wetland, the area and percentage of beach, grassland, and bushland kept decreasing from 1976 to 2014, while the river increased slightly. The total area of natural wetland decreased from 3488.2 km^2 in 1976 to 1120.9 km^2 in 2014, with the annual decreasing rate of 62.3 km^2/year. In contrast, both ditch and ponds (DP) and aquafarm and salt plains (AS) of artificial wetland kept increasing and showed a relative growth of 15 times of the original state. The areas of cultivated land and woodland both increased from 1976 to 2014 in spite of experiencing a decrease in the period 1995–2006. The area of construction land kept increasing from 1976 to 2014, and showed a relative growth of 274% from the original state.

Classes	1976		1984		1995		2006		2014	
	A (km²)	P (%)	A (km²)	P (%)	A (km²)	P (%)	A (km²)	P (%)	A (km²)	P (%)
Beach	1991.5	34.2	1402.6	23.5	852.9	14.2	813.6	13.7	480.6	8.1
Grassland	639.9	11.0	724.2	12.2	903.7	15.1	612.1	10.3	342.2	5.8
Bushland	747.4	12.8	891.2	15.0	362.3	6.0	238.0	4.0	169.6	2.9
River	109.4	1.9	136.8	2.3	129.3	2.2	129.4	2.2	128.5	2.2
Natural wetland	3488.2	59.9	3154.9	53.0	2248.2	37.5	1792.9	30.2	1120.9	18.9
Ditch and pond	75.6	1.3	112.6	1.9	194.1	3.2	238.8	4.0	250.1	4.2
Aquafarm and salt pan	17.7	0.3	37.7	0.6	364.3	6.1	717.8	12.1	1261.3	21.3
Artificial wetland	93.3	1.6	150.3	2.5	558.3	9.3	956.6	16.1	1511.4	25.5
Cultivated land	2030.2	34.9	2351.4	39.5	2636.1	44.0	2481.6	41.8	2499.7	42.1
Woodland	14.8	0.3	36.9	0.6	85.0	1.4	77.5	1.3	81.8	1.4
Construction land	192.8	3.3	264.8	4.4	467.2	7.8	628.9	10.6	722.6	12.2
Non-wetland	2237.9	38.5	2653.1	44.5	3188.2	53.2	3188.0	53.7	3304.2	55.7
YRD Area	5819.3	100.0	5958.2	100.0	5994.7	100.0	5937.5	100.0	5936.5	100.0

The analysis of land use areas also shows an abrupt or limited temporal transitional process where cultivated land and woodland decreased in the period 1995–2006. This is likely because of a lack of suitable lands for the development of agriculture and forest planting. Furthermore, the acceleration of construction expansion occupied more cultivated land and woodland in this period.

Figure 3 shows the annual changing rate of each category, revealing a diverse changing process for each land use category in different periods, thereby showing the systematic transitions in more detail: natural wetland increased in the former stages (1976–1984 and 1984–1995) and decreased in the recent stages (1995–2006, 2006–2015), while non-wetland and artificial wetland kept increasing from 1976 to 2014. Only the periods 1995–2006 and 1976–1984 have different trajectories, marked by a decrease in the percentage of woodland and cultivated land in the period 1995–2006, and a decrease in the percentage of beach in the period 1976–1984.

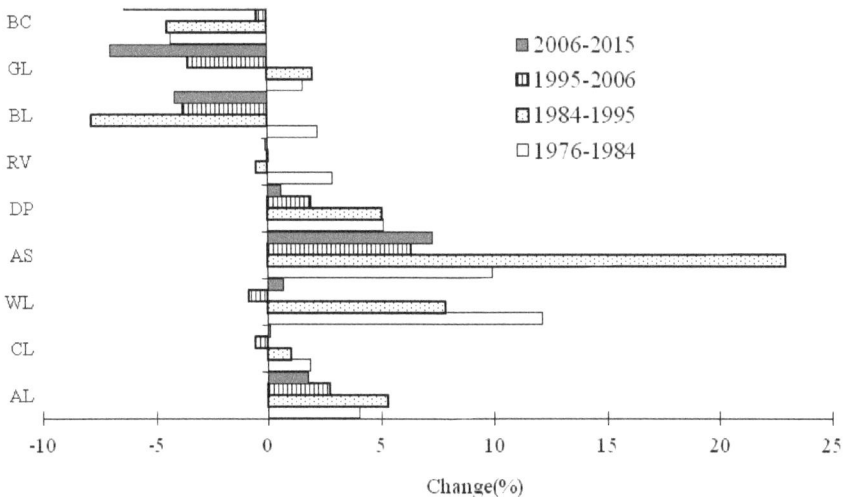

Figure 3. Annual changing rate of each category.

Although the increase of non-wetland and artificial wetland (on the left-hand side) in contrast to natural wetland areas (on the right-hand side) was clear, the annual changing rates appeared different in different periods from 1976 to 2014. During the period 1976–1984, woodland, cultivated land, and ditch and ponds had the biggest annual increasing rate, while the river, bushland, and grassland classifications all had a positive annual increasing rate. During the 1984–1995 period, construction land and aquafarm and salt pans had the biggest annual increasing rates, while bushland had the biggest annual decreasing rate, and grassland had a positive annual increasing rate. During the period 1995–2006, none of the categories had the biggest annual increasing or increasing rate, but the woodland and cultivated land annual changing rates were negative, showing a an opposite changing trend. During 2006–2014, beach land and grassland had the biggest annual increasing rate.

3.3. Dynamic Analysis of Changes in Land Use

3.3.1. Stability Grade

The transformation matrices for 1976–1984, 1984–1995, 1995–2006, and 2006–2014 subsequently made possible a detailed study of the dynamics of land use and occupation in five periods of analysis. The stability grade (SG) of the land cover was calculated for each transformation matrix to show the percentage of landscape that remained unchanged. About 73.2%, 62.3%, 81.2%, and 82.8% of the landscape persisted or 28.0%, 39.3%, 18.8%, and 17.2% of the landscape has changed during the period 1976–1984, 1984–1995, 1995–2006, and 2006–2014, respectively, indicating that persistence dominates in all periods. However, the stability during the period 1976–2014 was only 38.0%, and 62.1% of the YRD experienced transition from one category to a different category. The stability grade values for each land use type were calculated and are shown in Figure 4. There was a relative small stability grade in the period 1984–1995 for all the land use types in comparison with the other three periods. The stability grade values were bigger than 50% except those of grassland, bushland, woodland, and aquafarm and salt pans in the period 1984–1995, which also indicated that persistence dominates in all periods. Nevertheless, only cultivated land and river had stability grade values bigger than 60% in the whole period of 1976–2014. Therefore, the cumulative process of LUCC has resulted in the YRD having undergone significant land use/cover alterations over the 38 years considered. The analysis also showed that the land use class transfer does not take place all at once, but in a set of small sequential steps.

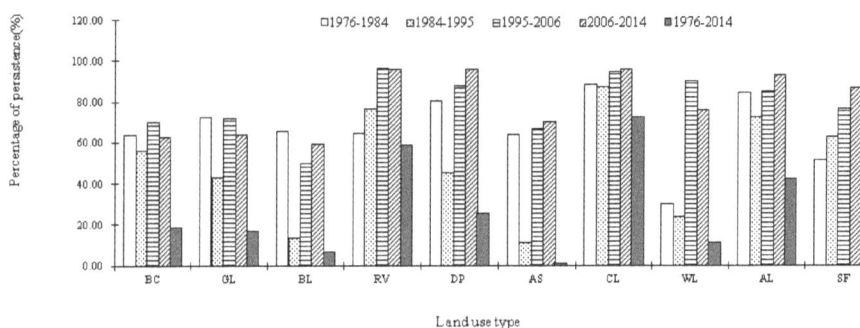

Figure 4. Stability grade values for each land use type in different periods.

3.3.2. Swap Change

Percentages of swap change for each land use type in different periods and the whole YRD were calculated and are shown in Figures 4 and 5. Swap change accounted for 57.6%, 53.4%, 46.0%, and 34.8% of total change during the periods 1976–1984, 1984–1995, 1995–2006, and 2006–2014, respectively, showing a LUCC evolution process from the change attributable to location to the change attributable to quantity. Swap change is greater than net change in the period before 1995, suggesting that the importance of the swapping component and common methods of land use/cover change study would miss these dynamics. During the period 1976–2014, changes in construction land experienced net change dynamics, whereas the changes in all other categories consisted of both swap and net changes. The river and sea surface both had relatively larger percentages of swap change, reflecting the transforming effects on the trail channels by the Yellow River Mouth and the coastline change caused by the accumulation of sediment and erosion. The type of change that each land use/cover experienced differs from period to period, but landscape types of natural wetland tend to have bigger swap changes than artificial land types such as aquafarms, salt pans, and construction lands.

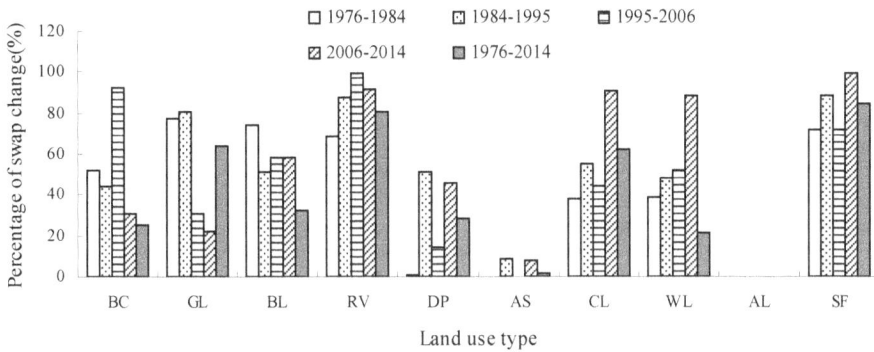

Figure 5. Percentages of swap change for each land use type in different periods.

3.3.3. Main Transitions

The landscapes of wetland, cultivated land, and construction land are closely related to ecology protection and human's survival and production depends. Based on the increase or decrease of these three landscapes, the main transitions were reclassified into seven categories, namely cultivated land to wetland, construction land expansion, wetland to cultivated land, internal transformation of natural wetland, natural wetland formation, artificialization of natural wetland, and coastline erosion.

Figure 6 shows the distributions of the main transition categories in different period from 1976 to 2014, and Table 7 shows the percentages and detailed compositions of theses main transitions. The transitions between sea surface and beach land were included in the main transitions in the four periods, showing that the YRD has been experiencing the process of erosion and sediment accumulation. The main transition categories of land use dynamics from 1976 to 2014 were the artificialization of natural wetland, transition from wetland to cultivated land, and construction land expansion, with the percentages of 32.6%, 20.3%, and 12.9%, respectively. During the periods 1976–1984 and 1984–1995, the main transition categories were characterized by internal transformation of natural wetland, transition from wetland to cultivated land, and natural wetland's formation and erosion. However, the main transition categories are predominantly natural wetland artificialization and construction expansion in the periods 1995–2006 and 2006–2014. The change of main transition category also revealed a continuous increase in artificial areas, indicating that land use trajectories were veering towards artificial surfaces.

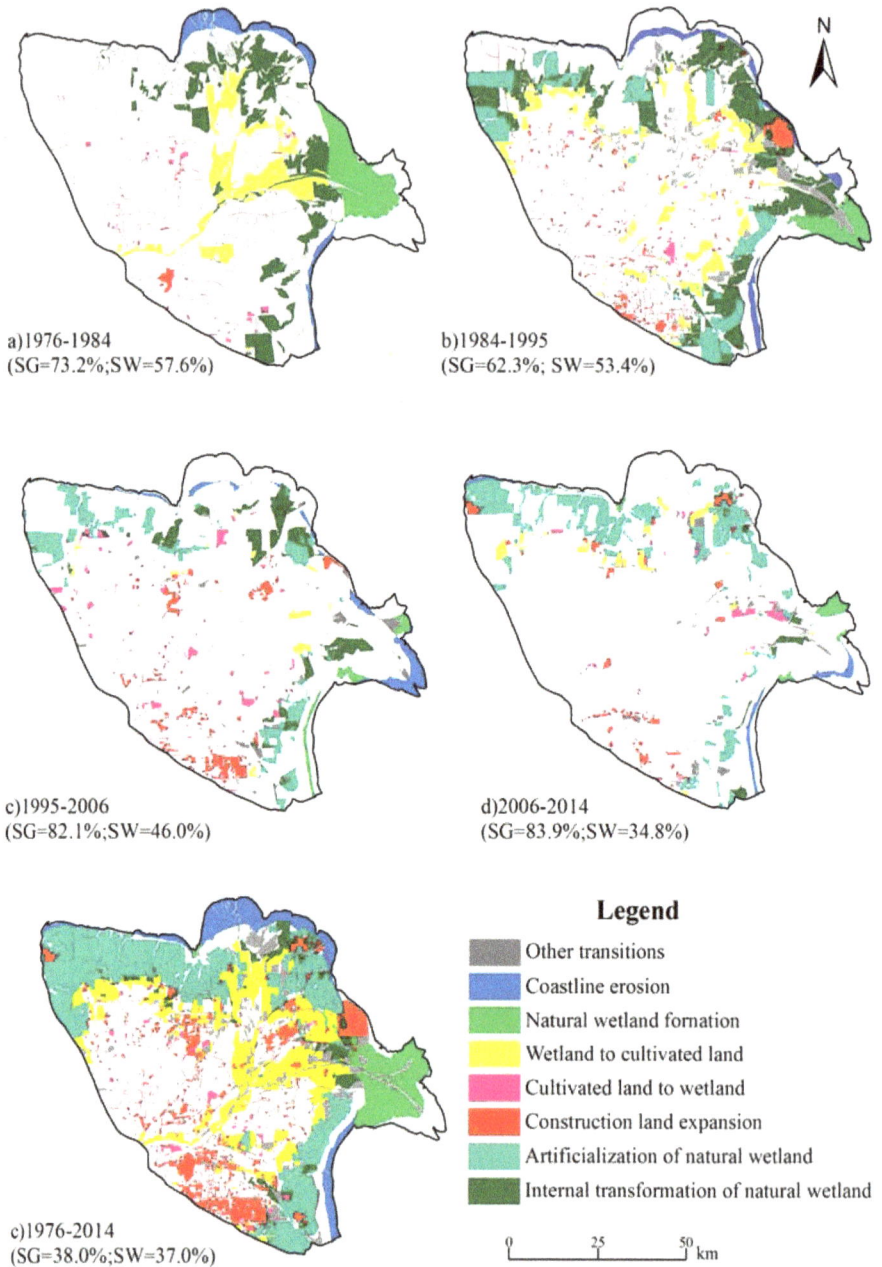

a)1976-1984
(SG=73.2%;SW=57.6%)

b)1984-1995
(SG=62.3%; SW=53.4%)

c)1995-2006
(SG=82.1%;SW=46.0%)

d)2006-2014
(SG=83.9%;SW=34.8%)

c)1976-2014
(SG=38.0%;SW=37.0%)

Legend

Other transitions
Coastline erosion
Natural wetland fornation
Wetland to cultivated land
Cultivated land to wetland
Construction land expansion
Artificialization of natural wetland
Internal transformation of natural wetland

0   25   50 km

Figure 6. Distributions of main transition categories in different periods from 1976 to 2014.

Table 7. The percentages (P/%) and detail compositions of main transitions in different period from 1976 to 2014.

Main TransitionType	1976–1984		1984–1995		1995–2006		2006–2014		1976–2014	
	P	Details	P	Details	P	Details	P	Details	P	Details
Cultivated land to wetland	1.4	CL-DP (1.4)	1.2	CL-DP (1.2)	6.5	CL-DP (3.2) CL-GL (1.8) CL-AS (1.5)	2.6	CL-GL (1.5) CL-AS (1.1)	1.4	CL-DP (1.4)
Construction land expansion	3.0	CL-AL (3.0)	6.2	CL-AL (3.9) BC-AL (2.3)	13.1	CL-AL (11.6) BC-AL (1.5)	8.1	CL-AL (3.1) GL-AL (2.6) BC-AL (2.4)	12.9	CL-AL (6.9) BC-AL (2.1) GL-AL (1.5) SF-AL (1.3) BL-AL (1.2)
Wetland to cultivated land	23.3	BC-CL (17.4) GL-CL (3.3) BL-CL (2.6) BC-BL (14.1)	17.9	GL-CL (8.6) BL-CL (8.0) DP-CL (1.3)	4.1	GL-CL (4.1)	9.4	GL-CL (7.9) AS-CL (1.5)	20.3	BC-CL (8.4) GL-CL (6.7) BL-CL (5.2)
Internal transformation of natural wetland	35.7	BC-GL (7.0) BL-GL (6.0) GL-BL (4.0) BL-GL (3.1) BC-RV (1.5)	33.0	BL-GL (13.6) BC-BL (10.5) BC-GL (7.1) BL-BC (1.8)	24.0	BL-BC (7.3) GL-BC (6.7) BC-BL (3.6) BL-GL (3.5) GL-BL (2.9)	5.1	BL-BC (2.0) BC-BL (1.8) BC-GL (1.3)	4.7	BC-GL (2.9) BL-GL (1.8)
Natural wetland formation	18.1	SF-BC (14.4) SF-BL (2.4) SF-GL (1.3)	6.2	SF-BC (6.2)	5.3	SF-BC (5.3)	5.0	SF-BC (3.9) SF-BL (1.1)	8.5	SF-BC (5.0) SF-BL (2.3) SF-GL (1.2)
Artificialization of natural wetland			17.8	BL-AS (5.8) BC-AS (5.5) GL-AS (2.8) BL-DP (2.5) GL-DP (1.2)	29.4	GL-AS (15.5) BC-AS (7.9) BL-AS (6.0)	52.9	BC-AS (27.6) GL-AS (16.1) BL-AS (9.2)	32.6	BC-AS (19.1) BL-AS (8.0) GL-AS (3.5) BC-DP (2.0)
Coastline erosion	10.1	BC-SF (10.1)	5.6	BC-SF (5.6)	11.4	BC-SF (11.4)	5.9	BC-SF (5.9)	8.0	BC-SF (8.0)
Total	91.6		87.9		93.8		89.0		88.4	

Although the main transition categories were characterized by artificialization, the composition details of main transition categories appeared different in different periods from 1976 to 2014. During 1976–2014, the form of natural wetland artificialization was mainly aquafarm and salt pan. The form of the transition from cultivated land to wetland was ditches and ponds in the period 1976–1984 and 1984–1995, The transitions of cultivated land to wetland were mainly in the form of the building of ditches and ponds in the periods 1976–1984 and 1984–1995, and there were small percentages of transitions from cultivated land to grassland and aquafarm land in the periods 1995–2006 and 2006–2014 which were caused by the policy of returning farmland to wetland and a new agriculture-fishery mode. The main source of construction land expansion was cultivated land in the four periods, but beach land and grassland also became part of the main sources in the last three periods. Both the composition types and percentages of natural wetland internal transformation decreased from early periods to late periods. The composition types only included the transition from beach to grassland and the transition from bushland to grassland. These results revealed that natural wetland internal transformation was the medium process of transition from the natural wetland to artificial land. The transitions from the beach, grassland, and bushland to cultivated land were the main transitions in the four periods, showing that the YRD had continuous agricultural development process from 1976 to 2014. The main land use types for the transitions from natural wetland to artificial wetland were aquafarm and salt pan which occurred in the last three periods. The main occupation land of construction land was cultivated land in the former periods (1976–1984 and 1984–1995), while grassland became the other main occupation source in the last period (2006–2014). The irrigation and water conservancy construction in the period 1995–2006 was one of the major types of transfer.

3.4. Trajectory of Wetland Change

The maps of six wetland trajectory types were shown in Figure 5, and areas and percentages of land use type in 1976 and 2014 for each wetland trajectory types were shown in Table 6. The results show that the classes of wetland changed greatly, even for the area percentages of wetland classes of stable wetland. The area of stable wetland was 746.7 km^2, accounting for 28.4% of the total wetland area in 2014. Wetland artificialization was mainly distributed beside stable wetland had an area of 1361.4km^2 and the main destination class was aquafarm and salt pan, accounting for 89.2%. Old degradation mainly distributed along the Yellow River and its old course was in the north (Diaokou Ditch). Recent degradations displayed scattered distribution except for a concentrated distribution outside the mouth of the Yellow River in the east. The area percentages for the old degraded and recently degraded wetlands were caused by coastline change (percentage to sea surface in 2014) which were 23.3% and 42.5%, respectively, which indicated that human activities other than coastline change were the main driving forces of wetland degradation. Wetland formation/restoration mainly came from the result of estuarine deposits in the mouth of the Yellow River (70.6%), followed by reservoir construction in the cultivated lands (13.2%).

During 1976–2014, the wetland change in the YRD included wetland formation/restoration, wetland degradation, and wetland artificialization, but the main wetland changes were wetland degradation and wetland artificialization (Figure 7 and Table 8). The areas of wetland degradation (1645.1 km^2) and wetland artificialization (1361.4 km^2) were much greater than that of the wetland formation/ restoration (524.1 km^2).

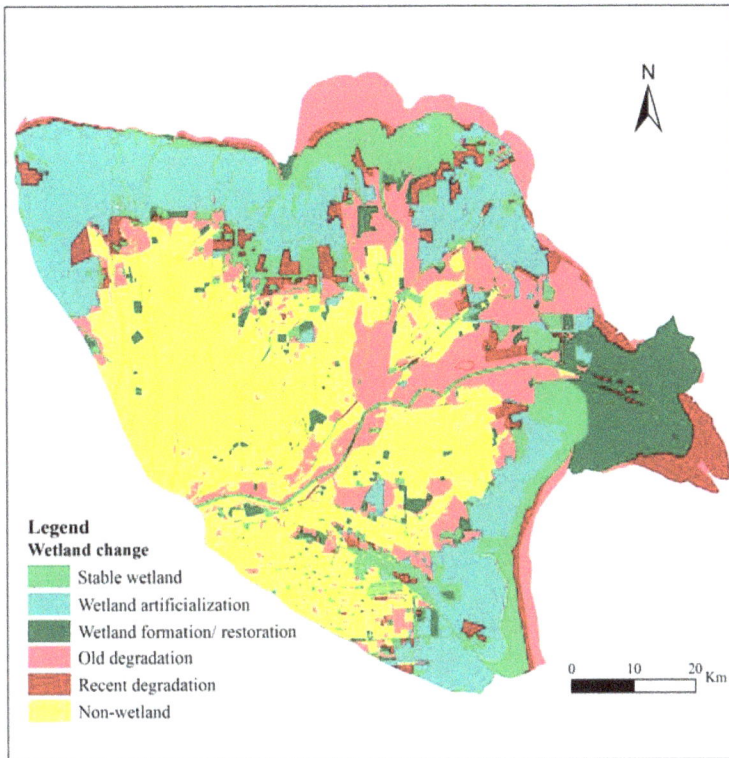

Figure 7. Wetland change trajectory map.

Table 8. Areas and percentages of land use type in 1976 and 2014 for each wetland change type.

Change Type	Time	Area (km²)	Percentage (%)									
			BC	GL	BL	RV	DP	AS	CL	WL	AL	SF
Non-wetland	1976	2120.5							90.2	0.6	9.1	0.1
	2014								77.4	1.0	21.7	0.0
Stable wetland	1976	746.7	49.5	16.7	17.1	9.8	5.7	1.2				
	2014		37.5	33.8	9.2	12.5	5.9	1.0				
Old degradation	1976	1234.4	51.9	20.0	16.7	2.2	2.5	0.7	0.7			5.4
	2014			0.0					59.1	3.6	14.1	23.2
Recent degradation	1976	410.6	32.3	21.8	13.0	0.5	0.6	0.1	1.1			30.5
	2014								31.7	4.1	21.7	42.5
Wetland formation/restoration	1976	524.1	2.4	2.5	4.7				19.5	0.3		70.6
	2014		38.3	17.1	19.2	6.7	13.2	5.5				
Wetland artificialization	1976	1361.4	61.4	12.2	24.7	0.5			0.2			1.0
	2014						10.0	90.0				

4. Discussion

The five cartographic outputs for changes in land use underline the major decrease in natural wetland areas, the increase in artificial wetland and non-wetland, especially in regards to the aquafarms, salt pans, and construction lands. The land use dynamics from 1976 to 2014 are similar with the observations before 2009 made by Zong [27], Zhang et al. [30], Sun et al. [28], Chen et al. [57], and Wang et al. [58]. About 28.0%, 39.3%, 18.8%, and 17.2% of the landscape in the YRD had experienced transition from one category to another category of land use/cover during the periods 1976–1984, 1984–1995, 1995–2006, and 2006–2014, respectively, indicating that persistence dominates in each period. However, the cumulative process of LUCCs had made the YRD undergo significant land use/cover alterations, and about 62.1% of the YRD experienced transition from one category to a different category of land use/cover over the 38 years considered. Therefore, the analysis also showed that the land use class transfer does not take place all at once, but in a set of small sequential steps.

Although an increase in all artificial land types (artificial wetland and non-wetland) was observed from 1976 to 2014, the aquafarm and salt pan land classification had the highest annual increasing area and rate, followed by construction land. The ditches and ponds, woodland, and cultivated land classifications mainly increased in the former two periods (1976–1984 and 1984 to 1995), and woodland and cultivated land decreased in the period 1995–2006, which was mainly caused by the occupation of construction land.

Swap change accounted for 57.6%, 53.4%, 46.0%, and 34.8% of total change during the period 1976–1984, 1984–1995, 1995–2006, and 2006–2014, respectively, showing a LUCC evolution process from the change attributable to location to that attributable to quantity. Swap change is greater than net change in the period before the 1995, demonstrating the importance of the swapping component and suggesting that common methods of land use/cover change study would miss these dynamics. The type of change that each land use/cover experienced differs from period to period, but landscape types of natural wetland tend to have bigger swap changes than artificial land types such as aquafarm, salt pan, and construction land classifications.

During the periods 1976–1984 and 1984–1995, the main transitions were characterized by internal transformation of natural wetland, transition from wetland to cultivated land, and natural wetland's formation and erosion. However, the main transitions are predominantly natural wetland artificialization and construction expansion in the period 1995–2006 and 2006–2014. The main transition valuation also revealed a continuous increase in artificial areas, indicating that land use trajectories were veering towards artificial surfaces. During 1976–2014, the destination of natural wetland artificialization was mainly aquafarm and salt pan, which will have likely exacerbated land subsidence, sea water invasion, and salinization [34,42,59].

Wetland change trajectory results demonstrate that the main wetland changes were wetland degradation and wetland artificialization. The percentages of old degradation and recent degradation transferred to sea surface were 23.2% and 42.5%, respectively. Meanwhile, the overlay analysis of wetland change trajectory map and coastline evaluation map shows 73.3% of old degradation wetland and 53.8% of recent degradation wetland are distributed in the stable land from 1976 to 2014. Therefore, coastline change is the subordinate effect for natural wetland degradation in comparison with human activities. The transitions of cultivated land to wetland were mainly a result of the building of ditches and ponds in the periods 1976–1984 and 1984–1995, and there were small percentages of transitions from cultivated land to grassland and aquafarm land in the periods 1995–2006 and 2006–2014, which were caused by the policy of returning farmland to wetland and a new agriculture-fishery mode.

The footprint of human disturbance on the YRD is becoming larger and larger, and the artificialization rate of the YRD increased from 40.1% in 1976 to 81.1% in 2014. The wetlands in the YRD are experiencing a continuous development and evolution process under the combined effects of natural factors and human factors. It is certain that the degree of human disturbance tended to increase with time, and the degree of influence has become deeper and deeper. Currently, the YRD is in a period of rapid development, and large-scale development and construction activities are inevitable. Although the establishment of Binzhou Shell Islands and Wetland Nature Reserve and the Yellow River Delta Nature Reserve and implementation of wetland restoration projects for the protection of coastal wetlands plays important role in preventing wetlands from loss and degradation, the overall loss and degradation trend is unlikely to change in the short-term. As such, the protection of coastal wetland ecological environment remains very difficult and long-term.

5. Conclusions

This research quantified the LUCC process, explored the systematic transitions of land cover, and identified wetland change trajectory for the period 1976–2014 in the Yellow River Delta through enhanced transition matrix and relevant quantitative indicators. This study provides reliable LUCC data, which is useful for the detection and refinement of conservation policies aimed at protecting estuarine wetland. The main wetland changes were wetland degradation and wetland artificialization, and anthropogenic activities were the major driving forces of wetland degradation. Our findings suggest that development of salt pan industry and the construction of built spaces occupying natural wetland needs to be controlled and well managed in order to help maintain the natural habits and mitigate seawater intrusion and soil salinization. Finally, this study highlighted that the identification of systematic transitions and their spatial statistical modeling under GIS environment enable researchers and planners to focus on the most important signals of systematic landscape transitions and allow a better understanding of the proximate causes of changes.

Acknowledgments: This research was supported by the National Natural Science Foundation of China (No. 41401663), the Natural Science Foundation of Shandong Province (No. ZR2016EEM18), the Science Foundation of Ministry of Education of China (No. 12YJC790254), and the Excellent Young Scholars Research Fund of Shandong Normal University. We would like to thank the Editor and the anonymous reviewers for providing constructive comments and suggestions.

Author Contributions: Baolei Zhang, Shumin Zhang, and Chaoyang Feng contributed to the research design. Qiaoyun Zhang contributed to data processing and analysis of datasets. Shumin Zhang and Baolei Zhang wrote the paper. Qingyu Feng and Chaoyang Feng reviewed the manuscript and offered useful insights for its improvement.

Conflicts of Interest: The authors declare no conflict of interest.

Abbreviations

The following abbreviations are used in this manuscript:

YRD	Yellow River Delta
LUCC	Land use/cover change
KC	Kappa coefficient
SG	Stability grade
SW	Swap change
BC	Beach
GL	Grassland
BL	Bushland
RV	River
DP	Ditch and ponds
AS	Aquafarm and salt pan
WL	Woodland
CL	Cultivated land
AL	Construction land
SF	Sea surface

References

1. Turner, B.L., II; Skole, D.; Sanderson, S.; Fischer, G.; Fresco, L.; Leemans, R. *Land-Use and Land-Cover Change-Science/Research Plan*; IGBP Report No.35 and HDP Report No.7; IGBP and HDP Secretariats: Stochkholm, Sweden; Geneva, Switzerland, 1995.
2. Houghton, R.A.; Hackler, J.L.; Lawrence, K.T. The US carbon budget: Contribution from land use change. *Science* **1999**, *285*, 574–578. [CrossRef] [PubMed]
3. Shi, P.J.; Gong, P.; Li, X.B. *Methods and Practice of Land Use Changes in Research*; Science Press: Beijing, China, 2000. (In Chinese)

4. Lawrence, P.L.; Chase, T.N. Investigating the climate impacts of global land cover change in the community climate system model. *Int. J. Clim.* **2010**, *30*, 2066–2087. [CrossRef]
5. Liu, J.Y.; Liu, M.L.; Deng, X.Z.; Zhuang, D.F.; Zhang, Z.X.; Luo, D. The land use and land cover change database and its relative studies in China. *J. Geogr. Sci.* **2002**, *12*, 275–282.
6. Dewan, A.M.; Yamaguchi, Y.; Rahman, M.Z. Dynamics of land use/cover changes and the Analysis of Landscape fragmentation in Dhaka Metropolitan, Bangladesh. *GeoJournal* **2012**, *77*, 315–330. [CrossRef]
7. Antrop, M. Why landscapes of the past are important for the future. *Landsc. Urban Plan.* **2005**, *70*, 21–34. [CrossRef]
8. Hasse, J.E.; Lathrop, R.G. Land resource impact indicators of urban sprawl. *Appl. Geogr.* **2003**, *23*, 159e175. [CrossRef]
9. Tavares, A.O.; Pato, R.L.; Magalhães, M.C. Spatial and temporal land use change and occupation over the last half century in a peri-urban area. *Appl. Geogr.* **2012**, *34*, 432–444. [CrossRef]
10. Li, S.N.; Wang, G.X.; Deng, W.; Hu, Y.M.; Hu, W.W. Influence of hydrology process on wetland landscape pattern: A case study in the Yellow River Delta. *Ecol. Eng.* **2009**, *35*, 1719–1726. [CrossRef]
11. Pontius, R.G., Jr.; Shusas, E.; McEachern, M. Detecting important categorical land changes while accounting for persistence. *Agric. Ecosyst. Environ.* **2004**, *101*, 251–268. [CrossRef]
12. Braimoh, A.K. Random and systematic land-cover transitions in northern Ghana. *Agric. Ecosyst. Environ.* **2006**, *113*, 254–263. [CrossRef]
13. Lambin, E.F.; Geist, H.J.; Lepers, E. Dynamics of land-use and land-cover change in tropical regions. *Ann. Rev. Environ. Resour.* **2003**, *28*, 205–241. [CrossRef]
14. Lu, D.; Mausel, P.; Brondízio, E.; Moran, E. Change detection techniques. *Int. J. Remote Sens.* **2004**, *25*, 2365–2407. [CrossRef]
15. Pontius, R.G.; Petrova, S.H. Assessing a predictive model of land change using uncertain data. *Environ. Model. Softw.* **2010**, *25*, 299–309. [CrossRef]
16. Manandhar, R.; Odeh, I.O.A.; Pontius, R.G. Analysis of twenty years of categorical land transitions in the lower hunter of new South Wales, Australia. *Agric. Ecosyst. Environ.* **2010**, *135*, 336–346. [CrossRef]
17. Isabel, P.; Cunha, M.; Pereira, L. Remote sensing based indicators of changes in a mountain rural landscape of northeast Portugal. *Appl. Geogr.* **2011**, *31*, 871–880.
18. Johnson, B.G.; Zuleta, G.A. Land-use land-cover change and ecosystem loss in the Espinal ecoregion, Argentina. *Agric. Ecosyst. Environ.* **2013**, *181*, 31–40. [CrossRef]
19. Liu, R.; Zhu, D.L. Methods for detecting land use changes based on the land use transition matrix. *Resour. Sci.* **2010**, *32*, 1544–1550.
20. Zhou, Q.; Li, B.; Kurban, A. Trajectory analysis of land cover change in arid environment of China. *Int. J. Remote Sens.* **2008**, *29*, 1093–1107. [CrossRef]
21. Wang, H.; Gao, Y.; Pu, R.L.; Ren, L.L.; Kong, Y.; Li, H.; Li, L. Natural and anthropogenic influences on a red-crowned crane habitat in the Yellow River Delta Natural Reserve, 1992–2008. *Environ. Monit. Assess.* **2014**, *186*, 4013–4028. [CrossRef] [PubMed]
22. Wang, H.; Gao, Y.; Pu, R.L.; Ren, L.L.; Kong, Y.; Li, H.; Li, L. Assessment of the red-crowned crane habitat in the Yellow River Delta Nature Reserve, East China. *Reg. Environ. Chang.* **2012**, *31*, 115–123. [CrossRef]
23. Bi, X.L.; Wang, B.; Lu, Q.S. Fragmentation effects of oil wells and roads on the Yellow River Delta, North China. *Ocean Coast Manag.* **2011**, *54*, 256–264. [CrossRef]
24. Fan, H.; Huang, H.J. Response of coastal marine eco-environment to river fluxes into the sea: A case study of the Huanghe (Yellow) River mouth and adjacent waters. *Mar. Environ. Res.* **2008**, *65*, 378–387. [CrossRef] [PubMed]
25. He, X.H.; Hörmann, G.; Strehmel, A.; Guo, H.L.; Fohrer, N. Natural and Anthropogenic Causes of Vegetation Changes in Riparian Wetlands Along the Lower Reaches of the Yellow River, China. *Wetlands* **2015**, *35*, 391–399. [CrossRef]
26. Kong, D.X.; Miao, C.Y.; Borthwick, A.G.L.; Duan, Q.Y.; Liu, H.; Sun, Q.H.; Ye, A.Z.; Di, Z.H.; Gong, W. Evolution of the Yellow River Delta and its relationship with runoff and sediment load from 1983 to 2011. *J. Hydrol.* **2015**, *520*, 157–167. [CrossRef]
27. Zong, X.Y. Study on Dynamic Changes of Wetland Landscape Pattern in Yellow River Delta. *J. Geo-inf. Sci.* **2009**, *11*, 91–97. [CrossRef]

28. Sun, X.Y.; Su, F.Z.; Lv, T.T. Analysis of temporal-spatial changes in wetlands over the Yellow River estuary. *Resour. Sci.* **2011**, *33*, 2277–2284.

29. Zhang, B.L.; Yin, L.; Zhang, S.M.; Feng, C.Y. Assessment on characteristics of LUCC process based on complex network in Modern Yellow River Delta, Shandong Province of China. *Earth Sci. Inform.* **2016**, *9*, 83–93. [CrossRef]

30. Zhang, X.L.; Ye, S.Y.; Yin, P. Characters and successions of natural wetland vegetation in Yellow River Delta. *Ecol. Environ. Sci.* **2009**, *18*, 292–298.

31. Li, S.N.; Wang, G.X.; Deng, W.; Hu, Y.M. Effects of runoff and sediment variation on landscape pattern in the Yellow River Delta of China. *Adv. Water Sci.* **2009**, *20*, 325–331.

32. Zhang, X.L.; Li, P.; Liu, L.J.; Li, P.Y. Degradation assessment of Modern Yellow River Delta coastal wetland. *Mar. Sci. Bull.* **2010**, *29*, 685–689.

33. Hong, J.; Lu, X.N.; Wang, L.L. Quantitative analysis of the factors driving evolution in the Yellow River Delta Wetand in the past 40 years. *Acta Ecol. Sin.* **2016**, *36*, 924–935.

34. Ottinger, M.; Kuenzer, C.; Liu, G.H.; Wang, S.Q.; Dech, S. Monitoring land cover dynamics in the Yellow River Delta from 1995 to 2010 based on Landsat 5 TM. *Appl. Geogr.* **2013**, *44*, 53–68. [CrossRef]

35. Fang, H.; Liu, G.; Kearney, M. Georelational analysis of soil type, soil salt content, landform, and land use in the Yellow River Delta, China. *Environ. Manag.* **2005**, *35*, 72–83. [CrossRef]

36. Xu, J.X. A study of thresholds of runoff and sediment for the land accretion of the Yellow River Delta. *Geogr. Res.* **2002**, *21*, 163–170.

37. Zhao, X.S.; Cui, B.S.; Sun, T.; He, Q. The relationship between the spatial distribution of vegetation and soil environmental factors in the tidal creek areas of the Yellow River Delta. *Ecol. Environ. Sci.* **2010**, *19*, 1855–1861.

38. Yu, J.B.; Chen, X.B.; Sun, Z.G.; Xie, W.J.; Mao, P.L.; Wu, C.F.; Dong, H.F.; Mu, X.J.; Li, Y.Z.; Guan, B.; et al. The spatial distribution characteristics of soil nutrients in new-born coastal wetland in Yellow River delta. *Acta Sci. Circumst.* **2010**, *30*, 855–861.

39. Kuenzer, C.; Ottinger, M.; Liu, G.H.; Sun, B.; Baumhauer, R.; Dech, S. Earth observation-based coastal zone monitoring of the Yellow River Delta: Dynamics in China's second largest oil producing region overfour decades. *Appl. Geogr.* **2014**, *55*, 92–107. [CrossRef]

40. Teferi, E.; Bewket, W.; Uhlenbrook, S.; Wenninger, J. Understanding recent land use and land cover dynamics in the source region of the Upper Blue Nile, Ethiopia: Spatially explicit statistical modeling of systematic transitions. *Agric. Ecosyst. Environ.* **2013**, *165*, 98–117. [CrossRef]

41. Wan, H. *Research on Wetland Information Extraction and Analysis of Yellow River Delta Based on RS and GIS*; China University of Petroleum (East China): Dongying, China, 2009.

42. Liu, G.L.; Zhang, L.C.; Zhang, Q.; Musyimi, Z.; Jiang, Q.H. Spatio–temporal dynamics of wetland landscape patterns based on remote sensing in Yellow River Delta,China. *Wetlands* **2014**, *34*, 787–801. [CrossRef]

43. Lavery, P.; Pattiaratchi, C.; Wyllie, A.; Hick, P. Water-Quality Monitoring in Estuarine Waters Using the Landsat Thematic Mapper. *Remote Sens. Environ.* **1993**, *46*, 268–280. [CrossRef]

44. Fan, Y.; Zhang, S.; Hou, C.; Zhang, L. Study on Method of Coastline Extraction from Remote Sensing—Taking Yellow River Mouth Reach and Diaokou Reach of Yellow River Delta Area as an Example. *Remote Sens. Inform.* **2009**, *8*, 67–70.

45. Piwowar, J.M. Digital Image Analysis. In *Remote Sensing for GIS Managers*; Aronoff, S., Ed.; ESRI Press: Redlands, CA, USA, 2005.

46. Ke, C.Q.; Zhang, D.; Wang, F.Q.; Chen, S.X.; Schmullius, C.; Boerner, W.M.; Wang, H. Analyzing coastal wetland change in the Yancheng National Nature Reserve, China. *Reg. Environ. Chang.* **2011**, *11*, 161–173. [CrossRef]

47. Liu, J.Y.; Liu, M.L.; Tian, H.Q.; Zhuang, D.F.; Zhang, Z.X.; Zhang, W.; Tang, X.M.; Deng, X.Z. Spatial and temporal patterns of China's cropland during 1990–2000: An analysis based on Landsat TM data. *Remote Sens. Environ.* **2005**, *98*, 442–456. [CrossRef]

48. Wang, W.J.; Zhang, C.R.; Allen, J.M.; Li, W.D.; Boyer, M.A.; Segerson, K.; Silander, J.A., Jr. Analysis and prediction of land use changes related to invasive species and major driving forces in the state of Connecticut. *Land* **2016**, *5*, 25. [CrossRef]

49. Ni, J.R.; Yin, K.Q.; Zhao, Z.J. Comprehensive classification for wetlands I. Classification. *J. Nat. Resour.* **1998**, *13*, 214–221.

50. Zhao, H.T.; Wang, L.R. Coastal wetland types in China. *Mar. Sci. Bull.* **2000**, *19*, 72–81.

51. Tang, X.P.; Huang, G.L. Study on classification system for wetland types in China. *For. Res.* **2003**, *16*, 531–539.

52. Foody, G.M. Status of land cover classification accuracy assessment. *Remote Sens. Environ.* **2002**, *80*, 185–201. [CrossRef]

53. Stehman, S.V. Selecting and interpreting measures of thematic classification accuracy. *Remote Sens. Environ.* **1997**, *62*, 77–89. [CrossRef]

54. Aldwaik, S.Z.; Pontius, R.G. Intensity analysis to unify measurements of size and stationarity of land changes by interval, category, and transition. *Landsc. Urban Plan.* **2012**, *106*, 103–114. [CrossRef]

55. Swetnam, R.D. Rural land use in England and Wales between 1930 and 1998: Mapping trajectories of change with a high resolution spatio-temporal dataset. *Landsc. Urban Plan.* **2007**, *81*, 91–103. [CrossRef]

56. Wang, D.C.; Gong, J.H.; Chen, L.D.; Zhang, L.H.; Song, Y.Q.; Yue, Y.J. Spatio-temporal pattern analysis of land use/cover change trajectories in Xihe watershed. *Int. J. Appl. Earth Obs. Geoinf.* **2012**, *4*, 12–21. [CrossRef]

57. Chen, J.; Wang, S.Y.; Mao, Z.P. Monitoring wetland changes in Yellow River Delta by remote sensing during 1976–2008. *Prog. Geogr.* **2011**, *30*, 585–592.

58. Wang, Y.L.; Yu, J.B.; Dong, H.F.; Li, Y.Z.; Zhou, D.; Fu, Y.Q.; Han, G.X.; Mao, P.L. Spatial Evolution of Landscape Pattern of Coastal Wetlands in Yellow River Delta. *Sci. Geogr. Sin.* **2012**, *32*, 717–724.

59. Higgins, S.; Overeem, I.; Tanaka, A.; Syvitski, J.P.M. Land subsidence at aquaculture facilities in the Yellow River delta, China. *Featur. Geophys. Res. Lett.* **2013**, *40*, 3898–3902. [CrossRef]

land

Article

Monitoring Urban Growth and the Nepal Earthquake 2015 for Sustainability of Kathmandu Valley, Nepal

Bhagawat Rimal [1], Lifu Zhang [1,*], Dongjie Fu [2], Ripu Kunwar [3] and Yongguang Zhai [1]

[1]   The State Key Laboratory of Remote Sensing Science, Institute of Remote Sensing and Digital Earth, Chinese Academy of Sciences, Beijing 100101, China; bhagawat@radi.ac.cn (B.R.); zhaiyg@radi.ac.cn (Y.Z.)

[2]   State Key Laboratory of Resource and Environmental Information System, Institute of Geographic Sciences and Natural Resources Research, Chinese Academy of Sciences, Beijing 100101, China; fudj@lreis.ac.cn

[3]   Cultural and Spatial Ecology, Department of Geosciences, Florida Atlantic University, Boca Raton, FL 33431, USA; rkunwar@fau.edu

*   Correspondence: zhanglf@radi.ac.cn; Tel.: +86-10-6483-9450

Academic Editors: Andrew Millington, Harini Nagendra, Monika Kopecka and Karen C. Seto
Received: 8 June 2017; Accepted: 14 June 2017; Published: 17 June 2017

Abstract: The exodus of people from rural areas to cities brings many detrimental environmental, social and cultural consequences. Monitoring spatiotemporal change by referencing the historical timeline or incidence has become an important way to analyze urbanization. This study has attempted to attain the cross-sectional analysis of Kathmandu valley that has been plagued by rampant urbanization over the last three decades. The research utilizes Landsat images of Kathmandu valley from 1976 to 2015 for the transition analysis of land use, land cover and urban sprawl for the last four decades. Results showed that the urban coverage of Kathmandu valley has tremendously increased from 20.19 km^2 in 1976 to 39.47 km^2 in 1989 to 78.96 km^2 in 2002 to 139.57 km^2 in 2015, at the cost of cultivated lands, with an average annual urban growth rate of 7.34%, 7.70% and 5.90% in each temporal interval, respectively. In addition, the urban expansion orientation analysis concludes the significant urban concentration in the eastern part, moderately medium in the southwest and relatively less in the western and northwest part of the valley. Urbanization was solely accountable for the exploitation of extant forests, fertile and arable lands and indigenous and cultural landscapes. Unattended fallow lands in suburban areas have compounded the problem by welcoming invasive alien species. Overlaying the highly affected geological formations within the major city centers displays that unless the trend of rapid, unplanned urbanization is discontinued, the future of Kathmandu is at the high risk. Since land use management is a fundamental part of development, we advocate for the appropriate land use planning and policies for sustainable and secure future development.

Keywords: land use/land cover change; urban expansion; geology; earthquake; GIS/RS; Kathmandu valley

1. Introduction

The terms land use and land cover are often used interchangeably, but are conceptually different. The former is anthropogenically manipulated and concerned with the purpose the land serves. It depends on the way humans manage the landscape. The latter is physically designed and related to vegetation (natural or artificially modified) and its structure, as well as the other land types covering terrestrial land. The relation between these two disciplines is such that land cover may have multiple land uses, and land use affects the land cover [1]. Global land use has been changed significantly in the past three decades due to population growth and its footprints. Population growth inevitably increases the urban footprint with significant consequences on biodiversity, climate and environmental resources [2,3]. The environmental implications of population and land use and land cover (LULC) changes have been extensively and empirically examined because the exponential population growth exerts outpaced pressure on the local environmental resources [4]. Rapid urbanization has also been a common phenomenon, especially in developing countries with an increasing desire for prosperity [5,6]. Urban growth, a complex phenomenon [7,8], and multiple driving forces are associated [9–11]. Urbanization is a feedback system that is economically motivated [12]. Urban areas have grown substantially in the last several decades worldwide, but urbanization varies considerably in different geographic regions and contexts [13].

Nepal is a country that has experienced population explosion and accelerated urbanization in the last six decades. The number of urban centers in Nepal rose between 1952 and 2015 [14]. Kathmandu, the bowl-shaped capital city of the country, has become densely populated and juxtaposes rapid urbanization, resulting in land use and socio-economic change [15,16]. High population growth, dramatic land use change and socioeconomic transformations have brought the inconsistency of rapid urbanization and environmental consequences in Kathmandu valley [16,17]. For the management of rapid urbanization, the development plan needs to be integrated in order to maintain the equilibrium amongst resources, users, managers and the environment [18,19]. We hypothesize that the ongoing urbanization in Kathmandu valley is discordant with its geology and geography, and the ongoing LULC change foments the jeopardy of the human-nature integrity. This study aims to analyze the impacts accompanied by the spatiotemporal patterns of urbanization and LULC changes in Kathmandu in the last 40 years (1976, 1989, 2002 and 2015) using GIS and remote sensing tools at the nexus of rapid land use change, population growth and urbanization and the 2015 earthquake impacts in the valley. More specific objectives of the study were to: (a) analyze the spatiotemporal dynamics of land use pattern from 1976 to 2015 of landscape change; (b) identify the urbanization pattern and dynamically map the urban expansion area; (c) explore urban expansion rate, urban expansion orientation and analysis of the significant urban concentration zone; (d) explain and map the 2015 earthquake impact area of the valley.

2. Materials and Methodology

2.1. Study Area

Kathmandu valley is an inter-montane lesser Himalaya bowl-shaped basin. Its geology mostly consists of fluvio-lacustrine sediments of Plio-Pleistocene age [20,21], which has characteristics of magnifying seismic waves and being prone to liquefaction [22]. The vulnerability of liquefaction is due to its high ground water level and potential strong earthquake motions in this area [23]. The valley is surrounded by Mahabharat mountain range associated with four hills namely Phulchowki (2762 m) in the southeast, Chandragiri/Champadevi in the southwest, Shivapuri (2762 m) in the northwest and Nagarkot in the northeast, formerly known as the forts of the valley [24]. The basement rocks of Phulchowki and the Bhimphedi groups of the Kathmandu complex are formed by Precambrian to Devonian rocks [25,26]. These groups together form the Kathmandu complex, which is interpreted tectonically as thrust mass (allochthonous) with the underlying Paraautochthonous Nuwakot Complex constituting the Mahabharat Synclinorium, the axis of which passes along the Phulchowki-Chandragiri range. The basement rocks are intersected by numerous faults systems; the geological formation of the valley has been divided into different groups [25] (Figure 1).

Administratively, the valley consists of three districts: Kathmandu, Lalitpur and Bhaktapur. Mainly five municipals areas, namely Kathmandu, Lalitpur Bhaktapur, Kirtipur and Madhyapur-Thimi, are exponentially developed and populated. Kathmandu valley, the selected study area, lies between 27°24′14″ and 27°49′10″ north latitude (the southern part of Thuladurlung village to the northern part of Gokarneshwor municipality) and 85°11′18″ and 85°33′56″ eastern longitude (the western part of Chandragiri municipality to the eastern part of Shankharapur municipality); however, nomenclature is being changed during the new local reconstruction process. The study area covers an area of 933.22 km^2 within the vertical span 410 m to 2831 m above sea level. Figure 2a indicates the location map of the study area, and Figure 2b represents the main part of Kathmandu valley, which is divided into eight main areas.

The Kathmandu valley is facing potentially insurmountable challenges (both demographic and physiographic) due to overpopulation led by immigration. Since 1971, the population of the valley has exploded from 0.6 million people to over 2.5 million (Table 1), with an annual growth rate of 2.3%–5.8% [27].

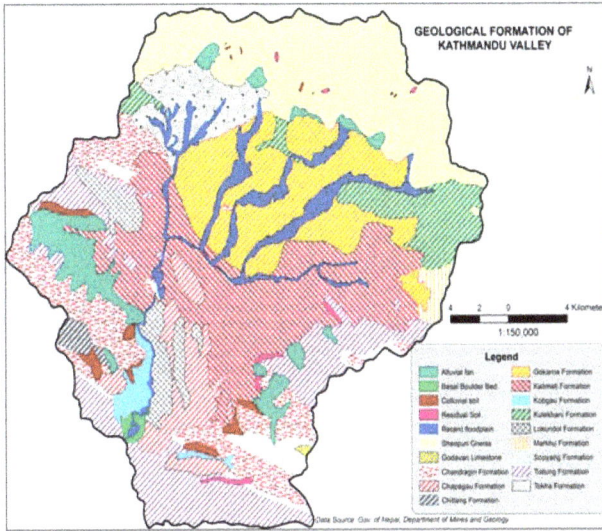

Figure 1. Geological formation map of Kathmandu valley.

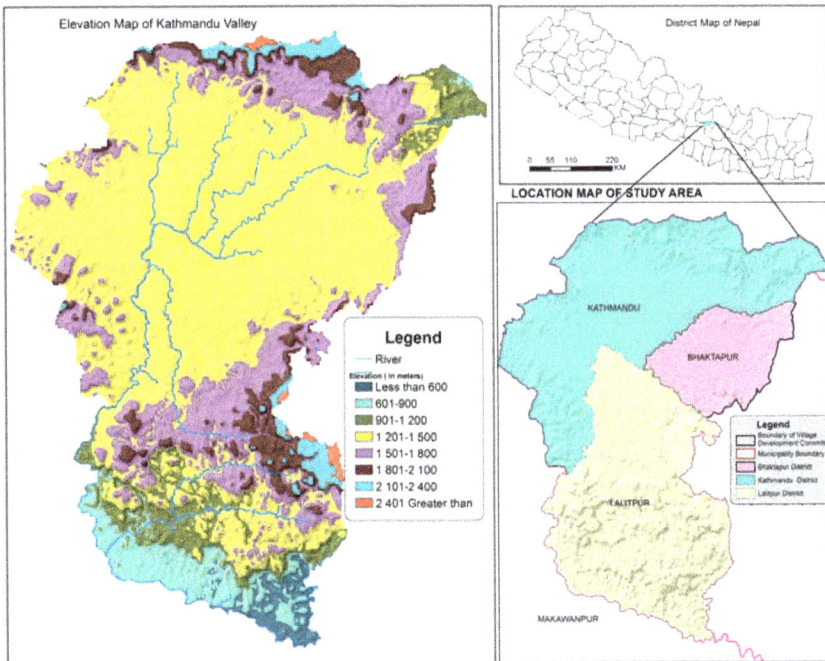

Figure 2. *Cont.*

59

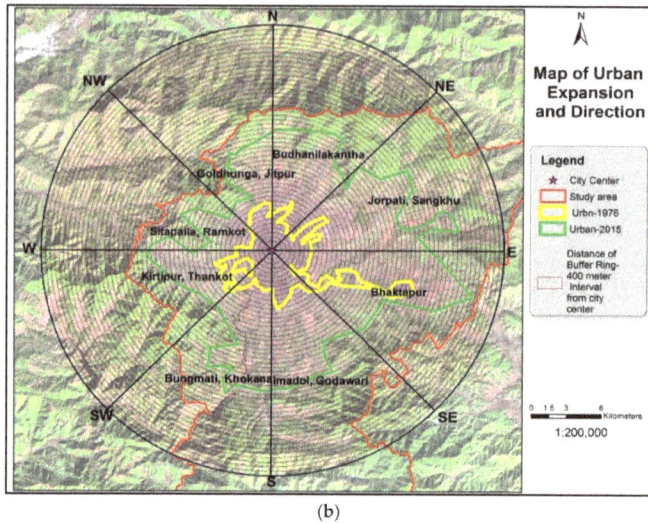

(b)

Figure 2. (a) Location map of the study area; (b) urban expansion of Kathmandu valley in eight main area.

Continuous haphazard urbanization has increased agricultural land loss and constrained the coping capacity in the course of disaster management [28]. The mushroomed concrete buildings constructed in the valley reduce the percolation of surface water, increase the demand for water and cause a fall in the water level each year, making the valley at the brink of land subsidence [25].

Table 1. Population distribution of Kathmandu valley.

Districts	1971	1981	1991	2001	2011
Kathmandu District	353,756	422,237	675,341	1,081,845	1,744,240
Lalitpur District	154,998	184,341	257,086	337,785	468,132
Bhaktapur District	110,157	159,767	172,952	225,461	304,651
Kathmandu valley *	620,882	768,326	1,107,370	1,647,092	2,519,034

Source: Central Bureau of Statistics (CBS 2011), * includes Kathmandu, Lalitpur and Bhaktapur districts.

2.2. Data

This study has analyzed the spatiotemporal dynamics of urbanization of LULC and identified the crosscutting impacts of the Nepal Earthquake, 2015. Four pairs of Landsat images were used for the classification: (i) Landsat Image 2 (Multi-spectral Scanner (MSS), image with Path/Row152/41) 28 October 1976; (ii) Landsat Image 5 image (Thematic Mapper (TM) with Path/Row 141/41) 31 October 1989; (iii) Landsat Image 7 (Thematic Mapper Plus (ETM+) with Path/Row 141/41) 27 October 2002; and (iv) Landsat Image 8 (Operational Land Image (OLI) with Path/Row 141/41) of 24 January 2015. All data were projected in the Universal Transverse Mercator World Geodetic System 1984. A topographical map published by the Survey Department, Government of Nepal, 1995 (scale 1:25,000), was used as the reference for image analyses. The Google Earth image was used for ground-truthing. A "region of interest" (ROI) boundary representing metropolitan, municipal and village level study area was delineated for remote sensing analysis. The study area boundary datasets were obtained from the Survey Department, Government of Nepal, and the administrative boundary of Kathmandu valley was imported in TerrSet.

2.3. Data Processing and Classification

All obtainable images, pre-processing, raster group (stacking), subset and classification were accomplished using Idrisi (TerrSet) developed by Clark Lab. Both unsupervised and supervised approaches with the maximum likelihood parameter (MLP) system were applied to improve the accuracy of land use classification. Since no single classification system is used by most of the scientific community [29,30], an adapted classification system recommended by Anderson et al., 1976 [31], was used for remotely-sensed data. In the study, land use is classified into six different classes: urban (built-up), water, open field, forest, bush and cultivated land. The Land Change Modeler (LCM) system was applied to analyze the changes and transitions matrix in LULC. The computed transition matrix consists of the row and column of landscape categories at times T1 and T2.

2.4. Urban Expansion Direction

For the detailed exploration of urban expansion and its direction, the major administrative center of the country, Singhadurbar is assumed as the central point, and the outward urban orientation of the valley is conducted through the lines created by utilizing ArcGIS 10.1. The outcomes should answer the questions: where and in which direction did the changes occur from the city core? The rays [30] consisting of eight subdivisions north, northeast, east, southeast, south, southwest, west, northwest and north (N-NE, NE-E, E-SE, SE-S, S-SW, SW-W, W-NW and WN-N), each representing 45°, were drawn, and 50 buffer zones were calculated at the regular intervals of 400 m.

2.5. Measuring Urban Expansion Rate

The urban expansion growth rate [32] of the study area is measured by calculating the total transformation of urban area. The urban expansion rate indicates the average annual urban area growth in the following years.

$$\text{MUER} = (U_{t1} - U_{t0})/(t_1 - t_0)$$

where MUER refers to measuring the urban expansion rate, U is urban area in km^2, t_1 is succeeding time and t_0 is preceding time.

2.6. Plotted of Earthquake Damage Area

Highly damaged localities were captured through GPS during the field observation, and the collected information has been overlaid in the geological map of the valley and analyzed accordingly (Figure 3). The study excludes the statistical information of loss and damage during the Nepal Earthquake, 2015.

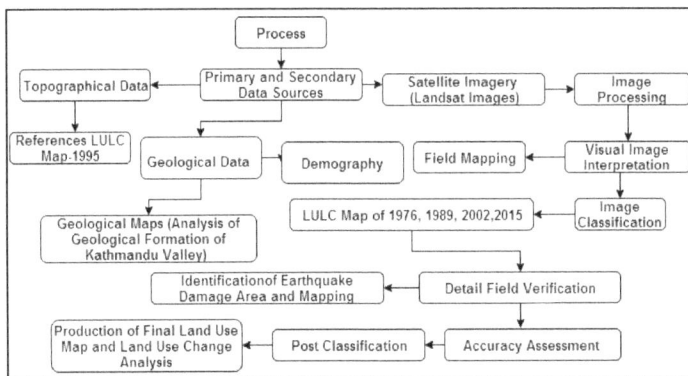

Figure 3. Methodological approach.

3. Results and Discussion

3.1. Urbanization in Kathmandu Valley

Random urbanization began to gain ground in the Kathmandu valley in the late 1950s [33], and accelerated population growth and migration complemented the process. In 1981, 93% of Nepal's population lived in rural areas, and the data came down to 83% in 2011 [27,34]. Since 2001, the average population density in the Kathmandu has vastly increased compared to the country as a whole. The population density of Kathmandu grew from 1837/km^2 in 2001 to 2699/km^2 in 2011, whereas Nepal's density of population only increased from 157/km^2 to 180/km^2 in the same time period. Consequently, the population density of Kathmandu was 1277/km^2. While Nepal's was 126/km^2 in 1991, in 1981, it was 963/km^2 in Kathmandu to 102/km^2 in Nepal and 623/km^2 against 79/km^2 in 1971. This shows that the population density in the Kathmandu valley was rather aggressively growing [24]. Urban population increased from 2.9% to 40.49% in a period of 63 years (1952–2015). The rural population of Nepal decreased from 97.1% down to 59.51% in the same period. The valley represents 22.77% of the total urban population of the country (Figure 4). Table 2 and Figure 5 depict the statistical comparison between national-level rural-urban population and the urban population proportion of Kathmandu valley.

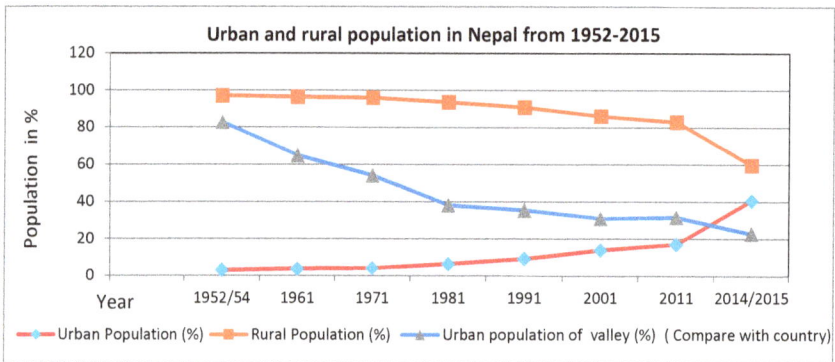

Figure 4. Urban and rural population of Nepal.

Table 2. Urban and rural population.

Year	Urban Centers	Urban Population (%)	Rural Population (%)	Urban Population of Valley (%) (of the Total Urban Population of the Country)
1952/1954	10	2.9	97.1	82.6
1961	16	3.6	96.4	64.9
1971	16	4	96	54
1981	23	6.4	93.6	38
1991	33	9.2	90.8	35.3
2001	58	13.9	86.1	30.9
2011	58	17.1	82.9	31.58
2014/2015	217	40.49	59.51	22.77

Source: Central Bureau of Statistics (CBS 2003), International Centre for Integrated Mountain Development (ICIMOD, 2007), (CBS, 2011), and (Ministry of Federal Affairs and Local Development | Nepal (MoFALD,2015). Note: The administrative boundary of the Village Development Committees and Municipality and the nomenclature are being changed due to the reconstruction and reclassification of the local-level administrative units. However, this study is based on local development construction before March 2017.

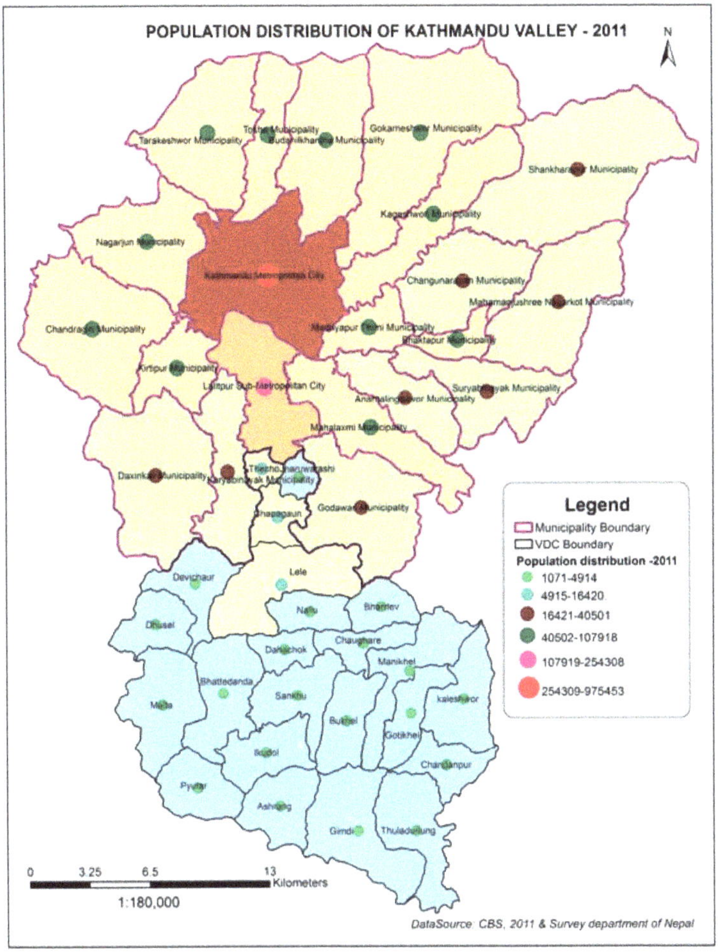

Figure 5. Population distribution in different parts of the Kathmandu valley. Note: The administrative boundary of Village Development Committees, Municipalities and their nomenclature are being changed due to the reconstruction and reclassification of the local-level administrative units, although this study is based on local development construction before March 2017.

3.2. Land-Use Land-Cover Dynamics

LULC change analysis of Kathmandu valley showed that 17.51 km^2 of cultivated land were converted into other land use classes between 1989 and 2002, whereas 55.01 km^2 of cultivated land were decreased between 2002 and 2015. Cultivated land was mostly transformed into urban/built up area and bush land, whereas forest areas were intensively degraded in rural areas. A total are of 11.84 km^2 was deforested between 1976 and 1989. The deforestation rate augmented (24.97 km^2) between 1989 and 2002 and is shown in Table 3 and Figure 6.

The land use pattern in urban areas had rapidly changed, and the urban development pattern is environmentally unmanageable in Kathmandu valley [35]. Rapid population growth and urbanization have caused much agricultural land transformation of anthropogenic structures and has intensified overexploitation of extant land covers. The escalated land use change is partially due to Nepal becoming a more service-based economy [36]; yet, 70% of Nepal's gross national product still depends on agricultural sector. Despite agricultural productivity, many people have migrated from rural areas for the quest of better living [37]. According to the statistics, Kathmandu's urban coverage totaled 20.19 km^2 in 1976 (Figure 7a), but increased to 39.47 km^2 in 1989 (Figure 7b), 78.96 km^2 in 2002 (Figure 7c) and 139.57 km^2 in 2015 (Figure 7d), which is considered as a rapid transition in the in the LULC analysis.

The effects of rapid urbanization were prevalent in cities and suburban areas of the valley. As the cities expanded, it directly impacted suburban areas, and those living in suburban faced many new challenges. A significant area (63.32 km^2) of cultivated land of Kathmandu valley was converted into other classes between 1976 and 2015, and 119 km^2 area appeared as urban in the same period, at the expense of agricultural lands, forest and bush areas. Much of the city's rapid growth in population has been accommodated in informal settlements, resulting in the destruction of natural systems. In order to raise the stewardship between humans and nature, more strict urban plans and policies that favor the protection of arable land are essential.

The spatial transition over a period of 13 years between 1976 and 1989 showed the pronounced conversion of forest into cultivated land uses, probably for agricultural uses such as rearing livestock, foraging and grass collection. Over 90% of the valley's population was agro-pastoral in the 1980s [33]. However, over time, urban sprawl became the second largest factor for converting forest into cultivated land, a precursor of urbanization (Table 4, Figure 8a).

Some extensive alterations in LULC were noticed between 1989 and 2002 (Table 5). The degeneration of forests continued with 20 km^2 being transformed into bush land and 17.62 km^2 into cultivated land. Urbanization was intense during this period with about 38 km^2 of land being converted from forest and cultivated lands to urban areas (Table 5, Figure 8b).

The conversion of a total of 59.74 km^2 of cultivated land into built up structures and the transformation of 13.72 km^2 of forest to cultivated land are the two remarkable changes in land use and land cover between 2002 and 2015. Forest land conversion was minimal because of increased forest and biodiversity conservation programs. Despite intensified urbanization, these recent conservation programs have helped emphasize the benefits of leaving greenery and forests intact. As a consequence, in the last decade, the change of forest into cultivation land (13.72 km^2) was surmounted by cultivated land into forest (Table 6, Figure 8c). Figure 8d demonstrates the spatiotemporal pattern of urbanization in the valley from 1976 to 2015.

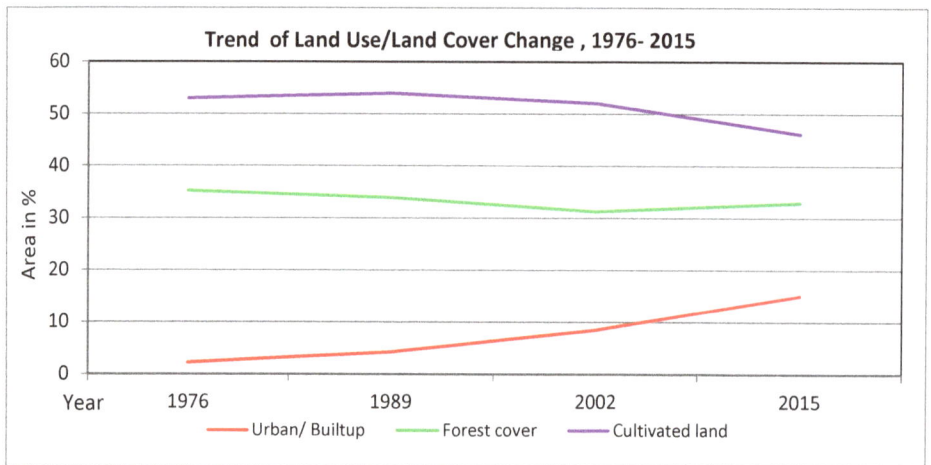

Figure 6. Trend of land use change in Kathmandu valley.

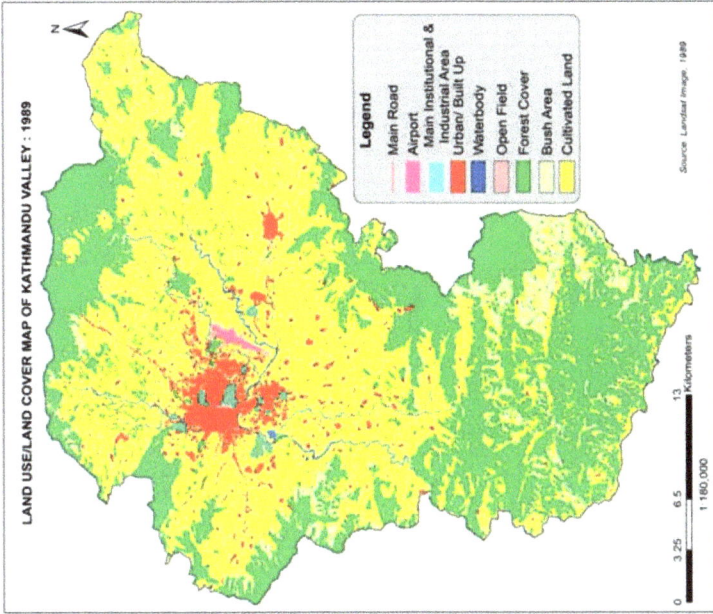

(b)

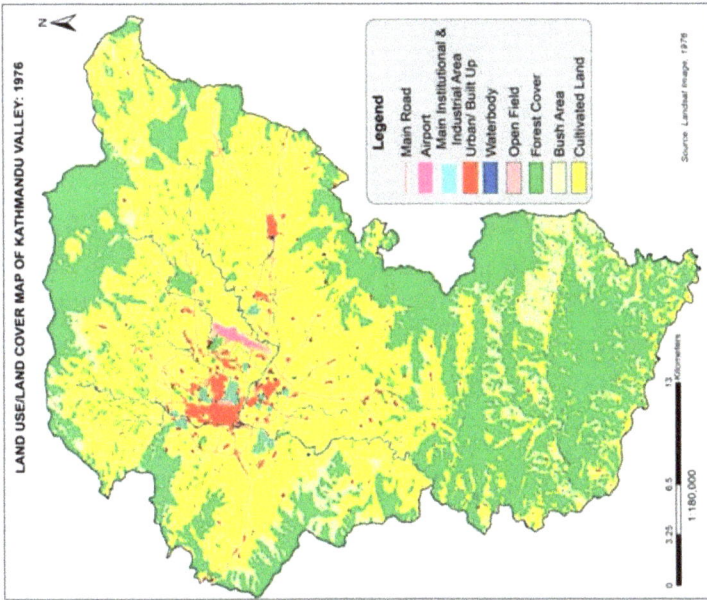

(a)

Figure 7. *Cont.*

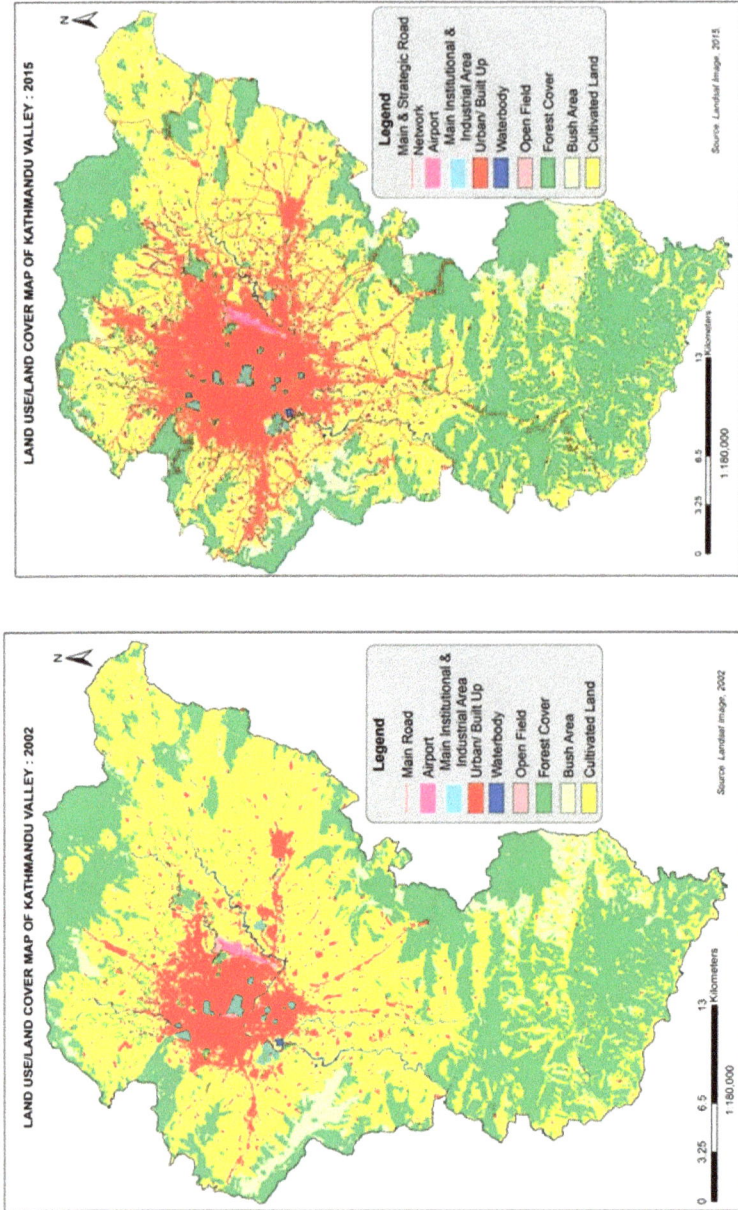

Figure 7. LULC map of Kathmandu valley: (**a**) 1976; (**b**) 1989; (**c**) 2002; and (**d**) 2015.

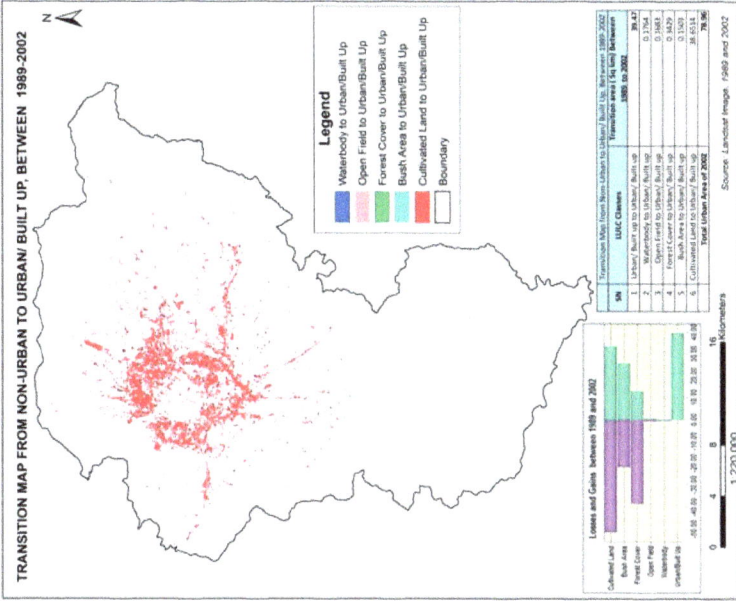

(b)

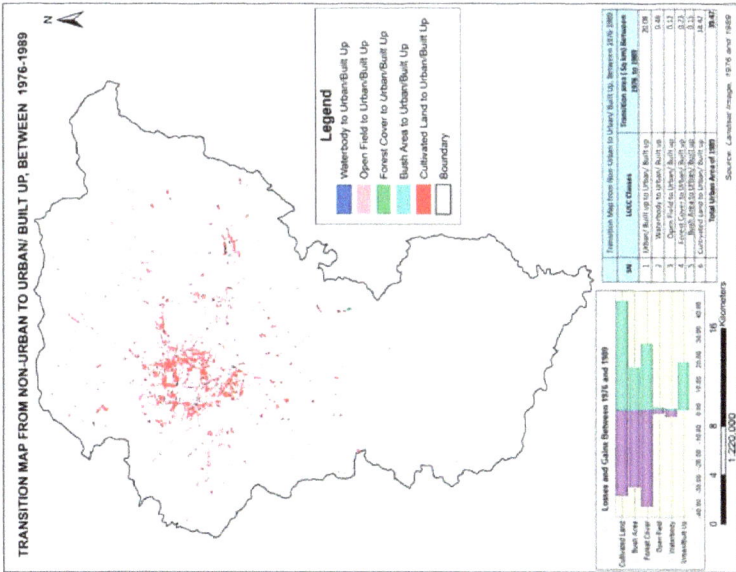

(a)

Figure 8. *Cont.*

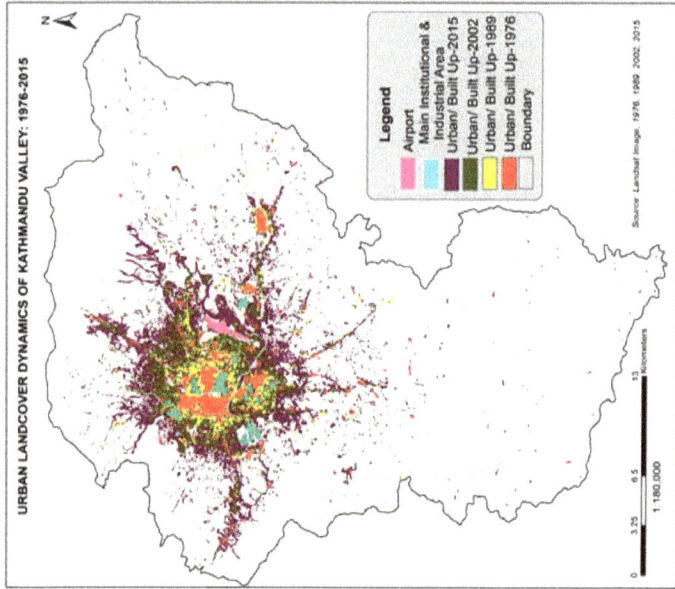

(d)

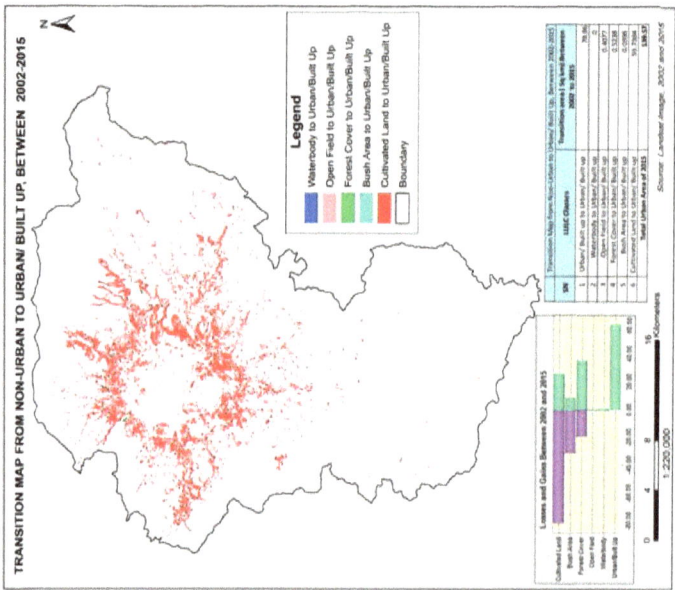

(c)

Figure 8. (**a**) Transition map from all land use/land cover classes to urban area 1976–1989; (**b**) transition map from all land use/land cover classes to urban area 1989–2002; (**c**) transition map from all land use/land cover classes to urban area 2002–2015; (**d**) urban land cover dynamics of the valley from 1976 to 2015.

Table 3. Land use/land cover statistics, 1976–2015.

Land Cover	1976 (Area) km²	%	1989 (Area) km²	%	Change between 1976 and 1989	2002 (Area) km²	%	Change between 1989 and 2002	2015 (Area) km²	%	Change between 2002 and 2015	Change between 1976 and 2015
Urban/built up	20.19	2.16	39.47	4.23	19.28	78.96	8.46	39.48	139.57	14.96	60.61	119.38
Water body	9.50	1.02	7.20	0.77	−2.30	6.59	0.71	−0.61	7.51	0.81	0.93	−1.98
Open field	3.23	0.35	2.49	0.27	−0.74	1.99	0.21	−0.50	1.75	0.19	−0.24	−1.48
Forest cover	327.60	35.10	315.76	33.84	−11.84	290.80	31.16	−24.97	306.73	32.87	15.93	−20.88
Bush area	78.94	8.46	65.33	7.00	−13.61	69.43	7.44	4.10	47.22	5.06	−22.22	−31.72
Cultivated land	493.77	52.91	502.97	53.90	9.20	485.45	52.02	−17.51	430.44	46.12	−55.01	−63.32
Total	933.22	100	933.22	100		933.22	100		933.22	100		

Table 4. Spatiotemporal transition of LULC, 1976–1989.

Year	LULC Classes	1989						km²
		Urban	Water	Open field	Forest	Bush	Cultivated	Total
1976	Urban	20.08	0	0.00	0.01	0.01	0.09	20.19
	Water	0.48	6.44	0.00	0.42	0.05	2.11	9.50
	Open field	0.12	0.00	1.59	0.03	0.10	1.39	3.23
	Forest	0.23	0.02	0.06	288.81	12.79	25.68	327.60
	Bush	0.15	0.00	0.14	16.19	47.92	14.53	78.94
	Cultivated	18.42	0.73	0.70	10.31	4.46	459.15	493.77
Total		39.47	7.20	2.49	315.76	65.33	502.97	933.22

Table 5. Spatiotemporal transition of LULC, 1989–2002.

Year	LULC Classes	2002						km²
		Urban	Water	Open field	Forest	Bush	Cultivated	Total
1989	Urban	39.47	0.00	0.00	0.01	0.00	0.00	39.47
	Water	0.18	6.54	0.03	0.02	0.00	0.43	7.20
	Open field	0.17	0.00	1.59	0.02	0.05	0.66	2.49
	Forest	0.34	0.01	0.01	277.27	20.52	17.62	315.76
	Bush	0.15	0.00	0.00	6.35	43.72	15.10	65.33
	Cultivated	38.65	0.03	0.37	7.14	5.14	451.64	502.97
Total		78.96	6.59	1.99	290.80	69.43	485.45	933.22

Table 6. Spatiotemporal transition of LULC, 2002–2015.

Year		2015						km²
	LULC Classes	Urban	Water	Open field	Forest	Bush	Cultivated	Urban
2002	Urban	78.86	0.01	0.00	0.04	0.00	0.05	78.96
	Water	0.00	6.56	0.00	0.03	0.00	0.00	6.59
	Open field	0.41	0.02	1.38	0.01	0.00	0.18	1.99
	Forest	0.52	0.13	0.05	271.64	4.74	13.72	290.80
	Bush	0.04	0.01	0.00	19.14	38.27	11.96	69.43
	Cultivated	59.74	0.79	0.32	15.86	4.20	404.54	485.45
Total		139.57	7.51	1.75	306.73	47.22	430.44	933.22

3.3. Spatial Orientation of Urban Land Expansion

The first decade between 1976 and 1989 accounted for limited urban enlargement throughout the entire study area, contrary to the large-scale urbanization between 1989–2002 and 2002–2015. Remarkable change of LULC (Figure 9a) and urban enlargement occurred in every direction from the city core (Figure 9b,c); yet, the extensions that took place in N-NE (Budhanilkantha area), NE-E (Jorpati, Sankhu, Gokarna area), E-SE (Bhaktapur area) and SE-S (Imadol, Godawari area of Lalitpur) are comparatively the highest. The Bhaktapur area is one of the important gateways to the valley and incorporates the large plain area where urban area expanded from 4.67 km² in 1976 to 8.3 km² in 1989. The dramatic increase from 15.2 km² in 2002 to 25.95 km² in 2015 is another large-scale expansion. Consequently, Jorpati, Sankhu area, experiences a similar trend where 2.6 km² urban area was augmented to 5.75 km² in 1989, which enlarged to 12.95 km² in 2002 and 23.75 km² in 2015 (Table 7). The areas adjacent to the major road networks that link the valley with the important hinterland villages (e.g., Lubhu, Lamatar, Chapagaun areas of Lalitpur) have gained a higher urban enlargement. S-SW (Bungmati, Khokana area) has moderately medium and SW-W (Kirtipur, Thankot area), W-NW (Sitapaila, Ramkot area) and NW-N (Goldhunga, Jitpur area) possess relatively less urban increase.

The first two areas share their western boundary with the hills, which has constrained the expansion, resulting in the comparatively lower expansion. Due to the on-going construction of the outer Ring road almost covering the whole valley, further expansion of the dense settlements in the remaining areas at the cost of cultivated land is predictable.

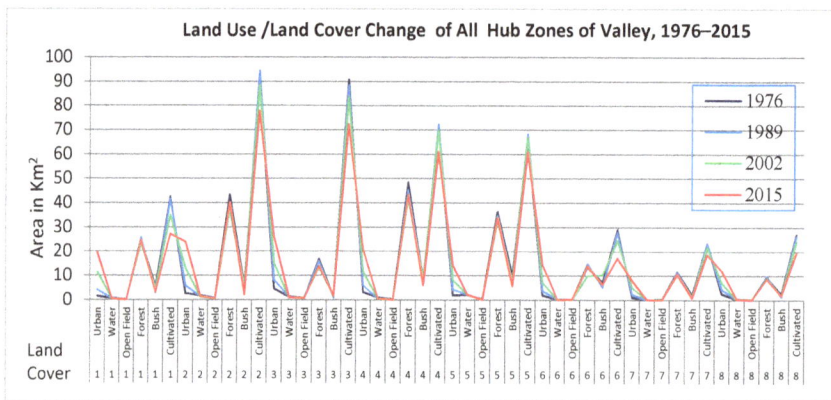

(**a**)

Figure 9. *Cont.*

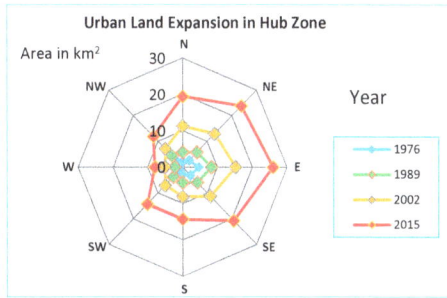

(b)

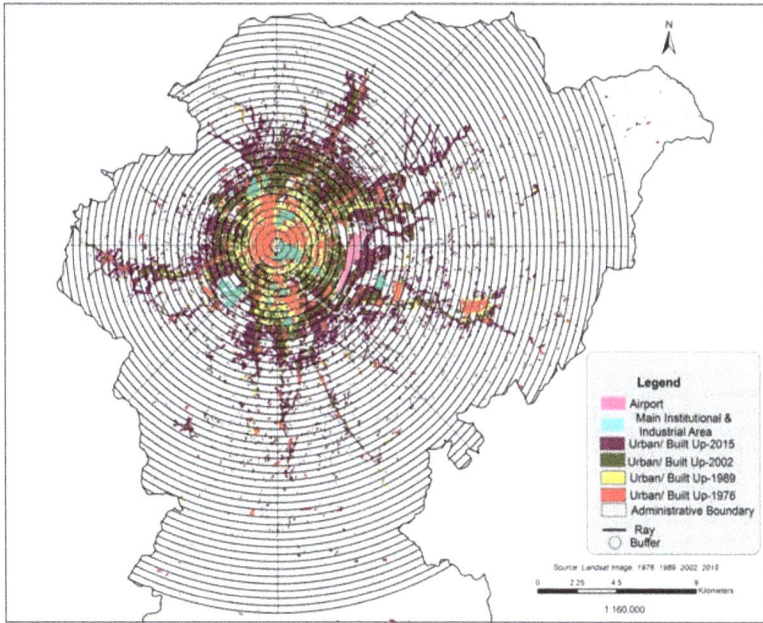

(c)

Figure 9. (**a**) Land cover change matric in all eight main zones in the valley, 1976–2015 (km²); (**b**) spider diagram of the urban expansion in eight directions in the valley, 1976–2015 (km²); (**c**) spatial orientation of urban land expansion in the valley, 1976–2015 (km²).

Table 7. Quantity of urban land expansion along 8 directions in the valley, 1976–2015 (km²).

Direction	1976	1989	2002	2015
N to NE	1.57	4.12	11.29	19.45
NE to E	2.63	5.76	12.96	23.75
E to SE	4.66	8.27	15.16	25.96
SE to South	3.16	5.93	11.28	20.69
S to SW	1.94	4.26	8.1	14.24
SW to W	1.93	3.78	7.11	14.35
W to NW	1.11	2.33	5.1	7.98
NW to N	2.58	4.36	7.07	11.9

3.4. Urban Expansion Rate

The total urban area of Kathmandu valley was 20.19 km^2 in 1976, which extended by almost two-fold in 1989 with 39.47 km^2, with an average growth rate of 7.34% per year in the following 13 years. In 2002 and 2015, the urban are reached 78.96 km^2 and 139.57 km^2, respectively. The total annual increase rate was 7.70% between 1989 and 2002 and 5.90% between 2002 and 2015. That is 6.91-times higher than that in 1976.

3.5. Socioeconomic Impacts

Kathmandu valley is the central hub for the political, educational and economic activities, employment opportunities, air transport, cultural and historical heritage. The availability of urban basic services such as roads, electricity and telephone has influenced the population acceleration in the last few decades. Thus, the fertile land has been confined due to the abrupt urbanization, and people's interest has shifted to commercial farming, mainly of horticulture and floriculture, from the substantial farming in the limited land within the valley [25].

Multiple driving forces are considered responsible for the urbanization process in Kathmandu; however, they were specific at spatial scales [9,37]. Economic opportunities in the city's core areas, population growth in the fringe and the political situation in the rural areas were identified as the major drivers of urbanization. A previous study reported that [37] physical condition was trivial in contributing to the urbanization process; nevertheless, its role greatly varied from 18% in the city, to 27% in the fringes, to 55% in the rural areas. Economic opportunities contributed 47%, 31% and 22% in urbanization in cities, fringes and rural areas, respectively. The land market business is highly concentrated in the rural and fringe in comparison to the city core due to the limited land availability. The land use-based approach is often regarded in sustainable urbanization [25].

3.6. Nepal Earthquake 2015

Nepal is listed as the 20th most disaster-prone country in the world, most vulnerable to the effects of climate change, 11th most at risk for earthquake damage and 30th most vulnerable to floods [38]. Kathmandu valley has the greatest earthquake risk among any of the 21 mega cities in the world because it lies in a very active seismic zone within a high hazard intensity (Figure 10). Seismic records of Nepal, which date back to 1255 A.D., show that destructive earthquakes happened in 1408, 1681, 1810, 1833, 1866, 1934, 1980, 1988 and 2011. The 1833 quake was the most destructive, and the 1934 earthquake impacted Kathmandu the most, resulting in more than 8500 deaths and destroying 38,000 buildings [38,39].

On Saturday, 25 April 2015, at 11:56 a.m. local time, a 7.6 magnitude (8.2 km depth) earthquake struck Barpak, Gorkha, about 76 km northwest of Kathmandu, followed by many subsequent aftershocks, the largest occurring at 12:50 local time on 12 May 2015 in Barpak, Gorkha, and Sunkhani, Dolakha, with a magnitude of 7.3. The 2015 earthquake was the most powerful jolt to hit Nepal since the quake of 1934 [40]. This earthquake killed 1746 people and severely injured 13,102 in Kathmandu valley alone. Altogether, 80,149 private properties and 321 government buildings were fully destroyed in Kathmandu alone; and 72,249 private houses and 545 government buildings partially crumbled [41]. The earthquake damaged (Figure 11a,d presents the view in Google Earth images before and after the earthquake of Kathmandu valley) many of the country's large administrative centers, including Singh durbar, Sheetal Niwas, the Supreme Court, etc. Seven UNESCO heritage sites were also damaged. The nine-story historical landmark tower Dharahara, in Nepali language, which was constructed during 1832 by the then Prime Minister Bhimsen Thapa, massively collapsed during the jolt, where a large number of visitors lost their lives (Figure 11e). Much of the damage occurred because many of the buildings across the country are not built to seismic standards, and so, any major jolt can cause them to crumble. Tens of thousands of houses, buildings and structures are in such a precarious state that a major jolt can result in harsh damage. Various structures are supported by bamboo, wooden or iron bastions to prevent from further crumbling. Now, even over a year after the quake, many structures still have not been rebuilt. Some people are living in homes with cracked foundations or holes in the stone walls. Countless numbers of people are also still reported to be living in temporary shelters in Kathmandu.

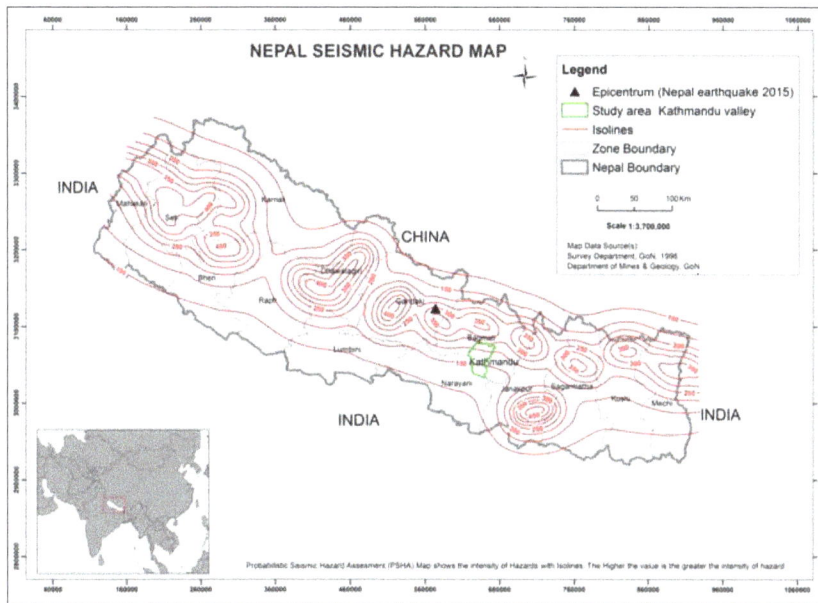

Figure 10. Seismic hazard map of Nepal.

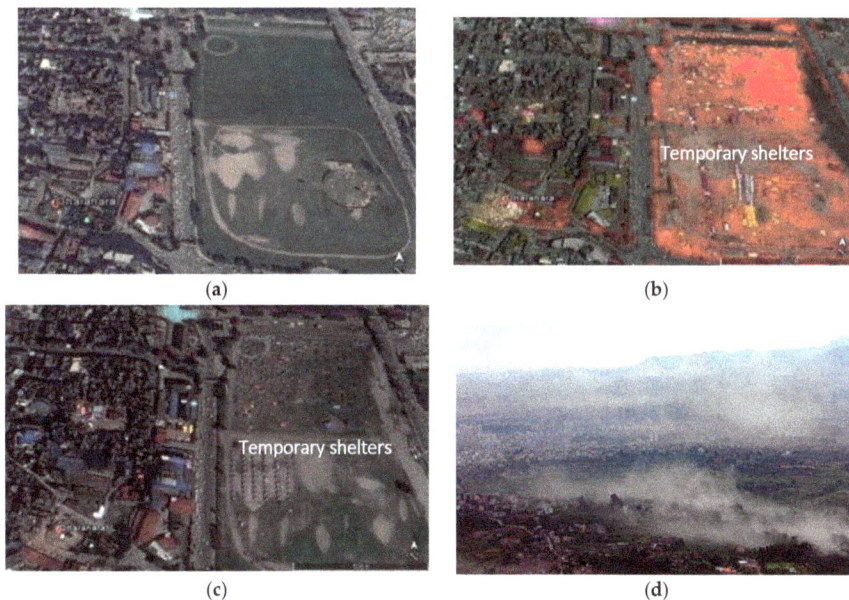

(a)

(b)

(c)

(d)

Figure 11. *Cont.*

Figure 11. Google Image of Kathmandu, City (focus on Dharahara and Tudhikhel. (**a**) Before the earthquake, 1 December 2014; (**b**) after the earthquake, 26 April 2015; (**c**) Google Image, after the earthquake, 12 May 2015; (**d**) dusty sky of Kathmandu a few minutes after the quake, 2015, photo, Bikram Rai; Bhimsen Tower, Dharahara, (**e**) before and (**f**) after the earthquake: Google search: 6 June 2016.

Due to the easy access of urban facilities, such as physical infrastructure, economic probability and advanced life, people are intending to move to city areas, resulting in villages being unattended and full of ruderals [36]. The investment of land markets and real estate has played a vital role in mushrooming the settlements in urban areas. The urban sprawl and LULC change of Kathmandu occurred without proper planning or policy, creating an area greatly at risk from earthquake damage. Limited understanding of earthquake disaster risk and lack of emergency preparedness are the major hurdles of the effective disaster risk reduction and mitigation process. Due to the lack of proper land use planning, unmanaged urban sprawl and risky settlements, many cultivable lands and forests are converted into built-up areas. Incomplete land use planning puts people at higher risk in natural disasters and increases food insecurity, as much of the area's cultivated land has been converted into shoddily-constructed homes. Due to the lack of seismic technologies, trained masons and engineers and the perceived cost of seismic-resilient construction, many buildings when constructed are seismically unsafe. To build seismically-resilient and LULC-friendly structures, effective disaster management and construction protocols must be implemented by the Nepalese government. These policies should consider mitigation and preparedness, emergency response and rehabilitation and reconstruction policies for post-disaster based on scientific criteria. Sustainable planning has the capacity to make this last disaster an opportunity to rebuild a better Nepal.

3.7. Earthquake, Land Use and Geological Formation

Spatial analysis of the geology in Kathmandu valley showed that the Gokarna formation, Kalimati formation and Chapagaun formation were all highly affected by the 2015 earthquake, while the Chandragiri formation and Kulekhani formation were both moderately affected (Figures 12 and 13). Mostly the areas enclosed within the Kalimati, Gokarna and Chapagaun formation, mainly Gongabu, Jorpati, Gokarna, Bhaktapur, Bungmati, Chapagaun and Kirtipur periphery, are the remarkable areas that obtain high urban orientation, as well as are highly impacted by the Nepal earthquake, 2015. The geological landscape affected the damage risk of Kathmandu valley, but housing construction also often determined how much damage a building experienced. Traditional homes constructed with bricks and mud mortar often suffered large amounts of damage, as well as the modern houses, which are made up of bricks and concrete, fail to meet the earthquake resilience criteria. Although concrete is a more modern material, it is heavy and often became dislocated during the quake. Additionally, poor equipment, unskilled labor, poor quality building materials and a lack of timely maintenance were risk factors for the quake damage result in fragile homes. Through spatial analysis and geological surveys, we conclude that the establishment of non-resilient, unregulated constructions on top of highly vulnerable geologic formations created the large amounts of loss and damage in Kathmandu valley. Further research is needed to fully understand different land uses and geological formations' vulnerability to earthquakes.

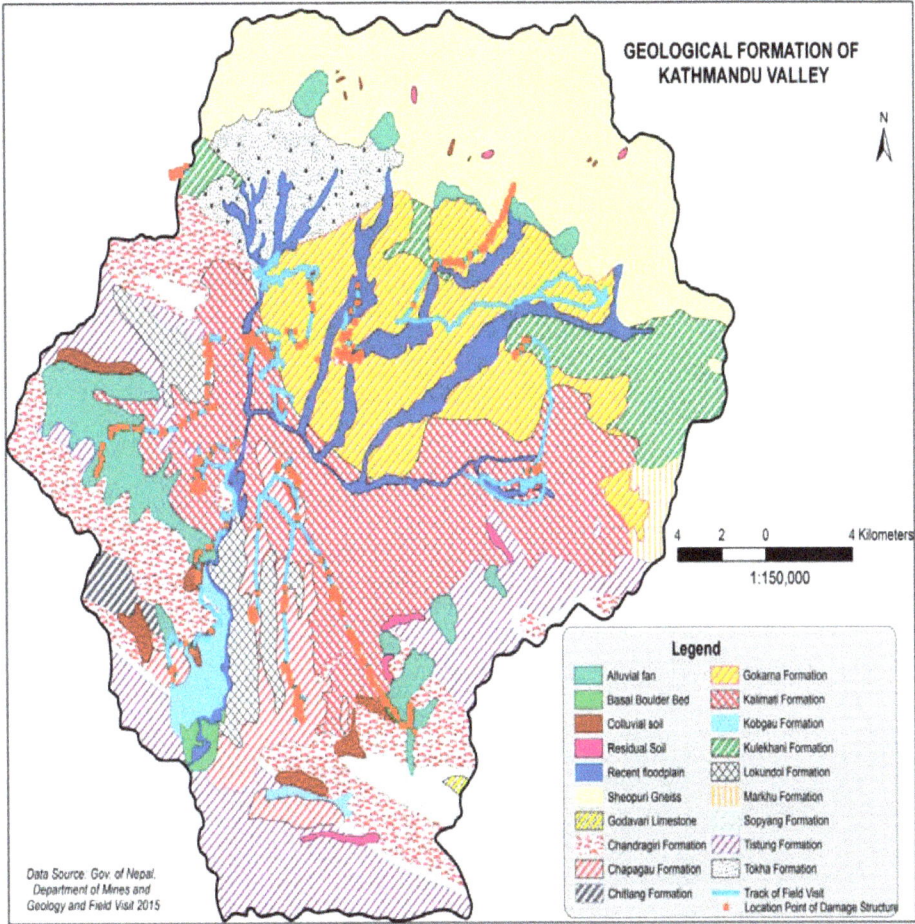

Figure 12. Overlay map of the earthquake damage structure and geological map of Kathmandu valley.

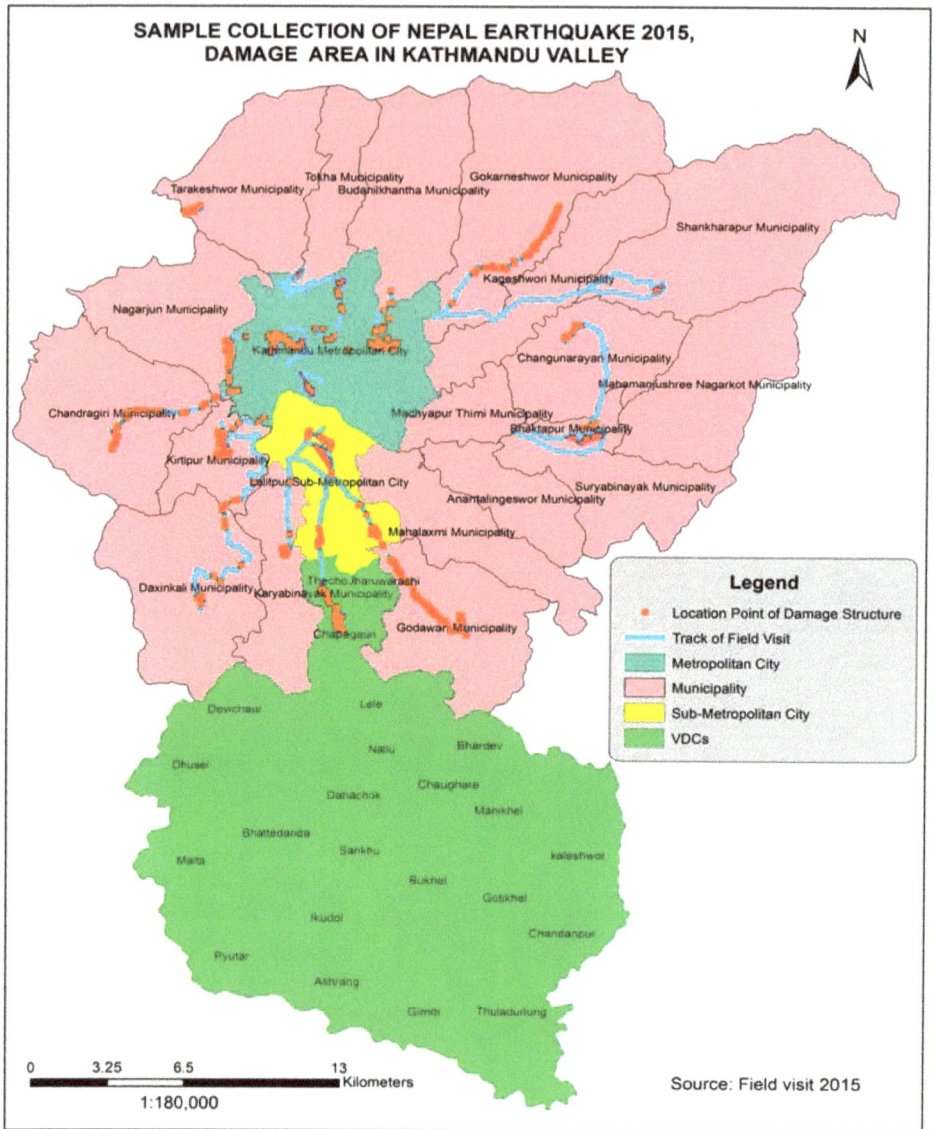

SAMPLE COLLECTION OF NEPAL EARTHQUAKE 2015, DAMAGE AREA IN KATHMANDU VALLEY

Legend

- Location Point of Damage Structure
- Track of Field Visit
- Metropolitan City
- Municipality
- Sub-Metropolitan City
- VDCs

Source: Field visit 2015

1:180,000

Figure 13. Overlay map of earthquake damage structure and administrative boundary map of Kathmandu valley (including all three district boundaries).

3.8. Planning and Policy

The government of Nepal (GoN) has introduced different land use acts and policies regarding the LULC management (e.g., Civil Code 1843, amended in 1963, Land Right Acquiring Act 1951, Survey and Measurement Act 1962, Land Act 1964, Land Administration Act and Land Revenue Act 1976, Land Acquisition Act 1977, Town Development Act 1989, National Agriculture Policy 2004, National Urban Policy 2007, Industrial Policy 2011, National Shelter Policy 2012, Agriculture Development Strategy (ADS) 2013, National Land Use Policy 2012 and 2015, Forest Policy 2015, Urban Planning and Building

Construction Guidelines, 2015); these are some of the noteworthy plans and policies promulgated by the government [42]. Similarly, several scattered laws and regulations in one way or other deal with land and its use. In 2002, the government approved a long-term development plan for Kathmandu valley prepared by the Kathmandu valley Town Development Committee. The plan was particularly based on land use criteria [25]. The plan conceptualized a vision for developing Kathmandu through the year 2020. It aimed at de-concentrating economic investments and employment opportunities in the valley core to promote natural, historical, cultural, touristic environments and delineating the urban growth boundaries to control urban growth and limit capital investments. However, the plan was jeopardized by a lack of long-term strategies, political commitment and the sustainability of growth in the valley. In 2015, Nepal adopted the Settlement Development, Urban Planning and Building Construction Guidelines. These guidelines are scientifically based, and their implementation is urgently required. Fixed scientific measures are considered to be strictly adopted in terms of infrastructure development in these newly-developing urban, fringe and rural areas. Geological studies and soil tests should be emphasized prior to construction of infrastructures. Specific road standards need to be maintained for road construction or the urban planning process. These guidelines must be followed closely for the protection of Kathmandu valley's high productivity agricultural land, green outskirts and open spaces and to smoothen the resilient urban development process.

4. Conclusions and Future Steps

The high rates of migration and population growth have directly contributed to rapid, often unmanaged urban growth in Kathmandu valley. Urbanization has occurred at the cost of fertile agricultural lands and cultural sites. Land use change in the valley has been aggravated through sporadic and persistent exploitations. Unattended fallow lands in suburban areas have worsened land degradation by welcoming invasive alien species and compromising the indigenous landscape and culture. The study area experienced rapid urbanization with the average annual urban growth rate of 7.34%, 7.70% and 5.90% between 1976, 1989, 2002 and 2015, respectively., The annual urban expansion growth rate indicates the level of urban area extension in the valley. Urban expansion direction indicates the value of socio-economic movement. Relatively the highest, moderate and the lowest urban concentration over the time period occurred in the east, southwest and west along with the northwest part of the valley, respectively.

This excessive trend of urban momentum in recent decades is being strengthened with the ongoing construction of the outer Ring-road and additional physical infrastructures, which are likely to convert the whole valley into a dense jungle of concrete in the coming decades. The government needs to introduce an effective urban plan to control the haphazard settlement practice in the city, fringe and rural areas. Geological studies, soil tests, building standards, road standards and wise land utilization practices should be conducted prior to the construction of private and public infrastructure

To this end, the monitoring of the urban expansion direction and variation through the application of multi-temporal images will be the crucial milestones, and geological information will remain as the stepping stones for the operative urban planning in the forthcoming days.

Since land use management is a fundamental part of development, our results advocate the essentiality of incorporating scientific urban planning and appropriate land use management practice for the sustainable urban development of Kathmandu valley, as well as other cities of Nepal.

Acknowledgments: The research was supported by the Chinese Academy of Sciences (CAS) President's International Fellowship Initiative (PIFI), Grant No. 2016PE022, and the National Natural Science Foundation of China (Grant No. 41471281, 41501396 and 41501473).The authors thank all of the scientists who have participated in the establishment of the database. Chris Lebos and Sushila Rijal are acknowledged for language editing. We would like to thank the Editor and the anonymous reviewers for their constructive comments and suggestions.

Author Contributions: Bhagawat Rimal designed the research idea for the paper and took overall responsibility for the study, including data collection and image analysis, preparation of figures, and finalization of the manuscript. Lifu Zhang supervised the research and provided useful suggestions throughout the preparation of the manuscript. Dongjie Fu, and Yongguage Zhai carried out the comprehensive revision of the article and gave useful advice. Ripu Kunwar contributed in putting the research in the local context. All authors revised and contributed to finalizing the manuscript. In addition, all authors approved the final manuscript.

Conflicts of Interest: The authors declare no conflict of interest.

References

1. Michetti, M.; Zampieri, M. Climate–human–land interactions: A review of major modeling approaches. *Land* **2014**, *3*, 793–833. [CrossRef]
2. Araya, Y.H.; Cabral, P. Analysis and modeling of urban land cover change in Setúbal and Sesimbra, Portugal. *Remote Sens.* **2010**, *2*, 1549–1563. [CrossRef]
3. Triantakonstantis, D.; Mountrakis, G. Urban growth prediction: A review of computational models and human perceptions. *J. Geogr. Inf. Syst.* **2012**, *4*, 555–587. [CrossRef]
4. Hunter, L.M.; de Manuel, J.; Gonzalez, G.; Stevenson, M.; Karish, K.S.; Toth, R.; Edwards, T.C., Jr.; Lilieholm, R.J.; Cablk, M. Population and land use change in the California Mojave: Natural habitat implications of alternative futures. *Popul. Res. Policy Rev.* **2003**, *22*, 373–397. [CrossRef]
5. Li, J.J.; Wang, X.R.; Wang, X.J.; Ma, W.C.; Zhang, H. Remote sensing evaluation of urban heat island and its spatial pattern of the Shanghai metropolitan area, China. *Ecol. Complex.* **2009**, *6*, 413–420. [CrossRef]
6. Zhang, H.; Zhou, L.G.; Chen, M.N.; Ma, W.C. Land Use Dynamics of the Fast-Growing Shanghai Metropolis, China (1979–2008) and its Implications for Land Use and Urban Planning Policy. *Sensors* **2011**, *11*, 1794–1809. [CrossRef] [PubMed]
7. Kanokporn, S.; Visut, I. Change of land use patterns in the areas close to the airport development area and some implicating factors. *Sustainability* **2011**, *3*, 1517–1530.
8. Kityuttachai, K.; Tripathi, N.K.; Tipdecho, T.; Shrestha, R. CA-Markov analysis of constrained coastal urban growth modeling: Hua Hin Seaside City, Thailand. *Sustainability* **2013**, *5*, 1480–1500. [CrossRef]
9. Cheng, J. Modelling Spatial and Temporal Urban Growth. Ph.D. Thesis, Faculty of Geographical Sciences Utrecht University, Utrecht, The Netherlands, 2003.
10. Yin, J.; Yin, Z.; Zhong, H.; Xu, S.; Hu, X.; Wang, J.; Wu, J. Monitoring urban expansion and land use/land coverchanges of Shanghai metropolitan area during the transitional economy (1979–2009) in China. *Environ. Monit. Assess.* **2011**, *177*, 609–621. [CrossRef] [PubMed]
11. Rimal, B.; Baral, H.; Stork, N.E.; Paudyal, K.; Rijal, S. Growing city and rapid land use transition: Assessing multiple hazards and risk in the Pokhara Valley Nepal. *Land* **2015**, *4*, 957–978. [CrossRef]
12. Sexto, J.O.; Song, X.P.; Huang, C.; Channa, S.; Baker, M.E.; Townshend, J.R. Urban growth of the Washington, D.C.–Baltimore, MD metropolitan region from 1984 to 2010 by annual, Landsat-based estimates of impervious cover. *Remote Sens. Environ.* **2013**, *129*, 42–53. [CrossRef]
13. Castrence, M.; Nong, D.H.; Tran, C.C.; Young, L.; Fox, J. Mapping urban transitions using multi-temporal Landsat and DMSP-OLS night-time lights imagery of the Red River Delta in Vietnam. *Land* **2014**, *3*, 148–166. [CrossRef]
14. Ministry of Federal Affairs and Local Development | Nepal Government. Available online: http://www.mofald.gov.np/en (accessed on 24 April 2017).
15. Sharma, P. Urbanization and development. In *Population Monograph of Nepal*; Central Bureau of Statistics: Kathmandu, Nepal, 2003; pp. 375–412.
16. Thapa, R.B.; Murayama, Y. Examining Spatiotemporal Urbanization Patterns in Kathmandu Valley, Nepal: Remote Sensing and Spatial Metrics Approaches. *Remote Sens.* **2009**, *1*, 534–556. [CrossRef]
17. Thapa, R.B.; Murayama, Y.; Ale, S. Kathmandu. *Cities* **2008**, *25*, 45–57. [CrossRef]
18. TV, R.; Aithal, B.H.; Sanna, D.D. Insights to urban dynamics through landscape spatial pattern analysis. *Int. J. Appl. Earth Obs.* **2012**, *18*, 329–343. [CrossRef]
19. Tsutsumida, N.; Saizen, I.; Matsuoka, M.; Ishii, R. Land cover change detection in Ulaanbaatar using the breaks for additive seasonal and trend method. *Land* **2013**, *2*, 534–549. [CrossRef]
20. Yoshida, M.; Gautam, P. Magnetostratigraphy of Plio-Pleistocene Lacustrine Deposits in the Kathmandu Valley Central Nepal. *Proc. Indian Natl. Sci. Acad.* **1988**, *54*, 410–417.
21. Sakai, T.; Gajurel, A.P.; Tabata, H.; Upreti, B.N. Small-amplitude lake-level fluctuations recorded in aggrading deltaic deposits of the Upper Pleistocene Thimi and Gokarna Formations, Kathmandu Valley, Nepal. *J. Geol. Soc.* **2001**, *25*, 43–51.

22. Botts, H.; Du, W.; Foust, B.; Ihinger, P.; Jeffery, T. Modeling Earthquake Risk-White Paper. Available online: http://www.corelogic.com/about-us/researchtrends/modeling-earthquake-risk-white-paper.aspx#.WUH9HNwlGpo (accessed on 24 April 2017).
23. Piya, B. Generation of Geological Database for the Liquefaction Hazard Analysis in Kathmandu Valley. Master's Thesis, International Institute for Geoinformation Science and the Earth Observation (ITC), Enschede, The Netherlands, 2004.
24. Sakai, H. Stratigraphic division and sedimentary facies of the Kathmandu Basin sediments. *J. Nepal Geol. Soc.* **2001**, *25*, 19–32.
25. International Centre for Integrated Mountain Development (ICIMOD). *Kathmandu Valley Environment Outlook (KVEO)*; International Centre for Integrated Mountain Development (ICIMOD): Kathmanud, Nepal, 2007.
26. Stocklin, J.; Bhattarai, K.D. *Geological Map of Kathmandu Area and Central Mahabharat Range (1:250,000)*; Department of Mines and Geology: Kathmandu, Nepal, 1986.
27. Central Bureau of Statistics (CBS). *Environment Statistics of Nepal and Nepal Population Report*; National Planning Commission Secretariat, Government of Nepal: Kathmandu, Nepal, 2011.
28. Bhattarai, K.; Conway, D. Urban Vulnerabilities in the Kathmandu Valley, Nepal: Visualizations of Human/Hazard Interactions. *J. Geogr. Inf. Syst.* **2010**, *2*, 63–84. [CrossRef]
29. Zhang, Z.X.; Wang, X.; Zhao, X.L.; Liu, B.; Yi, L.; Zuo, L.J.; Wen, Q.K.; Liu, F.; Xu, J.Y.; Hu, S.G. A 2010 update of National Land Use/Cover Database of China at 1:100,000 scale using medium spatial resolution satellite images. *Remote Sens. Environ.* **2014**, *149*, 142–154. [CrossRef]
30. Zhang, Z.; Li, N.; Wang, X.; Liu, F.; Yang, L. A Comparative Study of Urban Expansion in Beijing, Tianjin and Tangshan from the 1970s to 2013. *Remote Sens.* **2016**, *8*, 1–22. [CrossRef]
31. Anderson, J.R.; Hardy, E.E.; Roach, J.T.; Witmer, R.E. *A Land Use and Land Cover Classification System for Use with Remote Sensor Data*; United States Government Printing Office: Washington, DC, USA, 1976.
32. Xiao, P.; Wang, X.; Feng, X.; Zhang, X.; Yang, Y. Decting China's urban expantion over the past three decades using nighttime light data. *IEEE J. Sel. Top. Appl. Earth Obs. Remote Sens.* **2014**, *7*, 4095–4106. [CrossRef]
33. Toffin, G. Urban fringes: Squatter and slum settlement in the Kathmandu valley, Nepal. *CNAS J.* **2010**, *372*, 151–168.
34. Goldstein, M.; Ross, J.; Schuler, S. From a mountain/rural to a plains/urban society: Implications of the 1981 Nepalese consensus. *Mt. Res. Dev.* **1983**, *3*, 61–64. [CrossRef] [PubMed]
35. Muzzini, E.; Aparicio, G. *Urban Growth and Spatial Transition in Nepal, An Initial Assessment*; The World Bank: Washington, DC, USA, 2013.
36. Kunwar, R.M.; Baral, K.; Poudel, P.; Acharya, R.P.; Thapa-Magar, K.; Cameron, M.; Bussmann, R.W. Land-use and socioeconomic change, medicinal plant selection and biodiversity resilience in Far Western Nepal. *PLoS ONE* **2016**, *11*, e0169447. [CrossRef] [PubMed]
37. Thapa, R.B. Spatial Process of Urbanization in Kathmandu Valley, Nepal. Ph.D. Thesis, The Graduate School of Life and Environmental Sciences, The University of Tsukuba, Tsukuba, Japan, 2009.
38. Ministry of Home Affairs (MoHA). *Nepal Disaster Report*; Ministry of Home Affairs, Government of Nepal: Kathmandu, Nepal, 2011.
39. National Society for Earthquake Technology—Nepal (NSET). *The Kathmandu Valley Earthquake Risk Management Action Plan*; National Society for Earthquake Technology Nepal and GeoHazards International (GHI): Kathmandu, Nepal, 1999.
40. National Planning Commission. *Nepal Earthquake 2015, Post Disaster Needs Assessment*; Key Findings; Government of Nepal: Kathmandu, Nepal, 2015.
41. Government of Nepal. Nepal Disaster Risk Reduction Portal. Available online: https://drrportal.gov.np (accessed on 23 October 2015).
42. UN-Habitat. *National Land Use Plan*; Catalytic Support on Land Issues, Participatory Land Use Planning and Implementation; UN-Habitat: Pulchowk, Nepal, 2015.

land

MDPI

Article

Modeling Future Urban Sprawl and Landscape Change in the Laguna de Bay Area, Philippines

Kotaro Iizuka [1],*, Brian A. Johnson [2], Akio Onishi [3], Damasa B. Magcale-Macandog [4], Isao Endo [2] and Milben Bragais [4]

[1]   Center for Southeast Asian Studies (CSEAS), Kyoto University, 46, Yoshida Shimoadachicho, Sakyo-ku Kyoto-shi, Kyoto 606-8501, Japan
[2]   Institute for Global Environmental Strategies (IGES), 2108-11 Kamiyamaguchi, Hayama, Kanagawa 240-0115, Japan; johnson@iges.or.jp (B.A.J.); endo@iges.or.jp (I.E.)
[3]   Faculty of Environmental Studies, Tokyo City University, 3-3-1 Ushikubo-nishi, Tsuzuki-ku, Yokohama, Kanagawa 224-8551, Japan; onishi@tcu.ac.jp
[4]   Institute of Biological Sciences, University of the Philippines Los Baños, College, Laguna 4031, Philippines; demi_macandog@yahoo.com (D.B.M.-M.); mabragais@gmail.com (M.B.)
*   Correspondence: kotaro_iizuka@cseas.kyoto-u.ac.jp

Academic Editors: Andrew Millington, Harini Nagendra and Monika Kopecka
Received: 6 March 2017; Accepted: 11 April 2017; Published: 14 April 2017

Abstract: This study uses a spatially-explicit land-use/land-cover (LULC) modeling approach to model and map the future (2016–2030) LULC of the area surrounding the Laguna de Bay of Philippines under three different scenarios: 'business-as-usual', 'compact development', and 'high sprawl' scenarios. The Laguna de Bay is the largest lake in the Philippines and an important natural resource for the population in/around Metro Manila. The LULC around the lake is rapidly changing due to urban sprawl, so local and national government agencies situated in the area need an understanding of the future (likely) LULC changes and their associated hydrological impacts. The spatial modeling approach involved three main steps: (1) mapping the locations of past LULC changes; (2) identifying the drivers of these past changes; and (3) identifying where and when future LULC changes are likely to occur. Utilizing various publically-available spatial datasets representing potential drivers of LULC changes, a LULC change model was calibrated using the Multilayer Perceptron (MLP) neural network algorithm. After calibrating the model, future LULC changes were modeled and mapped up to the year 2030. Our modeling results showed that the 'built-up' LULC class is likely to experience the greatest increase in land area due to losses in 'crop/grass' (and to a lesser degree 'tree') LULC, and this is attributed to continued urban sprawl.

Keywords: landuse; change; open data; landscape; remote sensing; GIS; Markov Chain

---

1. Introduction

Urban sprawl is occurring at an accelerated pace in many developing countries worldwide due to rapid global economic and population growth coupled with globalization. Currently, 54% of the world's population lives in urban areas, and the United Nations has predicted that by 2050, 66% of the world's population will live in urban areas [1]. This rapid increase in urban population has forced nations to meet the changing demands for necessities such as food, energy, land, and water. A major concern related to this urban sprawl is land-use (LU)/land-cover (LC) change, which can dramatically alter the landscape in areas with high rates of urban expansion [2]. These LULC changes are often based on the plans of local governments to increase economic development and to support their growing populations. However, such plans may fail to consider other factors, including climate conditions, water resources, and food security [3,4]. Since population increases are expected to continue in many developing countries, governments need to take appropriate action to ensure that urbanization measures consider these factors. Thus, policymakers need to understand the historical trends in LULC change and must visualize future LULC scenarios to ensure the safety and standard of living of the residents [4]. LULC changes and related problems will likely continue to be major issues in the future [5,6], and, for governments to strategically plan future LULC development, they need to estimate the locations of LULC changes, the time scale of occurrence, and the factors driving these changes [7–9]. It is difficult to monitor LULC changes over large areas using field surveys or other ground-level data collection approaches, so many studies have instead detected and mapped LULC changes from above using satellite or aerial remote sensing data [2,10,11].

Nowadays, a great deal of attention is being paid in particular to rapidly growing cities in southeast Asia (and other regions), with the goal of understanding relationships between LULC changes and other factors such as climate change [12] and forestry [13]. The Metro Manila area of the Philippines is one of the most rapidly growing mega cities in the world [2], and, among other things, Manila's urban sprawl has threatened the ecology of the agricultural lands located along the urban fringe [14] and increased flood vulnerability in the area [15]. Other impacts related to food security are also likely to emerge in this area if agricultural lands keeps decreasing and the population keeps increasing [16]. Additionally, if Manila's urban sprawl continues into the remaining forest/agro-forest areas surrounding the city, various ecosystem services provided by these forest systems [17] will cease or decline, while, if the forested areas are instead converted to agricultural lands, it may lead to land degradations that also significantly impact the environment (e.g., increased erosion) [18,19]. Urbanization-related issues have large impacts on both human and environmental well-being, and it is clear that development in the region is causing various problems for both people and the environment. Thus, the LULC change trends in such a rapidly developing region should be monitored, and future LULC changes should be estimated to assess the impacts of future LULC changes.

The Laguna de Bay, located to the southeast of Metro Manila, is the largest lake in the Philippines and an important source of water for the population of Metro Manila and the surrounding cities and towns. Several river basins drain into this lake, and the LULC conditions of these drainage basins can have a significant effect on the lake water quality and quantity owing to different rainfall-runoff rates and pollutant loads of different LULC types [20]. Previous LULC change modeling studies in this region have focused mostly on the Metro Manila area [14,15,21,22], which does not give the full picture of the changes affecting the lake. Some past studies have also focused on specific drainage basins of the lake [19,23], but few works have studied the LULC changes of the entire surrounding landscape of the Laguna de Bay. According to the Global Footprint Network [24] Ecological Footprint Report, the Laguna lake watershed has undergone LULC changes during the last 30 years, wherein large rural areas have been converted into commercial, residential, and industrial areas. It was noted that the major LULC change occurred between 2003 and 2010, when the built up areas increased by 116%. During this period, the closed forests, observed mostly in the west, northwest, and southern parts of the lake, were reduced by 35% [25]. The lake has been the catch basin of much of the runoff from Metro Manila, so it has been heavily impacted by Metro Manila's urbanization and population increase. As Metro Manila has developed, the lake's water quality has deteriorated owing to increases in agricultural, industrial, and domestic pollution. Moreover, previous studies have reported that 66% of the lake watershed is vulnerable to erosion caused by urban growth, deforestation, and mining activities. Further, about four million tons of suspended solids flow into the lake annually [26]. From such perspectives, several local governments based along the lake as well as the national Laguna Lake Development Authority (a national government agency) have expressed interest in future LULC predictions and maps.

Using historical data, we can obtain information on past LULC changes that can be used to help model future LULC changes [11,27–29]. LULC change modeling for territorial planning has a long history, and a variety of modeling methods have been assessed at various locations, time periods, and spatial scales/spatial resolutions [30]. In the general sense, land change approaches can be roughly categorized into six types of approaches [31]; machine learning and statistical, cellular automata, sector-based economic, spatially disaggregated economic, agent-based, and hybrid approaches. In this work, we focused on the machine learning category, namely the Multilayer-Perceptron (MLP) Artificial Neural Network (ANN) approach, which is gaining attention for modeling LULC changes. ANN are powerful tools for modeling complex behaviors [32] (e.g, relationships between land transitions and their driving forces), and the usage of ANN for LULC change modeling has increased in recent years. As one example, Grekousis et al. [33] demonstrated the use of ANN to model future urban growth based on demographic time-series data. Triantakonstantis and Stathakis [32] used MLP for modeling future LULC transition probabilities based on information on past LULC changes and geomorphic drivers such as elevation, slope, and distance variables from specific land features, etc. Similar methods can be seen in a number of studies, although the number and types of driver variables utilized vary [27,34–38].

Past studies have been conducted at various geographical locations and spatial extents, e.g., the city level or provincial level. However, not many works have used the MLP method for modeling complex larger regions. In our case, we have focused on a larger ecological scale that encompasses multiple municipalities and provinces, each of which have different development policies, infrastructure, topography, climatic conditions, etc. Therefore, the MLP method for modeling the LULC transitions throughout the region becomes a strong decision tool, even when prior knowledge is lacking [31,37], and this is one of its advantages compared to other methods like SLEUTH (Slope, Land use map, Excluded area, Urban area, Transportation map, Hillside area), which require coefficient values to be set [39], and other cellular automata methods that require a suitability map [40] based on prior knowledge of change behavior. MLP can also handle a large number of data sets, so it could be a good predictor for recognizing the patterns of the changes in the area.

Our main objective in this study is to model the future LULC changes in the river basins that drain into the Laguna de Bay up to year 2030. This work implements the MLP with the Markov Chain method embedded in the Land Change Modeler (LCM) of the TerrSet software package [41]. This model is based on transition probabilities calculated using historical LULC change data and other freely/openly available geospatial datasets.

2. Study Area

The Philippines is one of the most rapidly developing countries in Asia. In particular, the Metropolitan Manila area has experienced large and rapid LULC changes owing to urban area expansion [2,6]. The Laguna de Bay, located just southeast of Manila, is the largest and most important and dynamic lake in the Philippines owing to its vital economic, political, and socio-cultural significance. With a surface area of 900 km^2, this lake is also one of the largest in Southeast Asia. From the 21 major river systems, more than 100 rivers that traverse the 292,000 ha watershed flow into the lake. The study area includes the municipalities surrounding the Laguna de Bay, located southeast of Metro Manila. Figure 1 shows that Metro Manila, situated to the northwest, has a denser urban concentration; while smaller, less densely populated cities and municipalities can be seen in the west and southwest areas surrounding the Laguna de Bay. The population of this area was estimated to be about 15 million in 2010 [42]. Agricultural lands are distributed mainly along the southwestern, southeastern, and northeastern shores of the lake. Large forest areas are located on the east side of the lake with mountain ranges (300–600 m), the highest peak of which, Mt. Banahao (2170 m), is located southeast of the lake. Just south of Laguna de Bay, Mt. Makilling (1090 m) is seen. The area is broad and the climate varies along different provinces of the area. Average annual temperatures are cooler in the mountainous areas (23 °C) than in the lower altitude plains and cities (25 °C and 27 °C respectively). Annual precipitation ranges from over 3000 mm in the east mountainous areas to 1900 mm in the western area around Manila bay. The area in focus is approximately 60 km × 80 km in the north-south and east-west directions, respectively, and it was chosen to visualize how urban sprawl will affect the LULC in the areas surrounding the Laguna de Bay. Therefore, we have ignored the center and northern areas of Metro Manila, even though it also experienced urban sprawl.

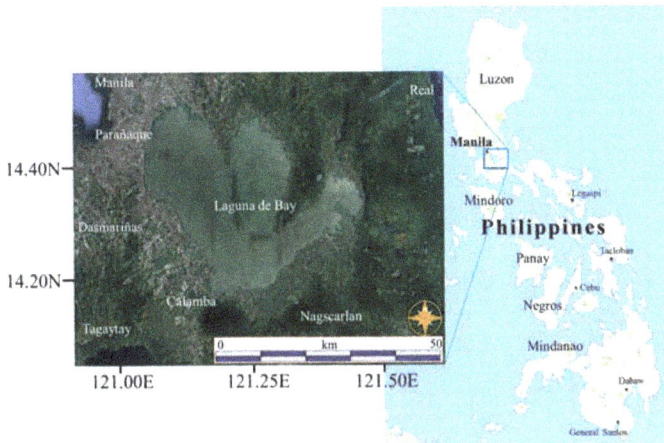

Figure 1. Overview of the study area, the Laguna de Bay district, and the surrounding environment in the Philippines. Metro Manila is shown in the northwest corner. Referenced from Google Earth [43].

3. Materials and Methods

3.1. Overview

As shown in Figure 2, the overall flowchart of the study details three main steps in generating maps of future LULC. The first step is to gather evidence of past LULC transitions by identifying the historical LULC changes in the area. For this, a map of recent LULC change from 2007 to 2015 was generated by using optical and synthetic aperture radar (SAR) satellite images from 2007 and 2015 and automated image classification techniques. Four LULC classes, including Built-up, Forests, Crop-Grass, and Water Bodies, were mapped for each year, and the LULC changes between 2007 and 2015 were identified by overlaying the two maps. The full details of the LULC change mapping methodology employed in this study are given in Johnson et al. [44], although brief information will be stated about the methods and the result of the developed LULC change map. In the second step, drivers of these historical LULC changes were identified by using various ancillary spatial datasets containing demographic, topographic, and climate information. In the third and final step, the future LULC of the area (2030) was modeled and mapped by using Markov Chain analysis.

In addition to the main procedure, the validity of the LULC change model was examined by comparing a 'simulated 2015 LULC map' with a 'reference 2015 LULC map'. The 'simulated 2015 LULC map' was generated using similar methods to those mentioned above. First, LULC maps from the years 2007 and 2010 were utilized along with the ancillary spatial datasets to model the LULC conditions in the year 2015 LULC (i.e., simulated 2015 LULC map). This map was then compared with the actual (i.e., reference) 2015 LULC map.

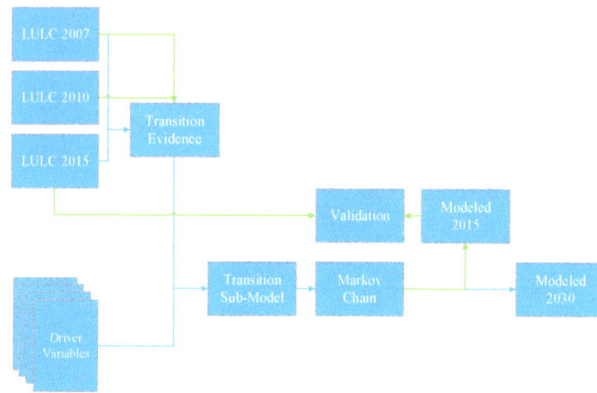

Figure 2. Overall flowchart of the methodology used in this study.

3.2. LULC Maps of 2007, 2010 and 2015

To develop a categorical LULC change map of the study area, we utilized optical (Landsat 5 and Landsat 8) and synthetic aperture radar (ALOS PALSAR-1 and PALSAR-2) satellite images from the years 2007, 2010, and 2015 and classified the pixels in the images from each year into one of four LULC classes (Built-up, Crop-Grass, Trees, and Water) using a semi-unsupervised classification approach [44]. Crop-Grass includes cropland, paddy fields, grassland, and pasture (however the majority lies within paddy and other agriculture). Trees includes forest and agroforestry plantations (e.g., coconut, banana, etc.). This four class LULC classification system, although relatively simple, was chosen because it was representative of the area and allowed us to maintain a relatively high LULC change mapping accuracy (adding more specific LULC classes usually decreases LULC mapping accuracy, and LULC change mapping accuracy even more so due to error propagation). The overall accuracy of the 2007–2015 LULC change map was estimated as 90.2%.

3.3. Evidence of LULC Change Transition

By using the two different periods of the LULC change maps, the net change in area of each LULC class was calculated, and the spatial distributions of all of the LULC changes were analyzed. The areas of transition and persistence of each LULC type within the 2007–2015 analysis was for both training and validation data with all of the driver variables when performing transition sub-modeling.

3.4. Collection and Processing of Data on Potential Driver Variables

Spatial data related to various potential drivers of LULC change were collected via the Internet. Only datasets which were openly available online were used, so our modeling approach can easily be replicated by other researchers. The drivers influencing LULC change processes are extremely diverse as well as highly variable from one location to another [45,46]. What is known is that the changes are typically the results of the local population's responses to economic opportunities [45], which gives relevance to various contextual features such as the distance from a location to nearby infrastructural features like major roads, town centers, and so forth, and many works show these kind of factors to be used as potential drivers for calibrating the change probability. These context features are also considered in this work for the model calibration. However, as mentioned before, we are not so aware of all the LULC change drivers in the area, so we want to shed some light on the question of 'What factors influence the LULC transitions?' Thus, data for a large number of possible drivers was collected to test which variables had the greatest levels of influence on LULC changes (a brief explanation is made later for why each variable was considered as potentially relevant). Table 1 shows the complete list of collected data related to these drivers; not all of these datasets were selected for the final LULC modeling.

Table 1. Complete list of all variables collected in this study.

Category	Driver	Abbreviation	Unit	Year	Data Source
Climate	Annual Mean Temperature Mean Diurnal Range	BIO1 BIO2	°C	1960–1990	PhilGIS [47]
	Isothermality	BIO3	%		
	Temperature Seasonality Max. Temperature of Warmest Month Min. Temperature of Coldest Month Temperature Annual Range Mean Temperature of Wettest Quarter Mean Temperature of Driest Quarter Mean Temperature of Warmest Quarter Mean Temperature of Coldest Quarter	BIO4 BIO5 BIO6 BIO7 BIO8 BIO9 BIO10 BIO11	°C		
	Annual Precipitation Precipitation of Wettest Month Precipitation of Driest Month	BIO12 BIO13 BIO14	mm		
	Precipitation Seasonality	BIO15	%		
	Precipitation of Wettest Quarter Precipitation of Driest Quarter Precipitation of Warmest Quarter Precipitation of Coldest Quarter	BIO16 BIO17 BIO18 BIO19	mm		
Topography	Elevation	DEM	m	2000	SRTM [48]
	Slope Aspect	Slope Aspect	degrees		
Spatial Context	Distance from Built-up Distance from Crop-Grass Distance from Trees Distance from Water	Dist_Built Dist_Crop Dist_Tree Dist_Water	Lat/Long degrees	2007	Classified LULC 2007
	Distance from Primary Road Distance from Secondary Road Distance from Tertiary Road Distance from Other Roads Distance from Canal Distance from River Distance from Stream	Dist_Road1 Dist_Road2 Dist_Road3 Dist_Road4 Dist_Canal Dist_River Dist_Stream		3 March 2016	OpenStreetMap [49]
	Distance from Golf Course Distance from Protected Area	Dist_Golf Dist_Protect		2004 2013	PhilGIS [47]
Nightlight Data	Night Light Data 2007 Night Light Data 2015 Night Light Change 2007 to 2015	NL_2007 NL_2015 NL_Ch	DN nanoWatts/cm^2/sr	2007 2015 -	NOAA Earth Observation Group [50]
Population	Population Map 2007 Population Map 2015 Population Change 2007 to 2015	Pop_2007 Pop_2015 Pop_Ch	People per hectare	2007 2015 -	WorldPop [51]

Gridded climate data with a 1 km resolution was obtained from the Philippine GIS Data Clearinghouse (PhilGIS) [47] (the original global dataset is distributed at WorldClim [52]). We included all of the climate data that was processed in the form of bioclimatic variables [53,54]. These climate variables can be considered drivers of LULC change because the agro-climatic zones with different climatic conditions can affect the suitability of agricultural lands for its productivity [55]; thus, no patterns of change in the preferred area might be identified relative to this factor. This is one challenge in our work since not many related studies implement climatic information in their calibration. The reason can be considered that, depending on the scale of the study area, the spatial variation of climatic factors might be too small to have any influence on LULC changes. However, our study area encompasses areas with different climatic conditions, which may allow us to identify if climate is affecting LULC change. Topographic data including elevation, slope, and aspect were obtained from the Shuttle Radar Topography Mission (SRTM) 1 s (30 m) digital elevation model (DEM) [48]. Topography is often a significant driver of LULC change [56] because areas with steep slopes are typically more difficult and thus less likely than flatter areas to be converted to built-up land or cropland. The SRTM DEM contains only elevation information; therefore, gridded slope and aspect data were generated by using the TerrSet software package [41]. Road and waterway data through March 03, 2016, were collected from OpenStreetMap (OSM) [49], and a 25 m grid map containing the Euclidean distance of each pixel to the nearest road/waterway was calculated by using TerrSet software [41]. The distance from roads in addition to other various LULC context features are often drivers of LULC change because more developed road networks are found with a greater rate of conversion [57]. This study focuses not only on the roads in general but also on different types/functions of roads by separating roads into detailed classes. It is expected that the importance of different road types could further distinguish the patterns of transitions throughout the study region. The case is similar for the waterways. Nightlight intensity information in the form of monthly average radiance composite images was obtained from the Visible Infrared Imaging Radiometer Suite (VIIRS) and Global Defense Meteorological Satellite Program-Operational Linescan System (DMSP-OLS) nighttime lights time series dataset [50]. The nightlight changes from 2007 to 2015 were computed by performing radiometric normalization of the images to ensure that the 2007 data had a radiance range similar to that of the 2015 data. The differences in radiance at each pixel location were then calculated. Nightlight intensity was considered a driver variable because it is strongly correlated with economic activity and the gross domestic product (GDP) [58,59]. Changes in nightlight intensity over time can also be considered an indicator of economic growth. Polygon data on the locations of protected areas were collected from the PhilGIS [47]. By using this dataset, the distance from each pixel to the nearest protected area was calculated in TerrSet [41], and all pixels located within a protected area were assumed to experience no LULC conversions in the LULC change modeling process. Unique LULC data such as golf course information were also collected and used to generate the distance information from those features. Compared to other unique LULC features such as markets and town centers, the location of a golf course would mostly not change over time, and there would be a lesser chance for new land areas to be developed as golf courses, keeping the consistency of the patterns; therefore we have used those data. Gridded population data were obtained from WorldPop [51]. This data is based on census population counts at the Barangay level but were downscaled to a 100 m × 100 m grid level utilizing various other spatial datasets, as outlined in Stevens et al. [60] and Linard et al. [61]. Population growth was calculated on the basis of population change between 2007 and 2015. All of these gridded datasets were resampled from their original resolutions to a 25 m resolution by using a cubic convolution resampling approach to match the resolution of the LULC change map (25 m) of the area. For the visual interpretation of the variables used in the model, summarized images are given in Figure S1 of the supplementary materials.

The explanatory power of all of these variables in relation to different LULC transitions was computed and examined by using Cramer's V [38]. Also known as Cramer's Coefficient (V), this method is used for quantifying the explanatory power of each variable, which is an optional quick test used to determine whether the variables are worthy of consideration in the model [29]. The final variables used for the modeling will be considered on certain criteria of this value.

3.5. Processing Transition Sub-Models (MLP)

MLP, a type of ANN method widely used for modeling complex behaviors and patterns, uses the back propagation algorithm to learn the characteristics of all the factors influencing the LULC transitions. Several studies show the advantages of MLP compared to logistic regression and other empirical models [62,63]. Further details of the MLP algorithm can be found in Riccioli et al. [37]. In our study, MLP's ability to handle a large number of input variables (some of which may be irrelevant and/or highly correlated with one another) in the model calibration process was very useful, as it allowed us to investigate over 20 explanatory variables. We focused on modeling the changes of three LULC classes; Built-up, Crop-Grass, and Trees. For all of the variables measuring 'distance from' a pixel to some geographic feature (e.g., road, built-up area, etc.), the distances were recalculated at a one year interval using that year's modeled LULC map. A random sample of 10,000 pixels from the 2007–2015 LULC change map was used for the building model. Of these, 50% were used for training and 50% were used for testing through a cross-validation process.

3.6. Change Modeling (Three Scenarios)

The probability of changes occurring in different years in the future was calculated using Markov Chain analysis, a technique for predictive change modeling that is able to model future changes based on past changes. On the basis of the observed data between the two periods (2007 and 2015 in our case), the Markov Chain computes the probability that a pixel will change from one LULC type to another within a specified period [64]. Table 2 shows the matrix of the probability that each LULC category will change to every other category (base rate), which is known as the transition probability. In this method, the probability is determined by the actual changes shown in the developed LULC map; further details have been reported by Takada et al. [65]. The target year of the modeling in the present study was set to 2030; the transition for each year was also produced to review the continuous dynamic changes in the study area. A total of three scenarios were output for the comparison. The base scenario considers the transition rate to be the same as the 2007 to 2015 change rate (i.e., a 'business-as-usual scenario'). The second scenario is if the development policy changes and the rate of LULC change reduces to half of the 2007–2015 transition rate (i.e., a 'compact development scenario'). The third scenario is if the rate of LULC change further accelerates to twice the 2007–2015 rate (i.e., a 'high sprawl scenario'). The deceleration and acceleration of the transition rates are controlled by simply half and double the values for each changing class in the transition probability matrix, respectively.

During the process of the simulation, the data of protected areas are used as constrain maps to control the process of transitions. The transition potentials associated with each transition are multiplied by the constraints map [64], so a value of 1 means unconstrained, while a values near to 0 acts as a disincentive and above 1 acts as an incentive. The protected area is given the value of very near 0 (i.e., 0.01).

Table 2. Markov transition probability matrix (business-as-usual scenario).

		Probability of Changing to (2030):			
		Built-Up	**Crop-Grass**	**Trees**	**Water**
LULC Given (2015)	Built-Up	1.0000	0.0000	0.0000	0.0000
	Crop-Grass	0.1137	0.5745	0.3118	0.0000
	Trees	0.0211	0.2372	0.7417	0.0000
	Water	0.0000	0.0000	0.0000	1.0000

3.7. Validation of Modeled Map

We assessed the validity of the model by comparing the simulated 2015 LULC (i.e., derived from LULC modeling using the 2007–2010 LULC change data) with the reference (i.e., mapped) 2015 LULC. Utilizing the 2007 and 2010 LULC maps, a LULC change model was calibrated using the same driver variables as used in the 2007–2015 calibration. Using the 2007–2010 model, the year 2015 LULC was simulated, and the projected 2015 map was compared with the reference 2015 map. Two statistical indexes were calculated for the validation; Figure of Merit (FoM) [66] and Kappa Index [67]. FoM determines the accuracy of LULC hits (model predicted change and actually observed change) compared with the sum of hits, misses (model predicted persistence but is observed change), and false alarms (model predicted change but it observed persistence), giving 0% for no match between the modeled and the reference map and 100% for a perfect match. Kappa Index is widely used in the remote sensing society for assessing the reliability of the map. Other than the standard Kappa ($K_{standard}$), Kappa for no ability (K_{no}) and Kappa for location ($K_{location}$) are computed. K_{no} considers and fixes the major problems of the standard Kappa, wherein it penalizes for large quantification error and fails to reward the simulation for specifying quantity [67]. $K_{location}$ indicates how well the grid cells are located on the landscape [64,67]. Kappa values range from -1 (no agreement) to 1 (perfect agreement). The water bodies were masked out for the computation.

4. Results

4.1. 2007–2015 LULC Changes

By using the Cross-Tabulation module, the transitions of each class from 2007 to 2015 were computed, as shown in Figure 3a. The gains and the losses for each LULC type, in ha, are also shown in Figure 3b. The transition map confirmed that the majority of changes in the Built-Up class are attributed to decreases in the Crop-Grass area, although a significant number of Trees areas are expected to become Built-Up land outside the protected forest areas.

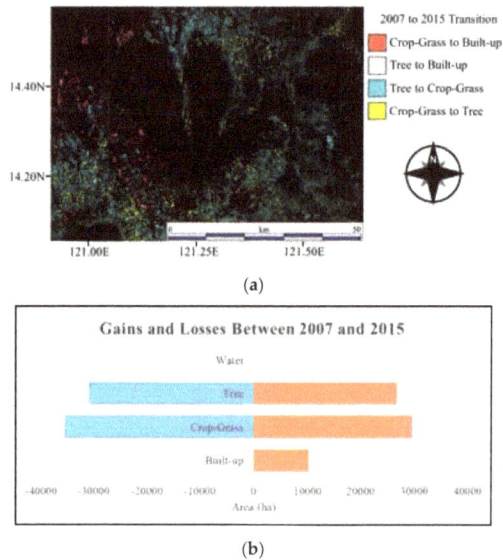

(a)

(b)

Figure 3. (a) Area of transitions occurring from 2007 to 2015 for each land-use/land-cover (LULC) type; (b) Total area of gains and losses computed within the study area for each LULC type.

For a visual understanding of the change patterns, Figure 4 shows the spatial variation of the different LULC transition trends. The spatial change pattern of the surface was created by coding areas of change with 1 and areas of no change with 0, while treating the values in a similar manner as that for quantitative values [64] and then interpolating them by using a 9th polynomial order function. The transition trend map is shown for each class transitioning to another including Crop-Grass to Built-Up, Trees to Built-Up, Trees to Crop-Grass, and Crop-Grass to Trees. This method enables identification and understanding of the spatial trends of the transition, which can provide a better comprehension of the sites of different changes at different spatial locations. The assumption shows the change patterns that occurred in the area from 2007 to 2015. For the Built-Up class, significant changes were detected from north to south along the west side of the Laguna de Bay. High probabilities of transition were detected from the Trees (Crop-Grass) class to Built-Up class north (west) of the center of the study area. The trends for the Trees class changing into Crop-Grass were most dominant north of the center of the study area, followed secondly by the southwest area and thirdly by the south to southeast area. The largest changes were recognized at the southwestern side;

however, smaller but important signs were detected at different parts of the surrounding environment, which possibly has relationships with the smaller cities located nearby. The changes of Crop-Grass to Trees were located mostly at the southern side of the study area, which is near protected areas. Thus, the Grass-Shrub-type LC is slowly changing into denser and taller vegetation due to the absence of effects from human activities.

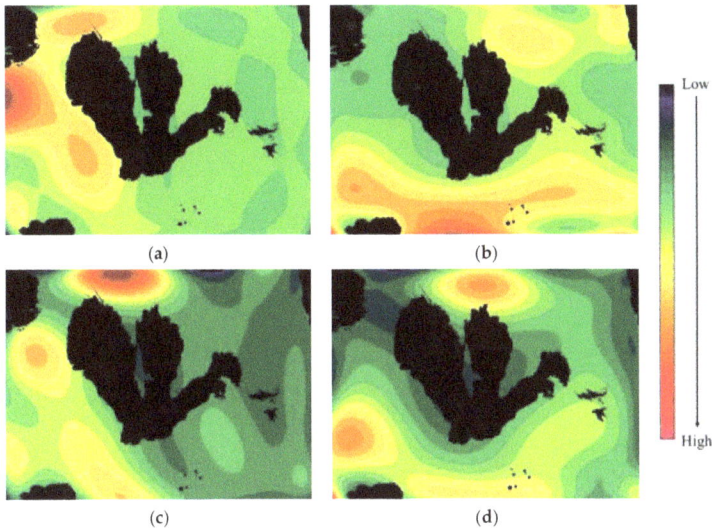

Figure 4. Patterns or trends of transition through space for (**a**) Crop-Grass to Built-up; (**b**) Crop-Grass to Trees; (**c**) Trees to Built-up; and (**d**) Trees to Crop-Grass. Red (blue) color indicates the higher (lower) chances of transition. The data contain no specific values because no mean is represented.

4.2. LULC Modeling

4.2.1. Potential Explanatory Power of Driver Variables

Table 3 shows the driver variables selected for modeling of the transition potential. Each variable has a potential explanatory power that describes the strength of its relationship to the actual transition of the classes and is computed by contingency table analysis. V takes values from 0 to 1. Values near 0 show little association between variables, whereas those near 1 indicate strong association. A value of about 0.15 contains little information for explanation; more than 0.4 is considered to be a good variable [64]. Only the final variables selected for the model are listed in the table. The selection criteria was that a variable had either (i) an overall V value greater than 0.3 or (ii) V values greater than 0.15 for all of the individual LULC classes. For example, although for *Dist_Protect* the overall V value is below 0.3, each class shows values above 0.15.

Table 4 shows the priority of the drivers compared with other variables, showing which are the most and least influential. Table 4 (a) to (d) show the accuracy and rankings of each variable that has the largest effect on the skill of the model for (a) Crop-Grass to Built-Up; (b) Crop-Grass to Trees; (c) Trees to Built-Up; and (d) Trees to Crop-Grass. The accuracy is based on the results of the 5000 testing pixels. The most (least) influential variable for model (a) was *NL_2015* (*BIO12*). For model (b), the most (least) influential was *DEM* (*NL_2015*). For model (c), the most (least) influential was *Pop_2007* (*Road_Dist2*). For model (d), the most (least) influential was *Slope* (*NL_ch*). The accuracy of each transition model in explaining its overall power for detecting the correct changes are Crop-Grass to Built-Up, 74.21%; Crop-Grass to Trees, 70.37%; Trees to Built-Up, 90.91%; and Trees to Crop-Grass, 74.31%.

Table 3. Test of the explanatory power (Cramer's V) of each variable.

	BIO3	BIO6	BIO7	BIO12	BIO15	BIO19	DEM	Slope	Dist-Built
Overall	0.3426	0.4680	0.4208	0.3250	0.4934	0.3757	0.6107	0.5649	0.3609
Built-up	0.2769	0.2908	0.2815	0.2803	0.5390	0.2748	0.4411	0.3142	0.3517
Crop-Grass	0.1242	0.1934	0.2315	0.1095	0.2223	0.1603	0.3273	0.3434	0.3583
Trees	0.4517	0.6166	0.6078	0.4476	0.6069	0.5680	0.6465	0.6220	0.2905
Water	0.4382	0.6450	0.5032	0.3886	0.5268	0.4256	0.8902	0.8400	0.4193

	Dist_Crop	Dist_Tree	Dist_Water	Road_Dist1	Road_Dist2	Road_Dist3	Road_Dist4	Road_River	Road_Canal
Overall	0.4469	0.5144	0.4352	0.3046	0.3767	0.3793	0.3921	0.3398	0.4165
Built-Up	0.1702	0.1664	0.2347	0.3634	0.4570	0.3979	0.3749	0.2757	0.4865
Crop-Grass	0.2912	0.2890	0.1899	0.2226	0.2815	0.3151	0.3746	0.2452	0.1678
Trees	0.2946	0.4073	0.6398	0.2488	0.3686	0.2974	0.2387	0.1875	0.5701
Water	0.7692	0.8848	0.5588	0.3390	0.3628	0.4630	0.5084	0.5261	0.3484

	Dist_Stream	Dist_Golf	Dist_Protect	Pop_2007	Pop_2015	P0p_Ch	NL_2007	NL_2015	NL_Ch
Overall	0.3612	0.3149	0.2484	0.4943	0.4910	0.5363	0.4659	0.4047	0.3218
Built-Up	0.3735	0.3941	0.2070	0.7183	0.7180	0.7132	0.6368	0.6597	0.4658
Crop-Grass	0.3159	0.1679	0.2147	0.3932	0.3838	0.4545	0.2989	0.1654	0.1922
Trees	0.3313	0.3677	0.2214	0.3731	0.3722	0.3477	0.5213	0.3142	0.3649
Water	0.4078	0.2746	0.3285	0.3308	0.3234	0.5002	0.2907	0.2046	0.1289

Table 4. The sensitivity of the model in maintaining selected inputs. The output shows the accuracy of the case in which all combinations of variables were used except for one to remain constant. Together it shows the ranking of variables from most to least influential given to the models for (a) Crop-Grass to Built-Up; (b) Crop-Grass to Trees; (c) Trees to Built-Up; and (d) Trees to Crop-Grass.

Variable Name	Model	Accuracy (%)				Influence Order			
		(a)	(b)	(c)	(d)	(a)	(b)	(c)	(d)
With all variables		74.21	70.37	90.91	74.31		N/A		
BIO3	Var.1 constant	74.00	70.39	90.90	74.35	13	19	15	24
BIO6	Var.2 constant	73.98	70.21	90.91	74.34	12	9	20	23
BIO7	Var.3 constant	74.16	70.38	90.91	74.29	21	18	21	20
BIO12	Var.4 constant	74.27	70.35	90.84	74.36	27	14	10	26
BIO15	Var.5 constant	73.62	70.39	90.99	74.26	5	20	24	18
BIO19	Var.6 constant	74.24	70.39	90.90	74.36	25	21	16	25
DEM	Var.7 constant	73.54	68.01	90.84	73.21	4	1	11	3
Slope	Var.8 constant	73.82	68.41	90.52	71.01	10	2	7	1
Dist_Built	Var.9 constant	74.16	70.39	90.92	74.14	20	22	22	15
Dist_Crop	Var.10 constant	74.21	70.37	90.91	74.30	23	17	18	21
Dist_Tree	Var.11 constant	74.21	70.37	90.91	74.31	22	16	17	22
Dist_Water	Var.12 constant	74.09	69.94	90.50	74.13	16	5	6	14
Road_Dist1	Var.13 constant	74.15	70.31	90.99	74.02	17	10	25	6
Road_Dist2	Var.14 constant	73.76	70.34	91.08	74.07	9	12	27	7
Road_Dist3	Var.15 constant	74.08	70.21	91.00	73.92	15	8	26	5
Road_Dist4	Var.16 constant	74.15	70.35	90.91	73.87	18	13	19	4
Dist_Canal	Var.17 constant	74.26	69.20	89.15	73.07	26	3	3	2
Dist_River	Var.18 constant	74.15	70.33	90.92	74.16	19	11	23	16
Dist_Stream	Var.19 constant	74.24	70.41	90.89	74.10	24	24	13	10
Dist_Golf	Var.20 constant	74.06	70.43	90.82	74.13	14	25	9	13
Dist_Protect	Var 21 constant	73.76	69.57	90.89	74.09	8	4	14	9
NL_2007	Var.22 constant	73.67	70.36	90.12	74.10	7	15	5	11
NL_2015	Var.23 constant	72.98	70.53	90.59	74.28	1	27	8	19
NL_Ch	Var.24 constant	73.38	70.47	90.85	74.37	3	26	12	27
Pop_2007	Var.25 constant	73.30	70.10	88.55	74.08	2	6	1	8
Pop_2015	Var.26 constant	73.64	70.12	88.74	74.11	6	7	2	12
Pop_Ch	Var.27 constant	73.82	70.39	89.90	74.22	11	23	4	17

4.2.2. LULC Change Modeling and Its Landscape

For visual interpretation of the dynamic changes in LULC classes for each stage of the modeling (each year of 2016–2030), Video S1 is provided in the supplementary material (for business-as-usual scenario). Here, we discuss only the beginning (2015), middle (2023), and end (2030) years of the LULC map. Figure 5 shows that changes occurred from the modeling at the Laguna de Bay region. A few significant characteristic trends of change depending on each LULC class can be identified. For the Built-Up classes, the first large change is the expansion of urban areas spreading more southward and expansion at the west side of the Laguna de Bay. This type of trend has occurred in the past. Google Earth images from the 1980s in those regions [68] show strong evidence of rapid LULC change in the southern part of Metro Manila and at the west side of the Laguna de Bay. Built-Up areas have expanded in the southwestern part of the study area near a smaller lake and show development along road infrastructures at the east side of this lake. At the east side of the Laguna de Bay, significantly fewer LULC changes have occurred compared with the west side because the east side of the lake consists mainly of rural areas with low population density and less infrastructure that are thus not affected by rapid development.

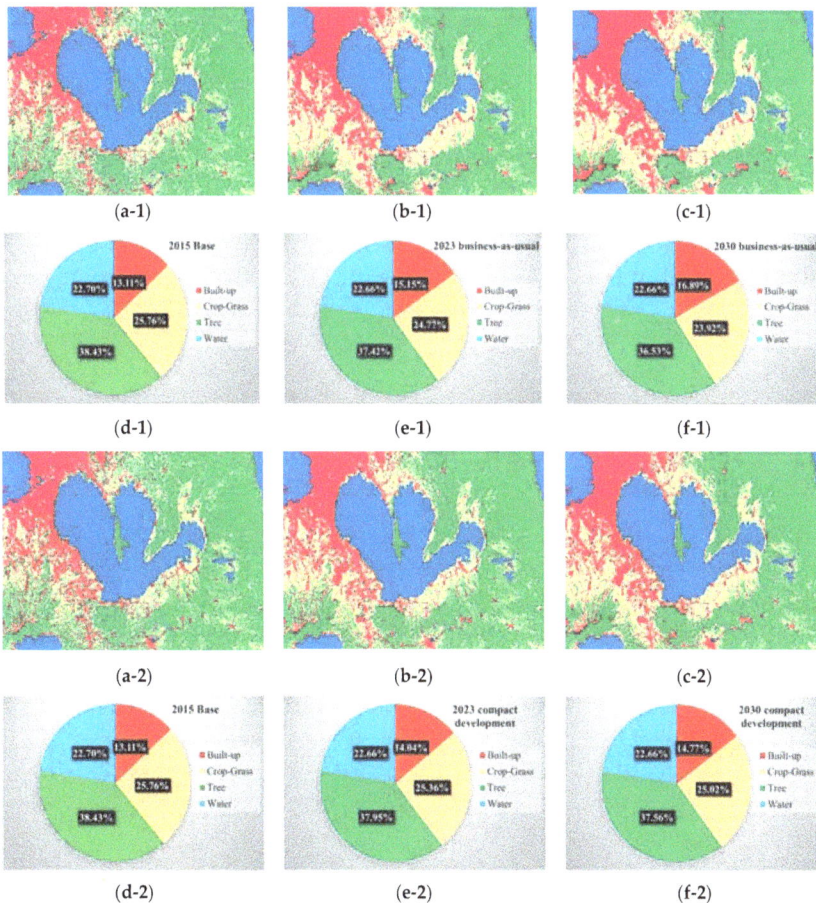

(a-1)   (b-1)   (c-1)

(d-1)   (e-1)   (f-1)

(a-2)   (b-2)   (c-2)

(d-2)   (e-2)   (f-2)

Figure 5. *Cont.*

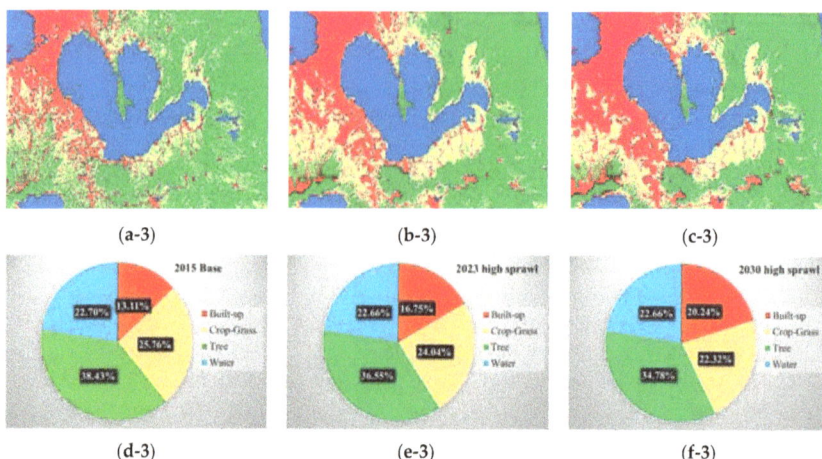

Figure 5. Land-use/land-cover (LULC) modeling of the Laguna de Bay environment for (a) 2015; (b) 2023; and (c) 2030. (a–c) shows the hard classified map; (d–f) shows the percentage of each LULC class within the study area. The numbers indicated after the alphabet represent the different modeled scenarios. Thus (a-1) is the business-as-usual scenario; (a-2) is the compact development scenario; and (a-3) is the high sprawl scenario. Legends in the pie-chart correspond to the colors of the LULC class in the hard classified map.

For the Crop-Grass class, the main changes were detected in three areas. The first includes changes along with the development of Built-Up areas at the west side of Laguna de Bay. In addition, the Crop-Grass class area has expanded and has ensured a considerable amount of area by invading the Trees class. The second site is at the east coast of the Laguna de Bay, where Built-Up areas have not changed significantly, although the Crop-Grass class has again ensured its area by invading large amounts of Trees class areas. The third location is at the center north of the Laguna de Bay, where a similar trend of Crop-Grass invading the Trees class was noted. Future scenarios can change depending on whether countermeasures for protecting the forests were planned and implemented. Similar to using the protected areas as constraining images for unchanges, probability maps for specific regions can be applied to reduce the transition probabilities of those areas. These types of planning scenarios for the forested areas can be considered a strong decision tool for planning REDD+ (Reducing Emissions from Deforestation and forest Degradation in developing countries) actions [69] because total changes in forests can be easily calculated quantitatively, enabling sound estimation of the CO_2 uptakes in those regions.

For the Trees class, the mountainous area at the east side of the Laguna de Bay and the protected areas show increments of Trees Class transition from the Crop-Grass. This occurrence depends on th ages of the people living in those rural areas, which have higher elevations and rugged terrain. When the residents become older, the land becomes more difficult to access. Agricultural lands would then begin to change into abandoned areas; thus, the trend in Crop-Grass LULC type would decrease. Other possible factors include the impact of the National Greening Program of the government [70]. In other regions, the majority of the Trees class trend shows a decrease in total area owing to its transition to the Built-Up and Crop-Grass classes.

4.2.3. LULC Change Statistics

Figure 6 shows the accumulated increase in land area of each transition class from the starting year of 2015 to 2030 in yearly increments for the business-as-usual scenario, compact development scenario, and high sprawl scenario. In the business-as-usual scenario, most of the LULC change classes showed increases in total area and different growing rates; however, the Crop-Grass to Trees class showed a decrease in area beginning in 2026. The Trees to Crop-Grass and Crop-Grass to Trees classes showed a polynomial-like increase, resulting in increases of 26,409 ha and 21,166 ha, respectively, in 2030. The Crop-Grass to Built-Up and Trees to Built-Up classes showed a linear increase of land area, resulting in increases of 14,137 ha and 3946 ha, respectively, in 2030. These changes are limited to the study area. More meaningful values might be extracted when areas are divided according to administrative boundaries. In-depth information on the LULC changes for each municipal boundary is given in Spreadsheet S1 in the supplementary material. For the compact development scenario, all of the classes show a linear-like increase compared to the business-as-usual scenario. The Trees to Crop-Grass and Crop-Grass to Trees classes shows increases of 18,274 and 16,036 ha respectively, which is about 70% of the business-as-usual scenario changes. The Crop-Grass to Built-up and Trees to Built-up showed 5783 and 2000 ha increases, indicating slightly less than 50% of their areas modeled in the business-as-usual scenario. These values are similar to the modeled 2021 LULC in the business-as-usual scenario. The high sprawl scenario shows a similar pattern to the business-as-usual scenario but with a steeper increase and faster point of decrease in increments for the Crop-Grass to Trees class. The Trees to Crop-Grass and Crop-Grass to Trees classes show increases of 30,021 and 20,182 ha, respectively. The former class shows a 113% increase, but the latter remains at 95% of that of the business-as-usual scenario, showing that more Crop-Grass conversions are occurring. Crop-Grass to Built-up and Trees to Built-up classes also show an increase compared to business-as-usual scenario (26,446 and 7836 ha respectively, approximate double the area of business-as-usual scenario).

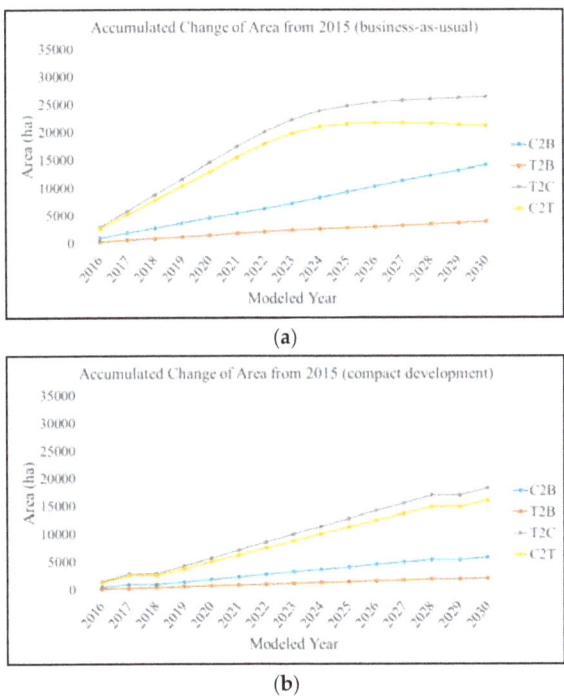

(a)

(b)

Figure 6. *Cont.*

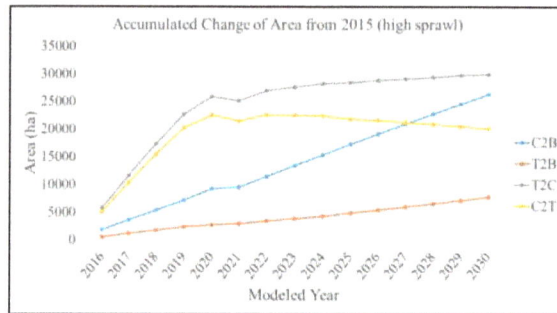

(c)

Figure 6. Trends of LULC class and net area of increase at each stage of the modeling. The base year is 2015. C2B, Crop-Grass to Built-up; T2B, Trees to Built-up; T2C, Trees to Crop-Grass; C2T, Crop-Grass to Trees. (**a**) Business-as-usual scenario; (**b**) compact development scenario; (**c**) high sprawl scenario.

4.3. 2007–2010 Model Validation

Firstly, FoM was carried out by computing hits, misses, and false alarms between the modeled and reference LULC maps of the year 2015. The total number of pixels with hits, misses, and false alarms was 517,671, 642,064, and 964,127 respectively, resulting in a FoM of 24.37%. Secondly, Kappa statistics were computed using the modeled and reference LULC maps. The calculated Kappa statistics were: $K_{standard} = 0.5825$; $K_{no} = 0.6620$; and $K_{location} = 0.6217$. A Kappa value of 1 indicates total agreement and 0 indicates totally by chance. This can be interpreted as for example using the $K_{standard}$ value that the modeled 2015 map is 58% better than a chance agreement.

5. Discussion

5.1. Influences of Driver Variables Overview

Generally, the topographical drivers such as *DEM* and *SLOPE* influenced the LULC transitions for all classes. As expected, areas of higher elevations and areas with steep slopes tended to experience higher rates of transition to trees, while lower and flatter areas were more likely to convert from tree-covered areas to built-up or agricultural lands. Context drivers showed a strong influence for all LULC change classes other than the Crop-Grass to Built-up class, with the *Dist_Canal* variable being the top candidate for the other three transition classes. These transitions were thus affected by water availability. Road infrastructure has shown also importance (and also road type), which is logical because they are highly related to people's mobility. The nightlight and population data showed a straightforward result of influencing changes to Built-up land. For example, Crop-Grass to Built-up transitions occurred more frequently in areas with higher populations, while Trees to Built-up transitions occurred more frequently in lower populated areas. Biophysical drivers (climatic variables) did not show as much influence as expected, although some relations were shown for variables such as *BIO15* (precipitation seasonality). Our primary expectation was that LULC changes to Crop-Grass might have some relationship with climate, but this was not the case in our study (the slope and population of the nearby area were much more significant drivers).

5.2. Scenarios of the Future

Comparing the three different scenarios of future changes that we modeled in this study, although the locations of the transitions did not change significantly (because all three scenarios were based on the same transition model), the quantity and rate of LULC changes differed. If we look at the compact development scenario, where the transition rate is half of business-as-usual scenario, the conversion to Built-up in 2030 does not catch up even to that of 2022 in the business-as-usual scenario. On the other hand, if we double the rate, the development of built-up area would be mostly completed at 2023 compared to 2030 for the business-as-usual scenario. Studies on the potential impacts of these development scenarios on the local environment are still under progress, although we have concerns related to crop production, biodiversity loss (including losses of patches and corridors for wildlife), changes in local climate due to increasing heat fluxes, flood vulnerability, and river and lake water quality. A work by Wijesekara et al. [71] shows good practice of how modeling results can be used as decision/planning tools.

To achieve higher accuracy for the modeling results, two possible factors can be investigated; improving the accuracy of the LULC change map of the region and/or incorporating the zoning policies and development plans of the local governments into the modeling process. The development plans can help determine better scenarios and model areas with higher chances for transitions and the manner of development. A limitation of future modeling, especially using Markov Chains that possess stationary distribution, is that the development rate is considered from past evidence of the changes. If this rate remains constant, the trend is closer to actual trend of future change. However, past information does not always explain future modeling, meaning this can vary depending on decisions by the government or local authorities.

5.3. Other Relating Works

In this section, we compare the results of our study with those of other similar studies. First of all, how does our LULC change modeling accuracy compare to that of other similar studies? As stated in the results section, the FoM was calculated in our study as 24.37%. If we look at other similar studies for comparison, the FoM values range from 1% to 59% depending on the spatial resolution, spatial extent, and number of LULC classes mapped [40]. To compare with our study, the Twin Cities or Detroit case [40] would be the most appropriate, as these studies used maps with a similar spatial resolution, number of pixels, and number of LULC classes to our study. These two studies had FoM values of 11% and 15% respectively; slightly lower than our FoM, even though we included one additional LULC class. The studies with much higher FoM values typically had access to site-specific driver variable information (local land allocation plans) and/or aggregated modeling results to coarser spatial resolutions (i.e., larger pixel sizes) to reduce the number of locational errors [40]. Although the LULC models implemented differ among studies and have different reference characteristics, this kind of comparison can give a general idea of the modeling accuracy of our study.

Comparing our study to others that used the same MLP Markov Chain modeling method that we used [32,34–38,72], we can find a few important aspects for discussion. One is that many previous studies did not provide reasoning for why certain driver variables were considered or discuss why certain factors were heavily influencing different types of LULC conversions. Although the number and types of driver variables investigated differ among studies, the level of importance of each variable and the reasons why the variable is affecting LULC changes also likely varies due to different social demands and issues in each study area. By understanding the factors driving LULC conversions, local governments can have a better idea of the issues that region that is facing. We addressed this in our study by first hypothesizing why each variable may be relevant and then by measuring the influence of each variable and discussing why certain variables were relevant for different types of LULC transitions. A second point that we would like to discuss is related to model behavior. Olmedo et al. [36] showed an important aspect of changes, which could not be simulated due to the acceleration of changes that occurred in the reference year, which did not show during the calibration time period. The MLP method is a stationary model, so it is determined that it will keep the same transition rate. This means, if the development scenario changes in the future, it could either increase or decrease its changing rates, resulting in an inaccurate projection. This issue is one reason why we simulated three scenarios of the future LULC in our study area. However, most other works only simulate one future scenario. Due to this model behavior issue and also to provide useful options to government agencies in the area, we recommend considering at least a few different scenarios in addition to the business-as-usual scenario. A third point we would like to emphasize is that there should be some general consensus for presenting model validation results. For instance, one issue with Kappa and other traditional LULC accuracy metrics is that they typically give high accuracy values in study areas with few LULC changes and much lower accuracy values in areas with many LULC changes [37,38]. On the other hand, because the FoM is a ratio metric, it is unaffected by the quantity of LULC changes, so it is a good practice to also report FoM in LULC change modeling studies (as we have done in this study).

5.4. Accomplished Tasks and Future Works

This work follows similar processes to other MLP Markov Chain studies to model the future landscape of Laguna de Bay. Using Landsat and SAR images, a spatial resolution of 25 m was achieved, showing finer information compared to using Landsat data alone. Developing LULC maps in tropical regions such as the Philippines, Indonesia, or other Southeast Asian countries is a difficult task due to frequent cloud cover (which obstructs the view of the land surface for optical satellite images). Therefore utilizing SAR data was an advantage in developing the historical LULC change data. We have worked on finding the characteristics of the changing pattern and compared these with other related studies. The 27 variables we considered was an enormous amount, around double the number of variables from the study that showed the highest number (14 drivers [72]). We have found explainable relations with the drivers used, and the study gave a clear idea of how LULC is likely developing in this area. Although we did use a large number of driver variables, we could not take into account the local zoning policies of the cities and towns in the study area (due to a lack of data availability), which also have a significant effect on LULC change. It is a very time-consuming and challenging task to visit all of the municipalities in the area to collect (and possibly digitize) their zoning and development plans, but this would probably further enhance the modeling results.

We found that urban sprawl, which was the focus of this study, is expected to continue occurring throughout the future timeframe that we considered (2016–2030). Looking at the surrounding area as a continuous landscape, we were concerned especially with how the existing agricultural and forested lands could be maintained in the future. We plan to continue the work to present a scientific standard for how these lands should be preserved according to various social-environmental impacts, which could be caused by the future LULC changes.

6. Conclusions

The objectives of this study were to model and map the future LULC of the Laguna de Bay area of the Philippines, which has a significant impact on the lake's water quantity and water quality. The study area is significant mainly because of the importance of the lake to the population of Metro Manila and the surrounding cities and towns. The future LULC was modeled and mapped by using Markov Chain analysis, and the transition probabilities were calculated by using historical LULC change data and freely available data related to the drivers of the LULC changes such as demographic, economic, topographic, infrastructure, and climate variables. Three questions are addressed in this work: (1) Where are different LULC changes taking place?; (2) What are the variables that best explain the changes (e.g., what are the drivers)?; and (3) What is the rate at which these changes are likely to take place? The major LULC changes in the area included an increase in Built-Up areas at the west side of the Laguna de Bay, south of Metro Manila, and changes of many areas between the Crop-Grass and Trees classes, owing to logging, cropland abandonment, reforestation, and other factors. In total, approximately 7800 to 44,000 ha of land within the study area are modeled to be converted to the Built-Up class by 2030, depending on the development scenario. We tested three scenarios: 'business-as-usual', 'compact development', and 'high urban sprawl'. The study has shown the extent of the changed areas in addition to the patterns and the locations of these future changes and identified the variables considered to be significant drivers of these changes (which varied for different types of LULC transitions). This information can be used by decision makers in deciding the necessary actions for preventing issues that might arise if such developments occur. The increase in Built-Up areas can affect local environmental factors such as temperature, local climatic conditions, and ecological effects [73,74]. These issues could also directly affect biodiversity and health by leading to an increase in the number of mosquitos carrying malaria and dengue fever [75,76]. Flooding is likely to become more frequent owing to the increase in urban areas; therefore, more strategic measures need to be developed to mitigate the impacts. The method implemented in this study can be used as a tool for making more informed decisions. Future work will assess the environmental impacts from future changes in the urban environments according to the various scenarios of development.

Supplementary Materials: The following are available online at www.mdpi.com/2073-445X/6/2/26/s1. **Figure S1**: Interpretation of all driver variables included in the model. (a) BIO3; (b) BIO6; (c) BIO7; (d) BIO12; (e) BIO15; (f) BIO19; (g) DEM; (h) SLOPE; (i) Dist_Built; (j); Dist_Crop; (k) Dist_Tree; (l) Dist_Water; (m) Road_Dist1; (n) Road_Dist2; (o) Road_Dist3; (p) Road_Dist4; (q) Dist_River; (r) Dist_Canal; (s) Dist_Stream; (t) Dist_Golf; (u) Dist_Protect; (v) Pop_2007; (w) Pop_2015; (x) Pop_Ch; (y) NL_2007; (z) NL_2015; (A) NL_Ch; **Spreadsheet S1**: Each LULC class total area (hectare) within each municipal boundaries for the 'business-as-usual', 'compact development', and 'high urban sprawl' scenarios; **Video S1**: LULC dynamics for Laguna de Bay region 2016–2030.

Author Contributions: K.I. designed, collected, and analyzed the data and took the lead in writing this article. B.A.J. collected and analyzed the data and edited the manuscript. A.O., D.B.M.-M., I.E., and M.B. gave critical information and revisions to this manuscript in all sections to improve in the quality of the work.

Conflicts of Interest: The authors declare no conflict of interest.

References

1. *World Urbanization Prospects: The 2014 Revision (ST/ESA/SER.A/352)*; Population Division, Department of Economic and Social Affairs, United Nations: New York, NY, USA, 2015.
2. Taubenböck, H.; Esch, T.; Felbier, A.; Wiesner, M.; Roth, A.; Dech, S. Monitoring urbanization in mega cities from space. *Remote Sens. Environ.* **2012**, *117*, 162–176. [CrossRef]
3. Gill, S.E.; Handley, J.F.; Ennos, A.R.; Pauleit, S. Adapting Cities for Climate Change: The Role of the Green Infrastructure. *Built Environ.* **2007**, *33*, 115–133. [CrossRef]
4. Satterthwaite, D.; McGranahan, G.; Tacoli, C. Urbanization and its implications for food and farming. *Philos. Trans. R. Soc. B* **2010**, *365*, 2809–2820. [CrossRef] [PubMed]
5. Oliver, T.H.; Morecroft, M.D. Interactions between climate change and land use change on biodiversity: Attribution problems, risks, and opportunities. *WIREs Clim. Chang.* **2014**, *5*, 317–335. [CrossRef]
6. Edelman, D.J. Managing the Urban Environment of Manila. *Adv. Appl. Sociol.* **2016**, *6*, 101–133. [CrossRef]
7. Bürgi, M.; Hersperger, A.M.; Schneeberger, N. Driving forces of landscape change—Current and new directions. *Landsc. Ecol.* **2004**, *19*, 857–868. [CrossRef]
8. Figueroa, F.; Sánchez–Cordero, V.; Meave, J.A.; Trejo, I. Socioeconomic context of land use and land cover change in Mexican biosphere reserves. *Environ. Conserv.* **2009**, *36*, 180–191. [CrossRef]
9. Kolb, M.; Mas, J.F.; Galicia, L. Evaluating drivers and transition potential models in a complex landscape in southern Mexico. *Int. J. Geogr. Inf. Sci.* **2013**, *27*, 1804–1827. [CrossRef]
10. Seto, K.; Güneralp, B.; Hutyra, L. Global forecasts of urban expansion to 2030 and direct impacts on biodiversity and carbon pools. *Proc. Natl. Acad. Sci. USA* **2012**, *109*, 16083–16088. [CrossRef] [PubMed]
11. Han, H.; Yang, C.; Song, J. Scenario Simulation and the Prediction of Land Use and Land Cover Change in Beijing, China. *Sustainability* **2015**, *7*, 4260–4279. [CrossRef]
12. Pitman, A.J.; de Noblet-Ducoudré, N.; Avila, F.B.; Alexander, L.V.; Boisier, J.-P.; Brovkin, V.; Delire, C.; Cruz, F.; Donat, M.G.; Gayler, V.; et al. Effects of land cover change on temperature and rainfall extremes in multi-model ensemble simulations. *Earth Syst. Dyn.* **2012**, *3*, 213–231.
13. Lasco, R.D.; Pulhin, F.B. Forest land use change in the Philippines and climate change mitigation. *Mitig. Adapt. Strateg. Glob. Chang.* **2000**, *5*, 81–97. [CrossRef]
14. Malaque, I.R.; Yokohari, M. Urbanization process and the changing agricultural landscape pattern in the urban fringe of Metro Manila, Philippines. *Environ. Urban* **2007**, *19*, 191–206. [CrossRef]
15. Murakami, A.; Palijon, A.M. Urban Sprawl and Land Use Characteristics in the Urban Fringe of Metro Manila, Philippines. *J. Asian Archit. Build. Eng.* **2005**, *1*, 177–183. [CrossRef]

16. Alexandratos, N.; Bruinsma, J. *World Agriculture towards 2030/2050: The 2012 Revision*; ESA Working Paper No. 12-03; FAO: Rome, Italy, 2012.

17. Perry, D.A.; Oren, R.; Hart, S.C. *Forest Ecosystems*, 2nd ed.; The Johns Hopkins University Press: Baltimore, MD, USA, 2008.

18. Rola, A.C.; Sajise, J.A.U.; Harder, D.; Alpuerto, J.M. Soil conservation decision and upland corn productivity: A Philippine case study. *Asian J. Agric. Dev.* **2009**, *6*, 1–19.

19. Briones, R.U.; Ella, V.B.; Bantayan, N.C. Hydrologic Impact Evaluation of Land Use and Land Cover Change in Palico Watershed, Batangas, Philippines Using the SWAT Model. *J. Environ. Sci. Manag.* **2016**, *19*, 96–107.

20. Tongson, E.E.; Faraon, A.A. *Hydrologic Atlas of Laguna de Bay 2012*; Laguna Lake Development Authority and WWF-Philippines: Quezon City, Philippines, 2012.

21. Murayama, Y.; Estoque, R.C.; Subasinghe, S.; Hou, H.; Gong, H. Land-Use/Land-Cover Changes in Major Asian and African Cities. *Ann. Rep. Multi Use Soc. Econ. Data Bank* **2015**, *92*, 11–58.

22. Boori, M.S.; Choudhary, K.; Kupriyanov, A.; Kovelskiy, V. Satellite data for Singapore, Manila and Kuala Lumpur city growth analysis. *Data Brief.* **2016**, *7*, 1576–1583. [CrossRef] [PubMed]

23. Abino, A.C.; Kim, S.Y.; Jang, M.N.; Lee, Y.J.; Chung, J.S. Assessing land use and land cover of the Marikina sub-watershed, Philippines. *Forest Sci. Technol.* **2015**, *11*, 65–75. [CrossRef]

24. Global Footprint Network. *Ecological Footprint Report: Restoring Balance in Laguna Lake Region*; Global Footprint Network: Oakland, CA, USA, 2013.

25. Wealth Accounting and the Valuation of Ecosystem Services (WAVES). Ecosystem Accounts Inform Policies for Better Resource Management of Laguna de Bay. Available online: https://www.wavespartnership.org/en/knowledge-center/ecosystem-accounts-inform-policies-better-resource-management-laguna-de-bay (accessed on 20 March 2017).

26. Rañola, R.F., Jr.; Rañola, F.M.; Casin, C.S.; Tan, M.F.O. *LakeHEAD Progress Report: The Social and Economic Basis for Managing Environmental Risk for Sustainable Food and Health in Watershed Planning: The Case of Silang-Sta; Rosa Sub-Watershed Communities in Lake Laguna Region*; Research Institute for Humanity and Nature: Kyoto, Japan, 2010–2011.

27. Mishra, V.N.; Rai, P.K.; Mohan, K. Prediction of land use changes based on land change modeler (LCM) using remote sensing: A case study of Muzaffarpur (Bihar), India. *J. Geogr. Inst. Jovan Cvijic* **2014**, *64*, 111–127. [CrossRef]

28. Fathizad, H.; Rostami, N.; Faramarzi, M. Detection and prediction of land cover changes using Markov chain model in semi-arid rangeland in western Iran. *Environ. Monit. Assess.* **2015**, *187*, 629. [CrossRef] [PubMed]

29. Megahed, Y.; Cabral, P.; Silva, J.; and Caetano, M. Land Cover Mapping Analysis and Urban Growth Modelling Using Remote Sensing Techniques in Greater Cairo Region—Egypt. *ISPRS Int. J. Geo Inf.* **2015**, *4*, 1750–1769. [CrossRef]

30. Agarwal, C.; Green, G.M.; Grove, J.M.; Evans, T.P.; Schweik, C.M. *A Review and Assessment of Land-Use Change Models Dynamics of Space, Time, and Human Choice*; U.S. Department of Agriculture: Quilcene, WA, USA; Forest Service: Washington, DC, USA; Northeastern Research Station: Newtown Square, PA, USA, 2002.

31. National Research Council. *Advancing Land Change Modeling: Opportunities and Research Requirements. Chapter: 2 Land Change Modeling Approaches*; The National Academies Press: Washington, DC, USA, 2014.

32. Triantakonstantis, D.; Stathakis, D. Urban growth prediction in Athens, Greece, using Artificial Neural Networks. *Int. J. Civil Environ. Struct. Construct. Archit. Eng.* **2015**, *9*, 234–238.

33. Grekousis, G.; Manetos, P.; Photis, Y.N. Modeling urban evolution using neural networks, fuzzy logic and GIS: The case of the Athens metropolitan area. *Cities* **2013**, *30*, 193–203. [CrossRef]

34. Zhai, R.; Zhang, C.; Li, W.; Boyer, M.A.; Hanink, D. Prediction of Land Use Change in Long Island Sound Watersheds Using Nighttime Light Data. *Land* **2016**, *5*, 44. [CrossRef]

35. Ahmed, B.; Ahmed, R. Modeling Urban Land Cover Growth Dynamics Using Multi-Temporal Satellite Images: A Case Study of Dhaka, Bangladesh. *ISPRS Int. J. Geo Inf.* **2012**, *1*, 3–31. [CrossRef]

36. Olmedo, M.T.C.; Pontius, R.G., Jr.; Paegelow, M.; Mas, J.-M. Comparison of simulation models in terms of quantity and allocation of land change. *Environ. Model. Softw.* **2015**, *69*, 214–221. [CrossRef]

37. Riccioli, F.; El Asmar, T.; El Asmar, J.P.; Fagarazzi, C.; Casini, L. Artificial neural network for multifunctional areas. *Environ. Monit. Assess.* **2016**, *188*, 67. [CrossRef] [PubMed]

38. Wang, W.; Zhang, C.; Allen, J.M.; Li, W.; Boyer, M.A.; Segerson, K.; Silander, J.A. Analysis and Prediction of Land Use Changes Related to Invasive Species and Major Driving Forces in the State of Connecticut. *Land* **2016**, *5*, 25. [CrossRef]
39. Chaudhuri, G.; Clarke, K. The SLEUTH and land use change model: A review. *Environ. Resour. Res.* **2013**, *1*, 88–105.
40. Pontius Jr, R.G.; Boersma, W.; Castella, J.-C.; Clarke, K.; de Nijs, T.; Dietzel, C.; Duan, Z.; Fotsing, E.; Goldstein, N.; Kok, K.; et al. Comparing the input, output, and validation maps for several models of land change. *Ann. Reg. Sci.* **2008**, *42*, 11–47. [CrossRef]
41. Clark Labs. *TerrSet Geospatial Monitoring and Modeling Software*; Clark Labs, Clark University: Worcester, MA, USA, 2015.
42. Laguna Lake Development Authority, Water Quality Report: Laguna de Bay and Its Tributaries. Available online: http://www.llda.gov.ph/index.php?option=com_content&view=article&id=218&Itemid=679 (accessed on 20 March 2017).
43. Google Earth, V 7.1.8.3036. (24 March 2016), Laguna de Bay, Philippines, 14.340535°N, 121.241762°E, Eye alt 62.90 km. DigitalGlobe 2016, Google 2016, CNES/Astrium 2016. Available online: http://www.earth.google.com (accessed on 20 March 2017).
44. Johnson, B.A.; Iizuka, K.; Bragais, M.; Endo, I.; Magcale-Macandog, D. Employing crowdsourced geographic data and multi-temporal/multi-sensor satellite imagery to monitor land cover change: A case study in an urbanizing region of the Philippines. *Comput. Environ. Urban Syst.* **2017**, *64*, 184–193. [CrossRef]
45. Lambin, E.F.; Turner, B.L.; Geist, H.J.; Agbola, S.B.; Angelsen, A.; Bruce, J.W.; Coomes, O.T.; Dirzo, R.; Fischer, G.; Folke, C.; et al. The causes of land-use and land-cover change: Moving beyond the myths. *Glob. Environ. Chang.* **2001**, *11*, 261–269. [CrossRef]
46. Lambin, E.F.; Geist, H.J.; Lepers, E. Dynamics of land-use and land-cover change in tropical regions. *Annu. Rev. Environ Resour.* **2003**, *28*, 205–241. [CrossRef]
47. PhilGIS—Philippines GIS Data Clearinghouse. Available online: https://www.philgis.org/ (accessed on 20 March 2017).
48. EarthExplorer (USGS). Available online: https://earthexplorer.usgs.gov/ (accessed on 20 March 2017).
49. OpenStreetMap—GEOFABRIK Downloads. Available online: http://download.geofabrik.de/asia/philippines.html (accessed on 3 March 2016).
50. NOAA Earth Observtion Group (EOG). Available online: https://ngdc.noaa.gov/eog/ (accessed on 20 March 2017).
51. WorldPop. Available online: http://www.worldpop.org.uk/ (accessed on 20 March 2017).
52. WorldClim—Global Climate Data. Available online: http://www.worldclim.org/ (accessed on 20 March 2017).
53. Hijmans, R.J.; Cameron, S.E.; Parra, J.L.; Jones, P.G.; Jarvis, A. Very high resolution interpolated climate surfaces for global land areas. *Int J Climatol.* **2005**, *25*, 1965–1978. [CrossRef]
54. O'Donnell, M.S.; Ignizio, D.A. *Bioclimatic Predictors for Supporting Ecological Applications in the Conterminous United States*; U.S. Geological Survey Data Series 691; U.S. Geological Survey: Reston, VA, USA, 2012.
55. Van Wart, J.; van Bussel, L.G.J.; Wolf, J.; Licker, R.; Grassini, P.; Nelson, A.; Boogaard, H.; Gerber, J.; Mueller, N.D.; Claessens, L.; et al. Use of agro-climatic zones to upscale simulated crop yield potential. *Field Crops Res.* **2013**, *143*, 44–55. [CrossRef]
56. Kim, I.; Le, Q.B.; Park, S.J.; Tenhunen, J.; Koellner, T. Driving Forces in Archetypical Land-Use Changes in a Mountainous Watershed in East Asia. *Land* **2014**, *3*, 957–980. [CrossRef]
57. Ruan, X.; Qiu, F.; Dyck, M. The effects of environmental and socioeconomic factors on land-use changes: A study of Alberta, Canada. *Environ. Monit. Assess.* **2016**, *188*, 446. [CrossRef] [PubMed]
58. Mellander, C.; Lobo, J.; Stolarick, K.; Matheson, Z. Night-Time Light Data: A Good Proxy Measure for Economic Activity? *PLoS ONE* **2015**, *10*, e0139779. [CrossRef] [PubMed]
59. Stathakis, D. Forecasting Urban Expansion Based on Night Lights. *Int. Arch. Photogramm. Remote Sens. Spat. Inf. Sci.* **2016**, *XLI-B8*, 1049–1054. [CrossRef]
60. Stevens, F.R.; Gaughan, A.E.; Linard, C.; Tatem, A.J. Disaggregating Census Data for Population Mapping Using Random Forests with Remotely-Sensed and Ancillary Data. *PLoS ONE* **2015**, *10*, e0107042. [CrossRef] [PubMed]
61. Linard, C.; Gilbert, M.; Snow, R.W.; Noor, A.M.; Tatem, A.J. Population Distribution, Settlement Patterns and Accessibility across Africa in 2010. *PLoS ONE* **2012**, *7*, e31743. [CrossRef] [PubMed]

62. Mas, J.F.; Puig, H.; Palacio, J.L.; Lopez, A.S. Modelling deforestation using GIS and artificial neural networks. *Environ. Model. Softw.* **2004**, *19*, 461–471. [CrossRef]

63. Yuan, H.; Van Der Wiele, C.F.; Khorram, S. An automated artificial neural network system for land use/land cover classification from Landsat TM imagery. *Remote Sens.* **2009**, *1*, 243–265. [CrossRef]

64. Eastman, J.R. *TerrSet Help File*; Clark University: Worcester, MA, USA, 2015.

65. Takada, T.; Miyamoto, A.; Hasegawa, F.S. Derivation of a yearly transition probability matrix for land-use dynamics and its applications. *Landsc Ecol.* **2010**, *24*, 561–572. [CrossRef]

66. Estoque, R.C.; Murayama, Y. Corrigendum to 'Examining the potential impact of land use/cover changes on the ecosystem services of Baguio city, the Philippines: A Scenario-based analysis'. *Appl. Geogr.* **2012**, *35*, 316–326.

67. Pontius, R.G. Quantification error versus location error in comparison of categorical maps. *Photogramm. Eng. Remote Sens.* **2000**, *66*, 1011–1016.

68. Google Earth, V 7.1.8.3036. (31 December 1984; 30 December 1990; 31 December 1996; 31 December 2002; 31 December 2008), Laguna de Bay, Philippines, 14.340535°N, 121.241762°E, Eye alt 62.90 km. Landsat/Copernicus. Available online: http://www.earth.google.com (accessed on 20 March 2017).

69. Iizuka, K.; Tateishi, R. Estimation of CO_2 Sequestration by the Forests in Japan by Discriminating Precise Tree Age Category using Remote Sensing Techniques. *Remote Sens.* **2015**, *7*, 15082–15113. [CrossRef]

70. National Greening Program, Department of Environment and Natural Resrouces. Available online: http://www.denr.gov.ph/priority-programs/national-greening-program.html (accessed on 20 March 2017).

71. Wijesekara, G.N.; Farjad, B.; Gupta, A.; Qiao, Y.; Delaney, P.; Marceau, D.J. A comprehensive land-use/hydrological modeling system for scenario simulations in the Elbow River Watershed, Alberta, Canada. *Environ Manag.* **2014**, *53*, 357–381. [CrossRef] [PubMed]

72. Losiri, C.; Nagai, M.; Ninsawat, S.; Shrestha, R.P. Modeling Urban Expansion in Bangkok Metropolitan Region Using Demographic–Economic Data through Cellular Automata-Markov Chain and Multi-Layer Perceptron-Markov Chain Models. *Sustainability* **2016**, *8*, 686. [CrossRef]

73. Dale, V. The Relationship between Land-Use Change and Climate Change. *Ecol. Appl.* **1997**, *7*, 753–769. [CrossRef]

74. Zhang, Y.; Smith, J.A.; Luo, L.; Wang, Z.; Baeck, M.L. Urbanization and Rainfall Variability in the Beijing Metropolitan Region. *J. Hydrometeorol.* **2014**, *15*, 2219–2235. [CrossRef]

75. Bravo, L.; Roque, V.G.; Brett, J.; Dizon, R.; L'Azou, M. Epidemiology of Dengue Disease in the Philippines (2000–2011): A Systematic Literature Review. *PLoS Negl. Trop. Dis.* **2014**, *8*, e3027. [CrossRef] [PubMed]

76. Su, G. Correlation of Climatic Factors and Dengue Incidence in Metro Manila, Philippines. *Ambio* **2008**, *37*, 292–294. [CrossRef]

Article

Analysis of Urban Green Spaces Based on Sentinel-2A: Case Studies from Slovakia †

Monika Kopecká [1],*, Daniel Szatmári [1] and Konštantín Rosina [2]

[1] Institute of Geography, Slovak Academy of Sciences, Štefánikova 49, 814 73 Bratislava, Slovakia;
    geogszat@savba.sk
[2] Joint Research Centre, European Commission, Via E. Fermi 2749, I-21027 Ispra, Italy;
    konstantin.rosina@ec.europa.eu
*   Correspondence: monika.kopecka@savba.sk; Tel.: +421-2-57510218
†   This paper is extended from the version presented at the 6th International Conference on Cartography and
    GIS, Albena, Bulgaria, 13–17 June 2016.

Academic Editors: Andrew Millington and Harini Nagendra
Received: 8 March 2017; Accepted: 10 April 2017; Published: 14 April 2017

Abstract: Urban expansion and its ecological footprint increases globally at an unprecedented scale and consequently, the importance of urban greenery assessment grows. The diversity and quality of urban green spaces (UGS) and human well-being are tightly linked, and UGS provide a wide range of ecosystem services (e.g., urban heat mitigation, stormwater infiltration, food security, physical recreation). Analyses and inter-city comparison of UGS patterns and their functions requires not only detailed information on their relative quantity but also a closer examination of UGS in terms of quality and land use, which can be derived from the land cover composition and spatial structure. In this study, we present an approach to UGS extraction from newly available Sentinel-2A satellite imagery, provided in the frame of the European Copernicus program. We investigate and map the spatial distribution of UGS in three cities in Slovakia: Bratislava, Žilina and Trnava. Supervised maximum likelihood classification was used to identify UGS polygons. Based on their function and physiognomy, each UGS polygon was assigned to one of the fifteen classes, and each class was further described by the proportion of tree canopy and its ecosystem services. Our results document that the substantial part of UGS is covered by the class *Urban greenery in family housing areas* (mainly including privately-owned gardens) with the class abundance between 17.7% and 42.2% of the total UGS area. The presented case studies showed the possibilities of semi-automatic extraction of UGS classes from Sentinel-2A data that may improve the transfer of scientific knowledge to local urban environmental monitoring and management.

Keywords: urban green spaces; Sentinel-2A; ecosystem services; Slovakia

---

1. Introduction

Urban expansion is occurring at an unprecedented rate. By 2020, approximately 80% of Europeans will be living in urban areas, while in seven countries, the proportion is expected to be 90% or more [1]. Although urban areas remain a relatively small fraction of the terrestrial surface (in 2012, artificial surfaces covered 4.4% of land in Europe [2]), the urban ecological footprint widely extends beyond city boundaries, and urban expansion is impacting heavily on ecological processes [3]. Ecologists have studied the relationship between urban biodiversity and socioeconomic patterns in cities since the 1970s [4]. For example, a study of vegetation in neighbourhoods in Chicago, Illinois, USA, related patterns of tree species richness to census tract block data for the neighbourhoods [5]. Research on street and yard trees [6] aimed to identify census and other socioeconomic predictors of species richness. Two urban maps of the city of Osnabrück, Germany [7], represented the socio-economic distribution of the human population and the distribution of plant communities. The comparison reveals that both distributions are closely linked. These studies attempted to relate patterns of biodiversity to specific types of neighbourhoods, thus building on ideas that were linked to theories about differentiation and spatial patterns in cities [4].

The diversity and quality of urban green spaces (UGS) such as parks, forests, green roofs, or community gardens are tightly linked to human well-being as UGS provide a number of ecosystem services for people. In recent years, many studies advanced our understanding of UGS in their biophysical, economic and socio-cultural dimensions. The crucial ecosystem service of urban vegetation is its regulatory effect on the urban microclimate [8]; other relevant benefits are stormwater infiltration, food security, physical recreation, and psychological well-being of residents [9–11]. It is also known that the percentage of green space in people's living environment has a positive association with the perceived general health of residents [12].

By definition, ecosystem services have societal relevance: they provide benefits that humans want or need. The Economics of Ecosystem Services and Biodiversity—TEEB Manual for Cities [13] grouped ecosystem services in four major categories: provisioning, regulating, habitat and cultural and amenity services. In the process of spatial planning, all these desired outcomes must be identified and compared to the current potential of the UGS. Analyses and inter-city comparison of UGS patterns and their functions require not only detailed information on their relative quantity but also a closer examination of UGS in terms of quality and land use, which can be derived from the land cover composition and spatial structure.

The spatial structure of impervious-vegetated mix is heterogeneous at much finer scale in urban landscape than elsewhere. As a result, for a long time, conventional methods of mapping urban vegetation have relied on a visual interpretation of aerial images and fieldwork. More recently, developed very high resolution (VHR) satellite remote sensing systems (IKONOS, QuickBird, GeoEye, RapidEye, WorldView, Pleiades) are capable of providing imagery with similar detail to aerial photography, and they offer opportunities to overcome the lack of reliable and reproducible information on urban vegetation across large areas [14–16]. However, the disadvantage of VHR satellites is their narrow swath and therefore limited coverage of the Earth's surface. Besides, the VHR satellites are commercially oriented services, and the data cost is relatively high.

One of the most recent sources of information on land cover, including UGS, is Sentinel-2A (S2A), a high-resolution optical Earth observation mission. Although it has coarser spatial resolution than the VHR satellites, it offers higher spectral resolution and is provided free of charge. Sentinel missions are part of the Copernicus program (previously called GMES), a joint initiative of the European Commission and European Space Agency to establish a European capacity for the provisioning and use of information for environmental monitoring and security applications.

The objectives of this study are to:

1. Apply a methodical procedure of UGS extraction from S2A satellite imagery to selected study areas.
2. Analyse the UGS composition in the context of ecosystem services and identify UGS major components typical for cities in Slovakia.
3. Present the results of UGS comparative analysis in three regional cities in Slovakia: Bratislava, Trnava and Žilina.

2. Materials and Methods

2.1. Data

The S2A multispectral imager covers 13 spectral bands with a swath width of 290 km and spatial resolutions of 10 m (three visible and a near-infrared band), 20 m (6 red-edge/shortwave infrared bands) and 60 m (3 atmospheric correction bands). The mission is intended to monitor variability in land surface conditions, and its wide swath width and high revisit time (10 days with one satellite and five days in full constellation with twin satellite Sentinel-2B) will support monitoring of changes to vegetation within the growing season. It also provides data and applications for operational land monitoring, emergency response, and security services. The coverage limits are from between latitudes 56° south and 84° north.

The 100% cloud-free S2A scenes acquired in August 2015 (study area Bratislava) and in September 2016 (study areas Trnava and Žilina) were downloaded from Copernicus Sentinels Scientific Data Hub (https://scihub.copernicus.eu/dhus/). We used orthorectified and radiometrically corrected images (processing level 1C). Since the study areas represent only a small fraction of the respective scene's footprint, we have assumed constant atmospheric conditions and no atmospheric corrections were applied. Each scene contains 13 spectral bands with native spatial resolutions of 10 m (blue, green, red, and near-infrared bands), 20 m (red edge bands), or 60 m (short wave infrared bands); all bands were resampled to 10 m resolution for further processing.

2.2. Methodology

Since the proposed classification scheme of UGS is largely land use oriented, it is not viable to obtain the information by automatic methods. Therefore, aerial or very high resolution (VHR) satellite images are needed to perform on-screen interpretation and classification of individual UGS polygons extracted from the S2A data. Finer than 10 m spatial resolution imagery is useful also in the process of selecting the training samples for supervised automatic classification of the S2A imagery (Figure 1) [17].

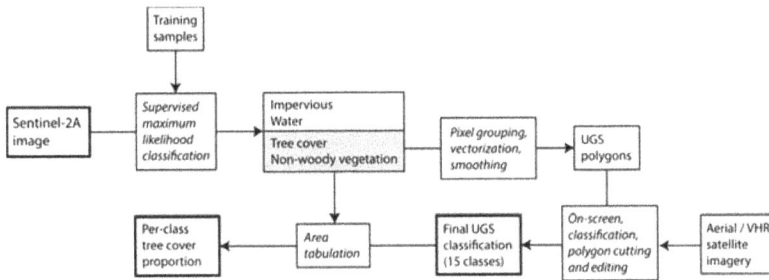

Figure 1. Scheme of the proposed methodology [17].

Given the spectral resolution and bandwidth of S2A data, we assumed that automatic classification methods have the potential to discriminate between a small number of spectrally different LC types with reasonable accuracy. We suggested a simple impervious-water-vegetation classification scheme; the vegetation was further divided into tree cover and non-woody classes. We preferred a supervised approach for higher accuracy. For each of the target land cover classes, a manually pre-classified training sample with a sufficient number of sample plots located evenly in the study area was created. The resulting set of total 435 sample plots (88 in Bratislava, 195 in Trnava and 152 in Žilina) was fed into the commonly used maximum likelihood classifier that was finally employed to perform per-pixel classification of the images. The S2A data were processed and classified using ESA SNAP 3.0 and ESRI ArcGIS Desktop 10 software, respectively. An example of true colour (red-green-blue) (RGB) composite is displayed in Figure 2a. Results of the initial classification are presented in Figure 2b. Using the tools for post-classification accuracy assessment in ArcGIS Desktop 10.4 we created 239 randomly sampled points. Reference values interpreted based on VHR satellite images and aerial images available in ArcGIS online Base map were compared with the classification results at the same locations. A summary of the accuracy assessment is presented in Table 1 in the form of a confusion matrix (per study area). The diagonal values for each city represent the correctly classified pixels. The ratio of their sum to the total number of tested pixels gives the total accuracy of the classification. User's accuracy (U Accuracy—rows of the table) shows pixels incorrectly classified as a given class.

Producer's accuracy (P Accuracy—columns of the table) shows pixels of a known class classified as something else. The Kappa coefficient (a real number from an interval [0,1]) is an overall assessment of the accuracy of the classification.

To extract the final UGS polygons from the classified images, these were reclassified into a binary form vegetation/non-vegetation. Contiguous pixels classified as vegetation were grouped (based on queen neighbourhood—each pixel can have maximum eight neighbours) and converted to vector polygons (Figure 2c). All polygon parts and holes smaller than 500 m^2 were removed. The remaining polygons were smoothed and generalised in order to remove pixelated borders, reduce the size, and improve the visual appearance (see Figure 2d).

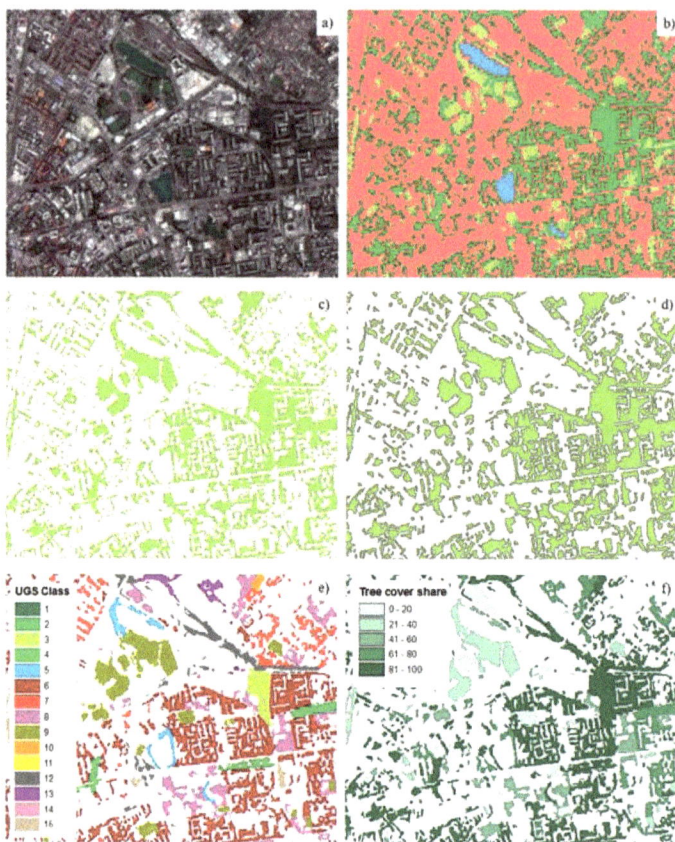

Figure 2. Six processing steps of the UGS extraction and classification (an example from study area Bratislava). (**a**) A true colour composite produced from S2A data; (**b**) result of the maximum likelihood supervised automatic classification; (**c**) binary map of vegetation/non-vegetation land cover; (**d**) vectorised and visually enhanced polygons with the minimum mapping unit of 500 m^2 applied; (**e**) result of UGS visual interpretation and polygon editing; (**f**) tree cover share estimated using (**b,e**) on a per-polygon basis [17].

From the perspective of ecosystem services and urban planning, it is important to consider how the identified UGS are utilised by city residents, the degree of human cultivation/intervention, and location relative to the prevalent use of urban land (residential, public, industrial). Considering these requirements, we recognised 15 classes presented and defined in Table 2. UGS polygons extracted in the previous step were overlaid on top of recent aerial orthophotos and visually classified at the scale range 1:10,000–1:5000. Most of the polygons were classified as such by filling the attribute values. In some cases (especially in places with abundant vegetation), the extracted polygons were spanning over larger areas and included multiple UGS classes. In such cases, the polygons were cut so that each polygon contains a single UGS class (see Figure 2e). Additional information useful for decision making is the type of vegetation. We suggest that tree vegetation provides a wider range of ecosystem services compared to herbaceous vegetation. We have therefore estimated the share of tree cover for each polygon (Figure 2f) and UGS class (based on the initial automatic LC classification).

Table 1. Supervised classification accuracy assessment.

Class Name	Water	Impervious	Tree Cover	Non-Woody Vegetation	Total	U Accuracy (%)	P Accuracy (%)
Water	20	0	0	0	20	100.00	100.00
Bratislava	10	0	0	0	10	100.00	100.00
Trnava	0	0	0	0	0	-	-
Žilina	10	0	0	0	10	100.00	100.00
Impervious	0	107	1	5	113	94.69	94.69
Bratislava	0	51	0	1	52	98.08	96.23
Trnava	0	33	1	2	36	91.67	94.29
Žilina	0	23	0	2	25	92.00	92.00
Tree cover	0	2	39	4	45	86.67	84.78
Bratislava	0	2	19	4	25	76.00	95.00
Trnava	0	0	10	0	10	100.00	83.33
Žilina	0	0	10	0	10	100.00	71.43
Non-woody vegetation	0	4	6	51	61	83.61	85.00
Bratislava	0	0	1	20	21	95.24	80.00
Trnava	0	2	1	14	17	82.35	87.50
Žilina	0	2	4	17	23	73.91	89.47
Total	20	113	46	60	239		
Bratislava	10	53	20	25	108		
Trnava	0	35	12	16	63		
Žilina	10	25	14	19	68		
Overall classification Accuracy (%):					90.79		
Overall Kappa coefficient:					0.862		

Urban ecosystem services have been classified in a variety of ways; most commonly, they are divided into four categories: provisioning services, regulating services, habitat or supporting services, and cultural services [13].

1. Provisioning services are ecosystem services that describe the material or energy outputs from ecosystems. They include providing raw materials, fresh water, food and medicinal resources.

2. Regulating services are the services that ecosystems provide by acting as regulators of local climate and air quality, carbon sequestration and storage, moderation of extreme events, wastewater treatment, erosion prevention and maintenance of soil fertility, pollination and biological control.

3. Cultural services include the nonmaterial, socio-ecological benefits (including psychological and cognitive benefits) people obtain from contact with the environment. They include recreation, physical and mental health (for example walking or playing sports in green areas), tourism, aesthetic appreciation and inspiration for culture, art and design, spiritual experience and sense of place (different sacred places or places with a religious meaning).

4. Habitat and supporting services underpin almost all other services by providing living spaces for organisms. They provide habitats for species and contribute to the maintenance of genetic diversity.

Table 2. UGS classification and definition of UGS classes.

No.	UGS Class	Description	Example	Ecosystem Services
1.	Urban forest/ Uncultivated park	Areas characterised by more than 50% woody vegetation with no signs of cultivation and without paved roads/paths		Habitat
2.	Cultivated park	Areas characterised by more than 50% woody vegetation with paved paths and scattered lawns		Cultural
3.	Cemetery	Areas of cemeteries with dominant vegetation		Cultural
4.	Urban public garden	Areas characterised by the regular shape of lawns, flowerbeds, shrubs, paths and scattered trees		Cultural
5.	Stream bank/lake shore vegetation	Green areas adjacent to ponds, lakes, rivers, or canals		Habitat/Regulating
6.	Urban greenery in apartment housing areas	Public greenery in residential zones between multi-flat houses and/or small commercial buildings		Regulating/Cultural
7.	Urban greenery in family housing areas	Greenery in residential zones between family houses, mostly comprising private horticultural gardens		Provisioning/Cultural
8.	Urban greenery in public facilities	Greenery in compact areas with particular public services, like hospitals, universities, school campus, ZOO etc. Sports facilities are not included in this class		Cultural
9.	Greenery in sports facilities	Green areas used for sports and leisure mainly covered by grass, such as football field, golf course, playground, and horse race circuit		Cultural
10.	Allotments	Area with small parcels of annual crops, pastures, fallow land and/or permanent crops, with scattered garden cabins		Provisioning/Cultural
11.	Cropland/pastures	Agricultural areas with signs of cultivation (e.g., tracks from ploughing or tractor use). This class contains both cropped areas and areas with grass in rotation, as well as orchards and vineyards		Provisioning
12.	Railway and roadside greenery	Verge with grass or other vegetation accompanying a railway, road or motorway		Regulating
13.	Green areas in industrial units	Areas covered by vegetation in factories with industrial production, storage facilities, logistic centres, etc.		Regulating
14.	Airport greenery	Grass areas of airports associated with runways		Regulating
15.	Ruderal vegetation	Areas with grass, herbaceous, shrub and/or scattered woody vegetation with no signs of recent cultivation. Usually heterogeneous in texture and colour. Fallow land and brownfields can also be part of this class		Regulating

Patterns of urban areas have significant implications for biodiversity and ecosystem functions. In this context, each UGS class was categorised according to their expected ecosystem services (Table 2).

Urban forests or uncultivated parks (class 1) as well as stream banks and lake shores (class 5) have become a refuge for many species, especially birds and amphibians. Cultivated parks, cemeteries and urban gardens (classes 2, 3 and 4) play an important role as providers of aesthetic and psychological benefits that enrich human life, reducing stress and increasing physical and mental health [18,19]. They are considered to be habitats in a city that truly demonstrate human expression and creativity [4]. Some of them (e.g., botanical gardens) are used also for environmental education purposes. Greenery in sports facilities, such as football pitches or aqua parks (class 9), increases the recreational potential of a city. However, the recreational opportunities of urban ecosystems also vary with social criteria, including accessibility, penetrability, safety, privacy and comfort [20].

Ecological infrastructure in cities regulates local temperatures and buffers the effects of urban heat islands [21]. Water from plants absorbs heat as it evaporates, thus cooling the air in the process [22]. Trees can also regulate local surface and air temperatures by reflecting solar radiation and shading surfaces, such as streets and sidewalks that would otherwise absorb heat. Decreasing the heat loading of the city is among the most important regulating ecosystem services trees provide to cities [23]. These regulating services are dominant in the classes *Urban greenery in apartment housing areas* and *Urban greenery in public facilities* (classes 6 and 8). Positive effects of vegetation on human health (cultural ecosystem services) are also important. For example, a view through a window looking out at green spaces could accelerate recovery from surgeries [24] and proximity of an individual's home to green spaces was correlated with fewer stress-related health problems and a higher general health perception [25]. Increasing the impervious surface area in cities leads to increased volumes of surface water runoff, and thus increases the vulnerability to water flooding. Green areas reduce the pressure on urban drainage systems by percolating water.

Urban greenery in family housing areas (class 7) prevailingly consists of greenery in residential zones between family houses, mostly comprising private horticultural gardens that offer multiple opportunities for family leisure activities. Traditionally, private gardens were important for the provisioning of food (usually some sorts of fruits and vegetables) for the city residents. However, in the recent two decades, a significant part of garden fruit trees was replaced by decorative conifers and only a small fraction of food consumed by families is home produced. Similar ecosystem services are provided by the class *Allotments* (10) represented by community gardens usually owned by people living in multi-flat houses. Despite our effort to exclude agricultural land from the study areas, some cultivated fields cannot be omitted as they are situated within the city area itself (see Chapter 2.3 Study area). They belong to the class 11 *Cropland/pastures* with provisioning ecosystem services. Air pollution from transportation and industry is one of the major problems for environmental quality and human health in the urban environment. Vegetation in industrial zones, railway, roadside and airport greenery (classes 12, 13 and 14) with dominant regulating ecosystem service improves air quality by removing pollutants from the atmosphere [26]. Vegetation in these localities is also an important factor for noise reduction. Ruderal vegetation (class 15) is typical for unused or abandoned localities. They are often covered by herbs and shrubs that also contribute to the reduction of soil sealing negative effects.

2.3. Study Areas

We have selected cities Bratislava, Trnava and Žilina, three of eight regional capitals of Slovakia's NUTS 3 administrative regions, as the study area (Figure 3).

The cities Bratislava and Žilina are located in large river valleys, and the diversified terrain is an important local climate factor; the city of Trnava is located in a flat area. The built-up area of the cities has increased significantly in recent twenty years due to the construction of new automobile plants (Volkswagen in Bratislava, Peugeot-Citroen in Trnava and Kia Motors in Žilina) and large shopping centres in suburban areas. As the rapid increase of impervious surfaces may worsen urban heat island effects, well-thought UGS management should play an important role in sustainable development of these cities.

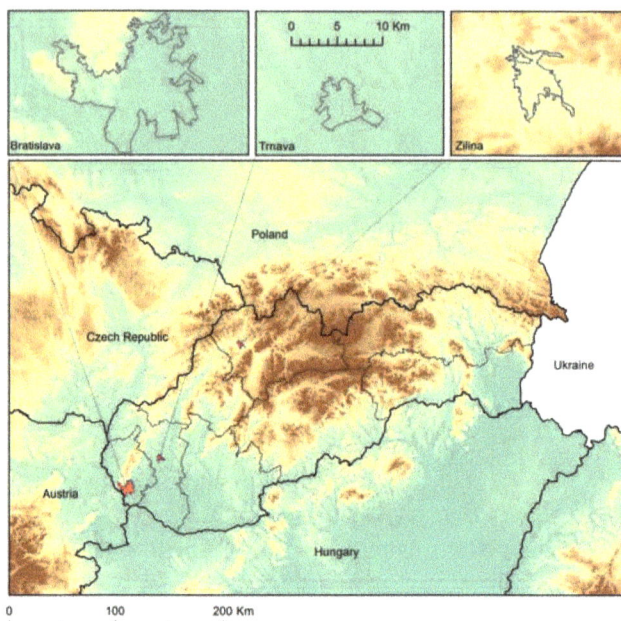

Figure 3. Location of the study areas Bratislava, Trnava and Žilina (Digital Elevation Model Over Europe—EU DEM—Source: http://land.copernicus.eu/pan-european).

The Bratislava study area is situated in the south-west part of Slovakia bordering Austria in the west and Hungary in the south. Bratislava is the capital of Slovakia, the country's largest city and it is the political, cultural and economic centre. Due to this fact and a good quality transport infrastructure, it is a territory with high potential for territorial development. The limiting factor for a further expansion of the city is the Little Carpathian mountain range located north of the city centre. Bratislava lies on the both banks of the Danube River, which crosses the city from the west to the south-east. The population of Bratislava at the end of 2015 was 422,453 inhabitants (7.78% of the population in Slovakia).

Trnava is a city in western Slovakia, 47 km north-east of Bratislava. It is located in the Danubian Lowland on the Trnávka River in the central part of Trnava Plain. Due to the character of relief, the close position to the capital city of Bratislava and good transport infrastructure, it has a great potential for territorial development. However, the presence of top quality soils in the hinterland of the city is actually a limiting factor for its further expansion. The population of Trnava at the end of 2015 was 65,596 inhabitants.

Žilina is a city in north-western Slovakia, around 200 km from the capital Bratislava, close to both the Czech and Polish borders. It is the fourth largest city of Slovakia with a population of 81,114, an important industrial centre and the largest city on the Váh River. The city is surrounded by the mountain ranges Malá Fatra, Súľovské vrchy, Javorníky and Kysucká vrchovina.

Definition of urban areas and their boundaries vary between countries and regions. Therefore, as the first step of any urban comparative analysis, a common definition of "urban" should be specified. Although the administrative definition of cities has the benefit of wide data availability, such boundaries often include large portions of rural landscape, mostly with agricultural and (semi-)

natural land cover, which cannot be considered as urban greenery. The focus of this study is on the services and benefits provided by urban ecosystems. Therefore, we suggest that for UGS comparison, a city should be defined rather by its continuously built-up area, where the concentration of people is the highest both during the daytime and night-time, and the density of buildings and other impervious surfaces is so high, that it can alter the microclimate significantly [27]. In the European context, the contiguously built-up area can be delineated based on open data from the Urban Atlas database that has harmonised definition, suitable spatial detail and is updated and validated regularly [28]. Particularly, we extracted all the artificial surfaces from Urban Atlas 2012 (code 1xxxx), excluding road and rail network (which is represented by a single extensive and complex polygon that spans the whole urban region including the commuting hinterland). Consequently, the polygons were buffered by 50 m and merged (to connect built-up blocks and join gaps in urban fabric up to 100 m wide). Holes in the resulting polygon were filled, and the result was buffered back by 50 m. This method produced an accurate picture of the selected cities (Figure 4), where only minor manual editing was needed. Additionally, this definition can be applied to any EU city and can be updated on a six-year basis to account for city expansion.

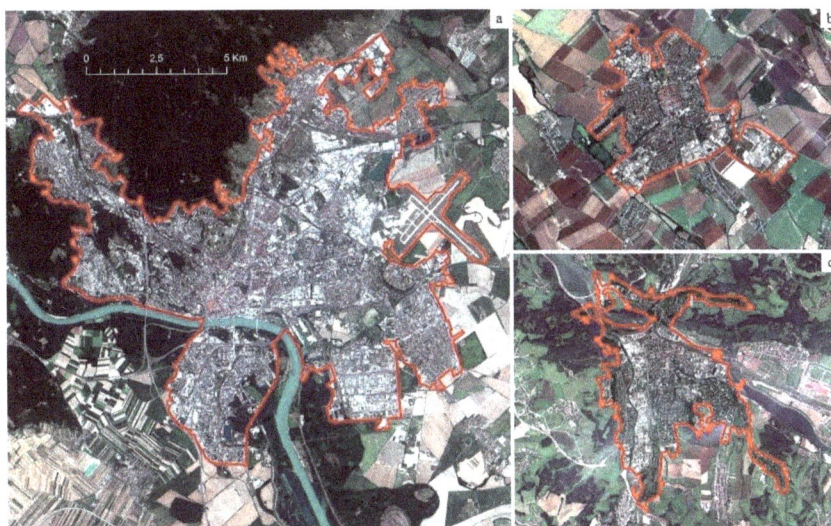

Figure 4. True colour composites (bands 4, 3, and 2 of cloud-free S2A images) of Bratislava (**a**), Trnava (**b**) and Žilina (**c**) study areas.

3. Results

The city of Bratislava is a densely populated capital city situated on both banks of a large river. According to the results of initial land cover classification, impervious surfaces covered 51.6% of the territory, urban vegetation covered 46.5% (out of which trees covered around 54%), and water covered 1.9%. Currently, the UGS area per capita is 121 m^2.

The largest part was covered by the class *Urban greenery in family housing areas* (7) with an area of more than 1000 ha, in total over 20% of the urban greenery (Table 3). UGS defined as *Urban greenery in apartment housing areas* (class 6) covered the second largest area of almost 680 ha, i.e., 13.2% of UGS. However, this class was represented by the highest number of patches (823 out of total number 2909 UGS polygons). On the other hand, the class *Urban public garden* (4) covered the smallest part of green areas within the city. Table 4 provides insight into the tree cover percentage for each class.

The highest proportions of woody vegetation were detected in class *Urban forest/Uncultivated park* (1) and *Cemeteries* (3). The lowest share of woody vegetation was found in class 14—*Airport greenery* (0.5%).

Table 3. UGS classification in Bratislava, Trnava and Žilina.

UGS Class	Polygon Count			Class Area (Ha)			Class Abundance (%)		
	Bratislava	Trnava	Žilina	Bratislava	Trnava	Žilina	Bratislava	Trnava	Žilina
1	36	6	13	469.55	29.71	67.06	9.2	4.0	6.1
2	39	19	3	74.63	43.94	9.82	1.5	6.0	0.9
3	10	7	12	58.57	6.51	11.97	1.1	0.9	1.1
4	7	3	2	12.43	2.54	1.22	0.2	0.3	0.1
5	64	5	29	149.54	8.29	59.87	2.9	1.1	5.4
6	823	151	112	679.34	73.96	139.27	13.2	10.0	12.6
7	589	215	120	1035.07	130.67	466.27	20.2	17.7	42.2
8	292	99	90	358.40	48.40	78.54	7.0	6.6	7.1
9	67	16	11	136.92	30.65	15.61	2.7	4.2	1.4
10	53	5	15	434.44	6.78	71.19	8.5	0.9	6.4
11	58	30	5	446.98	64.25	11.49	8.7	8.7	1.0
12	305	64	63	285.94	45.57	86.99	5.6	6.2	7.9
13	442	129	97	454.23	164.48	35.15	8.9	22.3	3.2
14	27	0	0	287.82	0.00	0.00	5.6	0.0	0.0
15	97	47	41	243.48	82.06	51.37	4.7	11.1	4.6
Total	2909	796	613	5127.34	737.82	1105.82	100.0	100.0	100.0

Table 4. Tree cover percentage and average patch size within UGS classes.

UGS Class	Tree Cover Percentage (%)			Average Patch Size (Ha)		
	Bratislava	Trnava	Žilina	Bratislava	Trnava	Žilina
1	92.9	74.6	65.7	13.04	4.95	5.16
2	84.9	59.0	49.9	1.91	2.31	3.27
3	88.4	58.9	19.0	5.86	0.93	1.00
4	71.5	42.4	0.0	1.78	0.85	0.61
5	81.7	22.4	8.7	2.34	1.66	2.06
6	67.2	38.0	2.4	0.83	0.49	1.24
7	57.5	19.3	3.2	1.76	0.61	3.89
8	57.5	34.4	2.9	1.23	0.49	0.87
9	33.4	7.9	2.4	2.04	1.92	1.42
10	51.4	2.7	23.3	8.20	1.36	4.75
11	13.0	3.4	1.6	7.71	2.14	2.30
12	66.0	29.2	1.9	0.94	0.71	1.38
13	36.7	8.0	0.3	1.03	1.28	0.36
14	0.5	0.0	0.0	10.66	0.00	0.00
15	43.0	8.7	8.4	2.51	1.75	1.25
Total	53.2	22.1	9.0	1.76	0.93	1.80

Trnava is a compact city with the highest share of impervious surfaces—over 59%. The city has the lowest UGS area per capita—112 m^2. The tree cover percentage is markedly lower than in Bratislava—22.1% of UGS. *Green areas in industrial units* (class 13) has the highest UGS class abundance (over 22%), which is affected mainly by large land areas used during the construction of the automobile factory in the south-east part of the city. The second largest area is covered by *Urban greenery in family housing areas* (class abundance is 17.7% and it has the highest number of polygons, see Table 3). However, private gardens in Trnava are relatively fragmented—the average patch size is only 0.61 ha (compared to 1.76 ha in Bratislava and 3.89 ha in Žilina).

The total green space within the city of Žilina amounts to 1120 ha, representing 50.64% of the city area. The city has the highest area of UGS per capita—136 m^2. However, UGS tree cover percentage is the lowest—only 9%. Residential gardens (class 7) represent as much as 42% of urban greenery area in the city. Although often ignored by ecologists as significant habitats in urban landscapes, gardens contribute to plant species richness and to insect and avian species diversity by providing critical habitat for nesting, food, and cover [4]. From this point of view, relatively compact class area plays the important role. As in Bratislava, the second largest area is covered by *Urban greenery in apartment housing area* with similar class abundance 12.6% (Figure 5). These two cities have also similar UGS average patch size—Žilina 1.80 ha and Bratislava 1.76 ha.

Woody vegetation is necessary for urban heat mitigation and improves the quality of life in residential areas. The share of tree cover in apartment housing areas (class 6) varies considerably in the cities—it is the highest in Bratislava (67.2%), much lower in Trnava (38%) and extremely low in Žilina (2.4%). Similar differences in tree cover percentage were recorded in family housing areas (class 7)—Bratislava 57.5 %, Trnava 19.3% and Žilina 3.2%; as well as in the surroundings of public facilities (class 8). The regulative function of woody vegetation is especially important along roads in the cities. While tree cover percentage in roadside greenery (class 12) in Bratislava is 66%, in Žilina it accounts only for 1.9% of the class area.

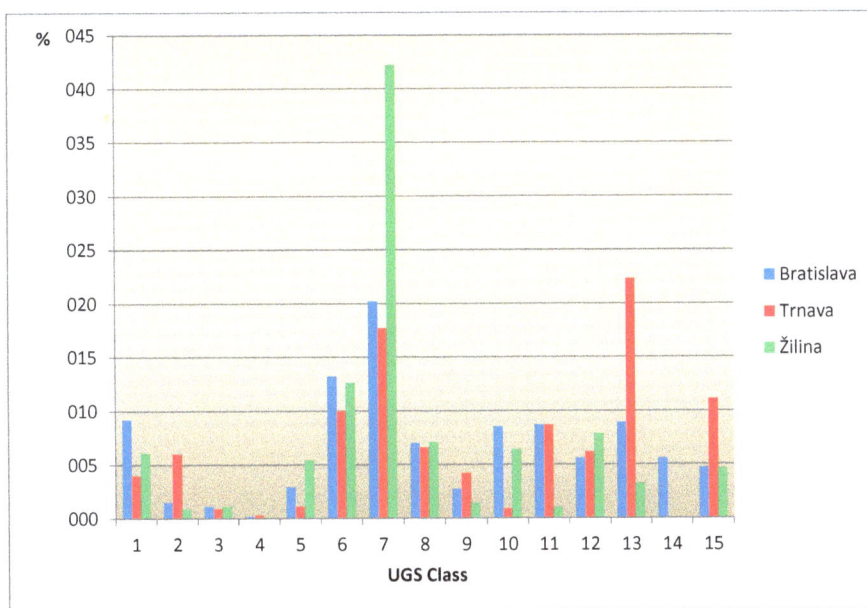

Figure 5. Proportion of UGS classes (%) in Bratislava, Trnava and Žilina.

The process of urbanisation transforms the original landscapes. It fragments natural vegetation resulting in habitat loss and isolation. Assessments of patches of remnant vegetation show that patch configuration plays a significant role in determining plant species [4]. In general, larger remnant patches contain more native species than smaller patches in urban landscapes. Consequently, conservation strategies in urban landscapes favour preserving larger patches over smaller ones. More than 12% of the green space in Bratislava corresponds to the areas with important habitat ecosystem services with a high proportion of natural vegetation. The average patch size for *Urban forest/Uncultivated park* and shore vegetation was 13.04 ha and 2.34 ha, respectively. The map of urban greenery presented in Figure 6 provides information about the spatial distribution of the ecosystem services within the city.

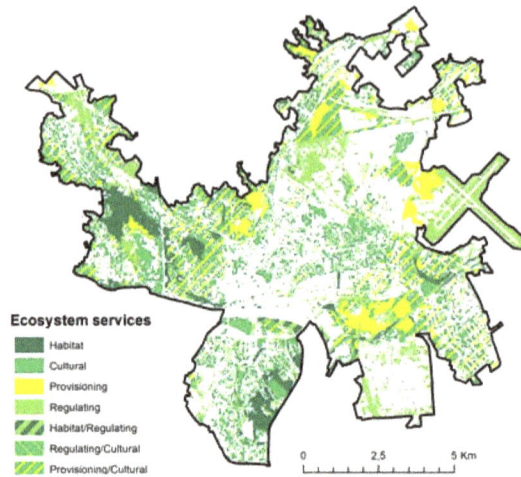

Figure 6. Ecosystem services in the city of Bratislava.

In Trnava, only limited amounts of the UGS correspond to urban forests or shore vegetation with the important habitat ecosystem services (ca. 5%). In Figure 7, areas with regulating ecosystem services dominate. In the city of Žilina (Figure 8), UGS with multifunctional ecosystem services cover the largest part of the territory.

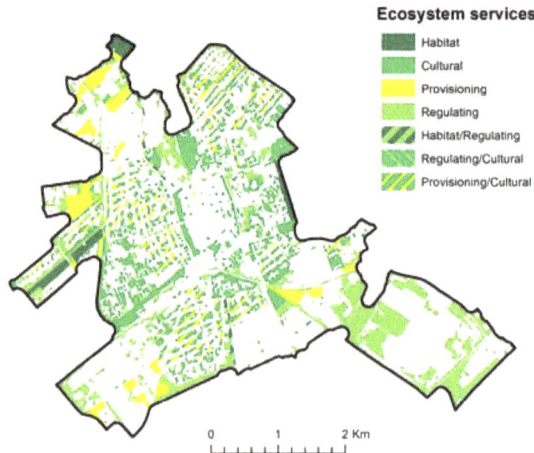

Figure 7. Ecosystem services in the city of Trnava.

As the central parts of the cities with densely built-up historical cores mostly lack UGS, the absence of ecosystem services is significant in the maps shown in Figures 6–8. Near the historical centres, small fragmented areas with cultural ecosystem services prevail. Multifunctional fragments of UGS are a typical feature of the urban landscape in residential zones adjacent to the city centre. Similar patterns can also be found in urban quarters that used to be separate villages in the past. Larger areas with cultural ecosystem services, such as parks, are located irregularly in different parts of the cities. Plots with production function are usually situated farther from the city centre and they often represent the localities of future development. UGS with regulating services are scattered along the roads throughout the cities; larger compact areas can be found in the periphery (industrial zones and the airport).

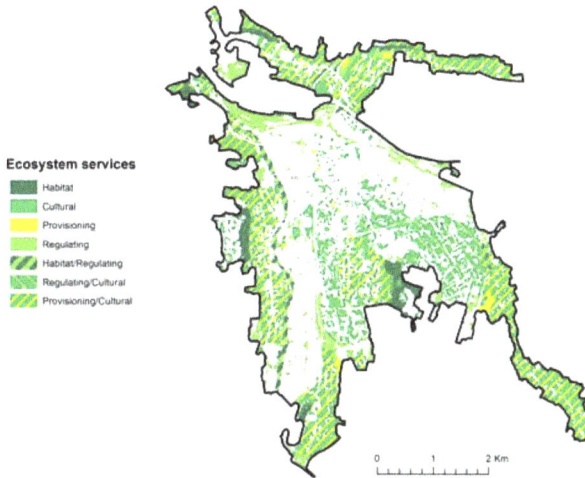

Figure 8. Ecosystem services in the city of Žilina.

4. Discussion

4.1. Processing of Sentinel-2A Data

The Sentinel-2 mission establishes a key European source of data for the Copernicus environmental monitoring program. Based on visual inspection of the results and comparison with VHR imagery, the overall quality of UGS extraction seems to be satisfactory. However, the distinction between tree and non-woody vegetation may be subject to commission and omission errors. The distinction is rather of continuous than binary nature since there are many tree, shrub, and other plant species of various ages and phenological phases. The amount of tree cover can also be biased by the number, location, and selection of training samples.

The results of post-classification accuracy assessment (Table 1) show good overall accuracy (over 85%). Although at the class level, as expected, there is confusion mainly between the two classes of vegetation. This has implications for the reported share of tree cover (Table 4). While in Bratislava there was commission error of 24%, in Trnava and Žilina, respective omission errors of 17% and 29% were estimated for the tree cover class. A likely source of the error is the temporal difference between the classified image and image used for finding the ground truth, as well as inaccuracies in delineation and classification of the training samples. Consequently, the actual cover tree cover percentage in the study areas may differ from the reported values, but we do not expect that ranking of the study areas would change (adjusted values of tree cover percentage: Bratislava 40%, Trnava 26% and Žilina 13%).

In addition, the morphology of the urban fabric may affect the results. In data derived from remote sensing, the classes that are dominant in a particular area tend to be overestimated and vice versa [29]. This effect is caused by cross-pixel spectral contamination (backscattered radiation from adjacent pixels influences the spectral response of a given pixel). Thus, cities with more fragmented UGS can be underestimated. With increasing fragmentation and heterogeneity of the urban landscape, the proportion of mixed pixels increases (when 10 or 20 m pixels are considered), and the accuracy of classification decreases. Five-metre resolution satellite data would be perhaps more suitable for UGS mapping. However, the wide swath, frequent revisit, spectral richness, and free availability of S2A data makes them very worthwhile to investigate their potential for monitoring built-up areas [30].

4.2. Assessment of Urban Ecosystem Services

Rapid urbanisation provides multiple opportunities to ensure basic human welfare and a viable global environment. According to some prognosis, more than 60 percent of the area projected to be urban in 2030 has yet to be built, and this presents opportunities to improve global sustainability, exploring how cities can be responsible stewards of biodiversity and ecosystem services within city boundaries [31]. With the growing awareness of the value of biodiversity and ecosystem services, cities should ensure that their biodiversity is conserved. Based on our results, specific management schemes and tools can be recommended for areas with various ecosystem services. For example, management policies could include leaving certain areas unmanaged (classes 1 and 5) while some areas could be managed lightly (class 3) and the others more intensively (classes 4 or 9). The spatially detailed and up to date information on the status of urban vegetation enables more efficient spending of limited resources by prioritisation and targeting neighbourhoods with least sufficient vegetation. Especially in combination with other data sources, it is possible to obtain valuable insights. For instance, together with demographic data, one can target vulnerable zones with a high proportion of elderly people sensitive to heat waves and an insufficient amount of tree coverage. Combined with noise exposure maps, the results could help improve roadside vegetation where most needed. Urban planners can also implement this information in planning frameworks including greenery coefficients (minimal share of vegetation per zone) or built-up indices.

Biodiversity studies in urban areas are often conducted in public spaces where access is not limited. However, comparative analysis of the three cities indicated the relatively high importance of the class *Urban greenery in family housing areas*, represented mainly by privately owned gardens. By controlling the vegetation structure in their properties, the private owners can have a major influence on urban biodiversity. Our results confirm the outputs of Goddard et al. [32] who point out the fact that in many countries, private gardens are a major component of urban green space and can provide considerable biodiversity benefits. They describe existing garden conservation strategies and the importance of gardens for raising awareness about biodiversity and public understanding of science. They also argue that citizens have a huge potential for enhancing urban environments by coordinating public management actions. The importance of privately managed UGS was confirmed also by Andersson et al. [33], who examined the importance of management scale on diversity and subsequently on ecosystem services in Stockholm, Sweden. They focused on three types of green spaces in Stockholm, Sweden: parks, managed by the city; cemeteries, generally managed by the Church of Sweden; and allotment gardens, managed by individuals. Those systems managed by individuals—bottom up management—had the greatest diversity and abundance of pollinators and a different suite of seed dispersers and insectivores than systems managed by the city and the Church of Sweden—top-down management. In this context, local authorities in the study areas should support environmental awareness and they can motivate local people to get involved in biodiversity measures. In Slovakia, all the garden parcels are subject to real estate tax defined by the municipality, so the tax reduction for the gardens with natural full-grown trees could be one of the suitable tools.

The presented maps in Figures 6–8 reflect dominant ecosystem services as defined in the TEEB Manual for Cities [13]. However, many UGS are multifunctional. Because city environments may be stressful for inhabitants, the recreational aspects of urban ecosystems are among the highest valued ecosystem services in cities. Nevertheless, cultivated parks are not only credited for their cultural ecosystem services and positive aesthetic and social values [34,35] but also act as hot spots of biodiversity in urban areas [36]. Similarly, together with cultural and provisioning ecosystem services, gardens provide habitat for many synanthropic species and support biodiversity [32]. Therefore, preservation of different types of UGS, including allotments and privately owned gardens, could be a target goal for urban planning initiatives in Slovak cities. Our recent land cover change analysis between the years 2002 and 2011 in Trnava based on VHR satellite data [37] documented not only intensive urban extension but also significant urban infill (UGS replaced by built-up areas). Presented results can serve as input data for the continuous monitoring of urban greenery in all three cities. Sentinel 2 imagery in the next years will allow us to monitor changes in the size and spatial structure of UGS classes and to evaluate the rate of urban infill. Although the initial classification of automatically extracted polygons is somewhat laborious, it is much less so for the future updates of the maps in which only a limited part of the polygons will need to be edited.

It is well known that vegetation abundance is negatively correlated to land surface temperature. Clustered or less fragmented UGS patterns lower surface temperature more effectively than dispersed patterns [38]. Based on our results, the UGS in Trnava study area are more fragmented than in Bratislava and Žilina. Another factor affecting the local microclimate is the tree cover proportion. While we have found that it is very low in Žilina, the city is also less prone to experiencing heatwaves compared to the other two study areas—thanks to its basin location and higher elevation, the average high temperatures in summer are around 2 °C lower. Presented results of UGS assessment will serve as the input data for the definition of the local climate zones [39], and the MUKLIMO model [40,41] will be tested in the study areas. Further research oriented on the relationship between the spatial pattern of urban features and land surface temperature in the study areas will bring deeper insight into the effects of UGS on local urban microclimate.

5. Conclusions

UGS developing and planning green spaces is of great importance in urban planning because it provides a wide range of ecosystem services, reduces the urban heat island, cleans the urban air, dampens noise, and absorbs CO_2. Land cover indices derived from remote sensing are crucial urban indicators, which contribute to sustainable urban planning and management. The recent availability of Sentinel-2 data is expected to bring land cover mapping and UGS monitoring to an unprecedented level. Among the main advantages of Sentinel-2, which make the satellite suitable for mapping and monitoring human settlements at a global level, is the combination of the wide swath and the frequent revisiting time at relatively high spatial resolutions. The results presented in this paper demonstrate the added-value of Sentinel-2 for mapping built-up areas in the context of ecosystem mapping and assessment. Current knowledge of ecosystem patterns and processes linked with landscape design enables not only planners and managers but also individual land owners to build sustainable landscapes for humans as well as urban flora and fauna. Presented comparative analyses of Bratislava, Trnava and Žilina can serve as a reference for decision- and policymakers in preserving urban biodiversity and improving life quality as well as the basis for other environmental research in these cities.

Acknowledgments: This paper is one of the outputs of the following projects: "Effect of impermeable soil cover on urban climate in the context of climate change" (Slovak Research and Development Agency—Grant Agency No. APVV-15-0136) and Changes in Agricultural Land Use: Assessment of the Dynamics and Causes Applying Land Cover Data and Selected Environmental Characteristics (VEGA—Grant Agency No. 2/0096/16) pursued at the Institute of Geography of the Slovak Academy of Sciences.

Author Contributions: Monika Kopecká conceived the UGS classification and wrote the paper; Konštantín Rosina proposed the methodology of UGS extraction and together with Monika Kopecká, analysed the data for the Bratislava study area; Daniel Szatmári analysed the data for Trnava and Žilina and prepared the figures.

Conflicts of Interest: The authors declare no conflict of interest.

References

1. European Environmental Agency. *Urban Sprawl in Europe. The Ignored Challenge*; EEA Report No. 10/2006; Office for Official Publications of the European Communities: Luxembourg, 2006; p. 56.
2. Soukup, T.; Feranec, J.; Hazeu, G.; Jaffrain, G.; Jindrova, M.; Kopecky, M.; Orlitova, E. Corine Land Cover 1990 (CLC 1990): Analysis and Assessment. In *European Landscape Dynamics: Corine Land Cover Data*; Feranec, J., Soukup, T., Hazeu, G., Jaffrain, G., Eds.; CRC Press: Boca Raton, FL, USA, 2016; pp. 69–76.
3. Grimm, N.B.; Faeth, S.H.; Golubiewski, N.E.; Redman, C.L.; Wu, J.; Bai, X.; Briggs, J.M. Global change and the ecology of cities. *Science* **2008**, *319*, 756–760. [CrossRef] [PubMed]

4. Müller, N.; Ignatieva, M.; Nilon, C.H.; Werner, P.; Zipperer, W.C. Patterns and Trends in Urban Biodiversity and Landscape Design. In *Urbanization, Biodiversity and Ecosystem Services: Challenges and Opportunities: A Global Assessment*; Elmqvist, T., Fragkias, M., Goodness, J., Güneralp, B., Marcotullio, P.J., McDonald, R.I., Parnell, S., Schewenius, M., Sendstad, M., Seto, K.C., et al., Eds.; Springer: Dordrecht, The Netherlands; Heidelberg, Germany; New York, NY, USA; London, UK, 2013; pp. 123–174.

5. Schmid, J.A. *Urban Vegetation: A Review and Chicago Case Study*; Department of Geography Research Paper No. 161; University of Chicago: Chicago, IL, USA, 1975; p. 266.

6. Whitney, G.G.; Adams, S.D. Man as a maker of new plant communities. *J. Appl. Ecol.* **1980**, *17*, 431–448. [CrossRef]

7. Hard, G. Vegetationsgeographie and Sozialökologie einer Stadt. Ein Vergleich zweier, Städtplane am Beispiel von Osnabrück. *Geographische Zeitschrift* **1985**, *73*, 125–144.

8. Lehman, I.; Mathey, J.; Rößler, S.; Bräuer, A.; Goldberg, V. Urban vegetation structure types as a methodological approach for identifying ecosystem services—Application to the analysis of micro-climatic effects. *Ecol. Indic.* **2014**, *42*, 58–72. [CrossRef]

9. Wolch, J.R.; Byrne, J.; Newell, J.P. Urban green space, public health and environmental justice: The challenge of making cities just green enough. *Landsc. Urban Plan.* **2014**, *125*, 234–244. [CrossRef]

10. Tzoulas, K.; Korpela, K.; Venn, S.; Yli-Pelkonen, V.; Kazmiercak, A.; Niemelä, J.; James, P. Promoting ecosystem and human health in urban areas using Green infrastructure. A literature review. *Landsc. Urban Plan.* **2007**, *81*, 167–178. [CrossRef]

11. Niemela, J. Ecology of urban green spaces: The way forward in answering major research questions. *Landsc. Urban Plan.* **2014**, *125*, 298–303. [CrossRef]

12. Maas, J.; Verheij, R.A.; Groenewegen, P.P.; Vries, S.; Spreeuwenberg, P. Green space, urbanity, and health: How strong is the relation? *J. Epidemiol. Community Health* **2006**, *60*, 587–592. [CrossRef] [PubMed]

13. TEEB—The Economics of Ecosystems and Biodiversity. *TEEB Manual for Cities: Ecosystem Services in Urban Management*; TEEB: Geneva, Switzerland, 2011; p. 43. Available online: www.teebweb.org (accessed on 20 February 2017).

14. Cheng, C.; Li, B.; Ma, T. The application of very high resolution satellite image in urban vegetation cover investigation: A case study of Xiamen City. *J. Geogr. Sci.* **2003**, *13*, 265–270.

15. Nichol, J.; Lee, C.M. Urban vegetation monitoring in Hong Kong using high resolution multispectral images. *Int. J. Remote Sens.* **2005**, *26*, 903–918. [CrossRef]

16. Tigges, J.; Lakes, T.; Hostert, P. Urban vegetation classification: Benefits of multitemporal RapidEye satellite data. *Remote Sens. Environ.* **2013**, *136*, 66–75. [CrossRef]

17. Rosina, K.; Kopecká, M. Mapping of urban green spaces using Sentinel-2A data: Methodical aspects. In Proceedings of the 6th International conference on cartography and GIS, Albena, Bulgaria, 13–17 June 2016.

18. Ulrich, R.S. Natural versus urban sciences: Some psycho-physiological effects. *Environ. Behav.* **1981**, *13*, 523–556. [CrossRef]

19. Kaplan, R. The analysis of perception via preference: A strategy for studying how the environment is experienced. *Landsc. Urban Plan.* **1983**, *12*, 161–176. [CrossRef]

20. Gómez-Baggethun, E.; Gren, A.; Barton, D.N.; Langemeyer, J.; McPhearson, T.; O'Farrell, P.; Andersson, E.; Hamstead, Z.; Kremer, P. Urban Ecosystem Services. In *Urbanization, Biodiversity and Ecosystem Services: Challenges and Opportunities: A Global Assessment*; Elmqvist, T., Fragkias, M., Goodness, J., Güneralp, B., Marcotullio, P.J., McDonald, R.I., Parnell, S., Schewenius, M., Sendstad, M., Seto, K.C., et al., Eds.; Springer: Dordrecht, The Netherlands; Heidelberg, Germany; New York, NY, USA; London, UK, 2013; pp. 175–252.

21. Moreno-Garcia, M.C. Intensity and form of the urban heat island in Barcelona. *Int. J. Climatol.* **1994**, *14*, 705–710. [CrossRef]

22. Nowak, D.J.; Crane, D.E. Carbon storage and sequestration by urban trees in the USA. *Environ. Pollut.* **2002**, *116*, 381–389. [CrossRef]

23. Wang, Y.; Bakker, F.; de Groot, R.; Wortche, H.; Leemans, R. Effects of urban trees on local outdoor microclimate: Synthesizing field measurements by numerical modelling. *Urban Ecosyst.* **2015**, *18*, 1305–1331. [CrossRef]

24. Ulrich, R.S. View through a window may influence recovery from surgery. *Science* **1984**, *224*, 420–421. [CrossRef] [PubMed]

25. Van den Berg, A.E.; van Winsum-Westra, M.; de Vries, S.; van Dillen, S.M.E. Allotment gardening and health: A comparative survey among allotment gardeners and their neighbors without an allotment. *Environ. Health* **2010**, *9*, 74. [CrossRef] [PubMed]

26. Kiss, M.; Takács, A.; Pogacsas, R.; Gulyas, A. The role of ecosystem services in climate and air quality in urban areas: Evaluating carbon sequestration and air pollution removal by street and park trees in Szeged (Hungary). *Morav. Geogr. Rep.* **2015**, *23*, 36–46. [CrossRef]

27. Vatseva, R.; Kopecká, M.; Otahel, J.; Rosina, K.; Kitev, A.; Genchev, S. Mapping urban green spaces based on remote sensing data: Case studies in Bulgaria and Slovakia. In Proceedings of the 6th International Conference on Cartography and GIS, Albena, Bulgaria, 13–17 June 2016.

28. European Commission: Mapping Guide for a European Urban Atlas. Available online: https://cws-download.eea.europa.eu/local/ua2006/Urban_Atlas_2006_mapping_guide_v2_final.pdf (accessed on 19 February 2017).

29. Hurbanek, P.; Atkinson, P.M.; Chockalingam, J.; Pazúr, R.; Rosina, K. Accuracy of Built-up Area Mapping in Europe at Varying Scales and Thresholds. In Proceedings of the Accuracy 2010: Ninth International Symposium on Spatial Accuracy Assessment in Natural Resources and Environmental Sciences, Leicester, UK, 20–23 July 2010; pp. 385–388.

30. Pesaresi, M.; Corbane, C.; Julea, A.; Florczyk, A.J.; Syrris, V.; Soille, P. Assessment of the added-value of Sentinel-2 for detecting built-up areas. *Remote Sens.* **2016**, *8*, 299. [CrossRef]

31. Elmqvist, T.; Fragkias, M.; Goodness, J.; Güneralp, B.; Marcotullio, P.J.; McDonald, R.I.; Parnell, S.; Schewenius, M.; Sendstad, M.; Seto, K.C.; et al. (Eds.) *Urbanization, Biodiversity and Ecosystem Services: Challenges and Opportunities: A Global Assessment*; Springer: Dordrecht, The Netherlands; Heidelberg, Germany; New York, NY, USA; London, UK, 2013; p. 755.

32. Goddard, M.A.; Dougill, A.J.; Benton, T.G. Scaling up from gardens: Biodiversity conservation in urban environments. *Trends Ecol. Evol.* **2009**, *25*, 90–98. [CrossRef] [PubMed]

33. Andersson, E.; Barthel, S.; Ahrné, K. Measuring social-ecological dynamics behind the generation of ecosystem services. *Ecol. Appl.* **2007**, *17*, 1267–1278. [CrossRef] [PubMed]

34. Bolund, P.; Hunhammar, S. Ecosystem services in urban areas. *Ecol. Econ.* **1999**, *29*, 293–301. [CrossRef]

35. Chiesura, A. The role of urban parks for the sustainable city. *Landsc. Urban Plan.* **2004**, *68*, 129–138. [CrossRef]

36. Cornelis, J.; Hermy, M. Biodiversity relationships in urban and suburban parks in Flanders. *Landsc. Urban Plan.* **2004**, *69*, 385–401. [CrossRef]

37. Kopecká, M.; Rosina, K. Identifikácia zmien urbanizovanej krajiny na báze satelitných dát s veľmi vysokým rozlíšením (VHR): Záujmové územie Trnava. *Geografický časopis* **2014**, *66*, 247–267.

38. Fan, C.; Myint, S.W.; Zheng, B. Measuring the spatial arrangement of urban vegetation and its impacts on seasonal surface temperatures. *Prog. Phys. Geogr.* **2015**, *39*, 199–219. [CrossRef]

39. Stewart, I.D.; Oke, T.R. Local Climate Zones for urban temperature studies. *Bull. Am. Meteorol. Soc.* **2012**, *93*, 1879–1900. [CrossRef]

40. Sievers, U.; Zdunkowski, W. A microscale urban climate model. *Contributions to atmospheric physics/Beiträge Zur Physik Der Atmosphäre* **1986**, *59*, 13–40.

41. Geletic, J.; Lehnert, M.; Dobrovolný, M. Modelled spatio-temporal variability of air temperature in an urban climate and its validation: A case study of Brno, Czech Republic. *Hung. Geogr. Bull.* **2016**, *65*, 169–180. [CrossRef]

land

MDPI

Article

Factors Influencing Perceptions and Use of Urban Nature: Surveys of Park Visitors in Delhi

Somajita Paul [1,2,]* and Harini Nagendra [3]

[1] Ashoka Trust for Research in Ecology and the Environment (ATREE), Royal Enclave, Jakkur Post, Srirampura, Bangalore 560064, India
[2] Manipal Academy of Higher Education, Manipal 576104, India
[3] School of Development, Azim Premji University, PES Institute of Technology Campus, Pixel Park, B Block, Electronics City, Hosur Road, Bangalore 560100, India; harini.nagendra@apu.edu.in
* Correspondence: somajitapaul@gmail.com; Tel.: +91-9312138242

Academic Editors: Andrew Millington and Monika Kopecka
Received: 5 February 2017; Accepted: 12 April 2017; Published: 18 April 2017

Abstract: Urban green spaces provide important recreational, social and psychological benefits to stressed city residents. This paper aims to understand the importance of parks for visitors. We focus on Delhi, the world's second most populous city, drawing on 123 interviews with park visitors in four prominent city parks. Almost all respondents expressed the need for more green spaces. Visitors valued parks primarily for environmental and psychological/health benefits. They had limited awareness of biodiversity, with one out of three visitors unable to identify tree species and one out of four visitors unable to identify animal species frequenting the park. Most of the daily visitors lived within 0.5 km of these parks, but a small fraction of visitors traveled over 10 km to visit these major parks, despite having smaller neighbourhood parks in their vicinity. This study demonstrates the importance of large, well-maintained, publicly accessible parks in a crowded city. The results can help to better plan and design urban green spaces, responding to the needs and preferences of urban communities. This research contributes to the severely limited information on people's perceptions of and requirements from urban nature in cities of the Global South.

Keywords: urban parks; public perception; recreation; visitors; environmental benefits

1. Introduction

With the increasing number of people living in urban areas, there are large-scale impacts on the sustainability of urban systems, impacting their biophysical and ecological components and eventually reducing human capacity for wellbeing. Increased urbanization and the consequent loss of green cover has been linked to reduced ground water recharge [1], degradation of water bodies [2], decreased biodiversity due to habitat loss and fragmentation [3–6], pollution [7], modification of rainfall [8–10] and urban warming [11–14].

Urban green spaces can increase resilience and reduce vulnerabilities to urbanization. Vegetation in urban areas contributes positively towards ecological heath in an urban system. Green spaces in urban areas provide ecosystem services [15] and recreational venues for diverse users [16,17]. Family recreation promotes the overall quality of family life and helps its members to develop life-long skills and values [18,19]. People staying close to nature are able to form stronger connections to nature, deriving both physical and psychological health benefits [20–24]. Urban green spaces facilitate social interaction and promote social cohesion, fostering a sense of place and belonging [25–27].

Various socio-demographic and environmental drivers of outdoor recreation have been identified by Bell et al. [28]. Proximity to recreational areas and parks is normally related to higher physical activity and healthier communities [29], and people derive health benefits [19]. The amount, quality and distance to urban recreation areas and green space affect the uses of green space by citizens to satisfy their daily recreational needs [30,31]. Proximity to urban green spaces thus tends to increase housing prices [32–34]. Some residents are also willing to pay for the use of urban green spaces for the derived benefits [35,36].

Thus, there are a growing number of studies on the environmental implications of urbanization and the benefits of urban nature. The greatest challenge lies in managing urban green spaces well, through the successful framing and implementation of environmental policies for sustainable urban nature. This would also help with the augmentation of public trust in the decision making process. Planning and management of urban nature is effective when it considers the diversity of knowledge of the public and stakeholders and the understanding and consideration of the user opinions, preferences, and attitudes towards conservation [37]. Peoples' perception and preferences for urban nature tend to vary from time to time and tend to be site specific. Hence, case studies are vital to bring out the local differences [38].

Most of the research on the use and importance of urban green spaces has been conducted in North America, Europe, and Australia [39–44]. There are also a growing number of studies from South East Asia [45–51]. A knowledge gap exists in terms of perception, provision and access to green space in Asia and specifically in India. This study attempts to fill this gap by understanding the relationship between park visitors and green spaces in the megapolis of Delhi. The objectives of the study are to (i) analyze the main uses of urban parks by different population groups; (ii) evaluate differences in the perception of different population groups of the quality of nature; and (iii) analyze distance to the green space and the relationship between distance to the green space and frequency of its use.

2. Materials and Methods

2.1. Study Area

The study sites are located in the heart of the National Capital Territory of Delhi (NCTD). Delhi is a rapidly expanding city with a high-density built-up area in the city centre and urban sprawl towards its periphery [52]. Delhi has a number of parks and gardens, spread over about 8000 hectares in various locations all over Delhi [53]. Administratively, the NCTD is divided into nine districts and 27 administrative sub-divisions or tehsils. The NCTD is administered by three local bodies; (i) the New Delhi Municipal Council (NDMC); (ii) the Delhi Cantonment Board (DCB); and (iii) the Municipal Corporation Delhi (MCD). New Delhi district has a population of 179,112 [54]. The British architect Edwin Lutyens designed the capital city of New Delhi, popularly known as Lutyens' Delhi, following a geometrical plan with large open green spaces and wide roads oriented along the main directions of the compass [55]. The NDMC administrative area corresponds to the New Delhi district. New Delhi district is the central and greenest part of the National Capital Territory of Delhi, which is now 'an oasis of nature in the midst of a vast urban desert' [56]. The large number of avenue trees, large parks, 'colony parks', green roundabouts, and bungalow gardens in Lutyens' Delhi shape the ecological and cultural character of this region, which 'nestles under a canopy of green' [56]. Parks and green spaces in New Delhi are maintained by different authorities, including the New Delhi Municipal Council (NDMC), Central Public Works Department (CPWD), Delhi Development Authority (DDA), and the Archaeological Survey of India (ASI). New Delhi is the most popular recreational area in Delhi [57]. Due to its historical and archaeological importance and it being the capital of India, New Delhi attracts people from diverse cultures and backgrounds from all over Delhi as well as India and abroad [57]. New Delhi thus represents an ideal locale for studying peoples' perceptions of urban green space in a crowded expanding city.

Within the New Delhi district, four large parks (Figure 1) managed by four different authorities were selected for study. The parks are (a) Buddha Jayanti Smarak Park (BJSP); (b) Lodhi Garden (LG); (c) Bhuli Bhatiyari Park (BBP), and (d) Safdarjung's Tomb (ST).

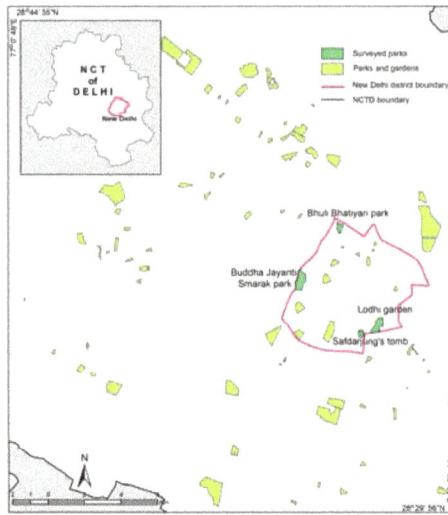

Figure 1. Study area.

Buddha Jayanti Smarak Park (Figure 2a) covers an area of 100 acres and is situated in the western part of the New Delhi district. The park forms a part of the well-known Delhi ridge forests, containing a mix of dry thorny native scrub with planted vegetation on a rocky, undulating, partially flat plain with high native biodiversity [58]. BJSP was established to commemorate the 2500th anniversary of the enlightenment of Gautama Buddha in the year 1957. It is free for public entry and remains open from 5:30 a.m. to 7:00 p.m. every day. It is managed by the Central Public Works Department of Government of India (CPWD).

Figure 2. Four parks: (**a**) Buddha Jayanti Smarak park; (**b**) Lodhi garden; (**c**) Safdarjung's tomb; and (**d**) Bhuli Bhatiyari park.

The Lodhi Garden (Figure 2b), having an area of 90 acres, is located in the southern part of the New Delhi district. This garden contains monuments established by the Sayyids and Lodhis between 15th and 16th century [59]. The park was developed during the British Period and was inaugurated by Lady Willingdon in the year 1936. The park was initially named Lady Willingdon Park. J. A. Stein and Garrett Eckbo redesigned the park in 1968 [59]. It is free for all visitors from 6:00 a.m. to 7:00 p.m. The monument is protected by the Archeological Survey of India (ASI) and the garden is maintained by the New Delhi Municipal Council (NDMC).

Safdarjung's Tomb (Figure 2c) has an enclosed garden around the tomb of Mirza Muqim Abul Mansur Khan, who was popularly known as Safadarjung. The monument and the garden premises are maintained by the ASI. The 32 acre garden is a Persian style or Charbagh Garden laid out in the form of four squares with wide foot paths and water tanks, which have been further subdivided into smaller squares. This is a historical funerary garden remodeled into a public park and the ASI took up the horticultural improvement of the tomb in 1918–19 [60]. There is a nominal entry fee for the garden and it remains open all days of the week from 7:00 a.m. to 5:00 pm

Bhuli Bhatiyari park (Figure 2d) is located in the northern part of the district and also forms a part of the ridge. Emperor Firuz Shah (1351–88) of the Tughlaq dynasty had built a hunting lodge named Bhuli-Bhatiyari-ka-Mahal (palace). In 1989, this place was developed as an ideal tourist location by the Delhi Tourism Development Corporation [61]. The remnants of the palace are protected by the ASI, and the park of 60 acres is maintained by the Delhi Development Authority (DDA). Entry is free, and the park remains open all days from sunrise to sunset.

2.2. Methodology

Except for Safdarjung's tomb, entry to the other three parks in the study area is free for visitors, hence the park administration does not maintain entry records of park visitors. Due to the non-availability of visitor records and to maintain parity in the sampling methodology amongst the parks, it was not possible to apply a simple random sampling technique to draw a true probability sample for the study [62]. Before beginning interviews, we conducted on-site observations of visitors to the park and activities taking place within the boundaries of the park. Due to security concerns, which are considerable in isolated locations in Delhi, we conducted interviews in frequently visited areas and not in the interior parts of the park, where visitors were fewer. We had an overall target of approximately 250 interviews across four parks and therefore decided to conduct interviews [63] with all visitors until the desired number of 60 per park was reached. Although the target of 60 interviews is admittedly not derived from a statistical estimate of sample size, we found that the responses did not vary appreciably after this point, which gave us confidence that the responses we received were representative of the majority of the visitors to the park. The interviews were conducted face to face, on weekends, weekday evenings, and other times when visitors were in large numbers, thereby sampling across the representation available of gender, education, and professional background. The survey was conducted on site by the lead author, both in English and Hindi. One hundred and twenty-three interviews were carried out in Buddha Jayanti Smarak park ($n = 26$), Lodhi garden ($n = 28$), Safdarjung's tomb ($n = 32$), and Bhuli Bhatiyari park ($n = 37$). Other researchers have used a similar or lesser number of interviews to study visitor perceptions and have drawn inferences from them. D'Souza and Nagendra [64] conducted 63 interviews of lake visitors, while Tucker et al. [63] and Krenichyn [65] interviewed 82 and 41 park visitors, respectively. The sample size of 123 had a maximum standard error of 0.045. The acceptable margin of error is 5%; thus a sample size that achieves standard errors lower than 0.05 is acceptable [66]. A total of nine visitors did not respond to the survey (Buddha Jayanti Smarak park = 2; Lodhi garden = 2; Safdarjung's tomb = 4, and Bhuli Bhatiyari park = 1). Amongst the non-respondents, three were female and six were male; the estimated age of four respondents was below 25, while two of them were above 60 years of age.

Visitors were approached for participation and informed that the purpose of the survey was to access the environmental awareness of the visitors and the distance to the park. It took five to seven minutes for the visitors to answer the questionnaire and ensure that it reflected their immediate experience. The surveys were conducted both on weekdays and weekends, in the mornings between 8:30 a.m. and 10 a.m. to collect views of morning walkers and in the evenings from 5 p.m. to 6 p.m. for other respondents. Five visits each were made to Buddha Jayanti Smarak park and Lodhi garden, followed by seven visits each to Safdarjung's tomb and Bhuli Bhatiyari park to conduct interviews.

The response formats were either open, in ranking scale, or closed (dichotomous, multiple choices, likert scale). Nineteen questions gauged the visitor's views on the quality of the park, uses of green space, environmental awareness, and distance to the green space (Table 1). Demographic and other information about the respondents were also collected, which included their name, age, gender, educational status, means of livelihood and information about companions. While it was a discretionary response, 'name' helped in developing a familiarity, rapport, and a humane connection with the respondent. The authors prepared a questionnaire containing a set of questions assessing park visitors' environmental awareness, main uses of green space, perception of the quality of the parks and distance to the parks, based in part on similar studies elsewhere [67–70]. The questionnaire was initially tested with a small group of park visitors before administering it to visitors to the four parks. This helped in further refining the questionnaire so that it ensured that the participants understood the questions uniformly and they were easily able to provide data for fulfilling the aim of the study. The visitor responses were further categorized for the ease of analysis. The distances of the closest park near the respondents' residences have been categorized into four classes (<0.5 km; 0.5–1 km; 2–5 km; and >5 km). The distances travelled by the respondents to visit the parks have been categorized into five (<0.5; 0.5–1 km; 1–4 km; 5–10 km; and >10 km) classes. The frequency of park visits has been grouped into six (everyday; several times a week; weekly; monthly; half yearly; yearly; and first time) classes. Thereafter, qualitative data analysis and interpretation of people's perception was carried out, which was then associated with the potential predictors or group dependent variables (gender, age, education, occupation, and companion). ANOVA and t tests [71] were used to examine the relationships between the variables.

Table 1. Green spaces and quality of urban life.

Criterion	Description	Groups of Questions on Survey	Type of Response	Role
Environmental Awareness	Pro-environmental attitude tends to shape ecological behaviour.	Do you have plants at home?	Dichotomous (Yes/No)	To assess the environmental awareness of the visitors.
		Do you feel the need for more green spaces/parks?	Dichotomous (Yes/No)	
		Do you take part in protecting nature and how?	Dichotomous (Yes/No) and open	
		What plant and animal species have you noticed in this park?	Open	
Main uses of green space	The use of green space reflects the benefits the visitors cherish from nature.	What are the uses of green spaces?	Open	What aspects the visitors value most in green space.
Quality of nature	Urban green spaces should approach levels of ecological and environmental quality desired by visitors	What is your assessment of the quality of this park?	Likert scale (i. Very good; ii. Good; iii. Satisfactory; iv. Bad; v. Very bad)	Satisfaction level of the visitors with the parks and their preferences.
		How do you think this park can be improved so that more people come here?	Open	
		What are the changes in plants and animal species over time?	Open	
Distance to the green space and frequency of use	The green space should be easily accessible, i.e., within walking distance of the communities.	How close do you stay to the park?	Open	To determine the distance to the green space.
		How often do you visit this park?	Open	
		How far is the closest park from your place of residence?	Open	

3. Results

3.1. Social Characteristics of the Visitors

As per the 2011 census, the density of population of the New Delhi district is 4057 persons/km^2. About 54% of the population is male and 46% is female [54]. Most respondents (N = 123) in the survey were male (57%; n = 70). The greatest (58%; n = 71) number of respondents were between 25 to 55 years old, followed by the age group above 55 years (25%; n = 31). About 77% (n = 95) have received university-level education. The majority was employed (44%; n = 54), and 22 % (n = 27) were self-employed in business. Homemakers, retired people, and unemployed people constitute the 'at home' category and account for 24% (n = 30) (Table 2). A majority (54%) of the respondents visit the park with family, followed by those visiting alone.

Table 2. Overall socio-demographic characteristics.

Socio-Demographic Characteristics	Percentage	Number
Gender		
Male	57%	70
Female	43%	53
Age		
<25 years	17%	21
25–55 years	58%	71
>55 years	25%	31
Occupation		
Self-employed/business	22%	27
Service	44%	54
Student	10%	12
At home	24%	30
Education		
Under graduates	23%	28
Graduates	56%	69
Post graduates	21%	26
Companion		
Alone	25%	31
Family	54%	66
Friends	21%	26

3.2. Environmental Awareness of the Users

A positive attitude of the respondents at all parks towards green spaces was found. The visitors were asked whether they had 'plants at home'. About 57% of the visitors had plants at home. More than 95% respondents felt that there is 'need for more green spaces/parks'. Environmental awareness and perception is also reflected in the level of participation in conserving nature. More than 63% of the respondents 'took part in protecting nature'. In comparison to the older age groups, fewer young adults (18%) took part in nature conservation measures (p = 0.043) (Table 3). The participants did it either by planting and nurturing trees at home or in the parks (43%), as a part of school/college curriculum (11%), making people aware of the usefulness of the green spaces (11%), asking people not to harm trees (13%), by being a member of some green groups or by making monetary contributions towards the protection and maintenance of the garden/parks (10%), or by not littering and segregating wastes, recycling, and refraining from using polyethylene carry bags (12%). Even though the visitors were not asked why they don't take part in the conservation of nature, 16% of the visitors spontaneously cited 'lack of time' as a reason. Amongst all age groups, a significant proportion

($p = 0.004$) of the younger (<25 years) visitors took part in nature conservation as a part of a school curriculum. The overall visitors on average could recognize 1.79 plant species and 1.80 animal species in the parks. Knowledge of plants and animal species was analysed with relation to socio-demographic factors, namely gender, age, education, occupation, and companion (Table 3). More than 30% of the visitors to the parks could not identify any plant species, while 21% of the visitors could not name any animal species in the park (Figure 3). It was also seen that 27% and 17% visitors could name at least two plant and animal species respectively. No significant variation was found between the socio-demographic variables in terms of (a) total number of identified species; (b) plant species; and (c) animal species. Even though insignificant, the average number of species identified by graduates and undergraduates is 3.5, but postgraduates tend to identify a higher number (4.3) of species in the parks. Visitors taking part in activities related to the protection of nature tend to identify more species in the park ($p < 0.005$).

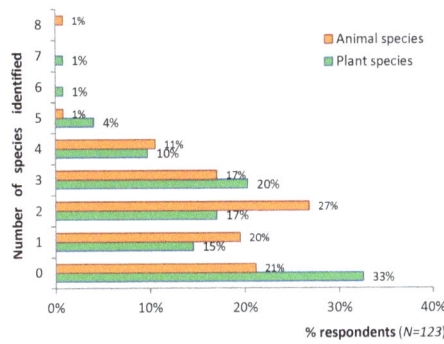

Figure 3. Number of plant and animal species identified by the visitors.

Table 3. Differences in environmental awareness among different socio-demographic groups.

Socio-Demographic Variables		Took Part in Nature Conservation	Average Number of Species Identified		
			Plants	Animal	Total
Gender	Male ($n = 70$)	53% (41)	1.54	1.76	3.30
	Female ($n = 53$)	47% (37)	2.11	1.84	3.96
	P (T-test)	0.30	0.07	0.23	0.32
Age groups	<25 years ($n = 21$)	18% (14)	1.71	1.48	3.19
	25 to 55 years ($n = 71$)	50% (39)	1.82	1.90	3.72
	>55 years ($n = 31$)	32% (25)	1.77	1.77	3.55
	P (ANOVA)	0.04 *	0.97	0.51	0.71
Education	Under graduation ($n = 28$)	22% (17)	1.75	1.79	3.54
	Graduation ($n = 69$)	51% (40)	1.67	1.68	3.35
	Post-graduation ($n = 26$)	27% (21)	2.15	2.12	4.27
	P (ANOVA)	0.12	0.43	0.44	0.28
Occupation	Business/Self-employed ($n = 27$)	23% (18)	1.93	2.04	3.96
	Service ($n = 54$)	36% (28)	1.48	1.69	3.17
	Students ($n = 12$)	10% (8)	2.58	1.75	4.33
	At home ($n = 30$)	31% (24)	1.90	1.80	3.70
	P (ANOVA)	0.08	0.16	0.80	0.36
Companion	Alone ($n = 31$)	26% (20)	1.77	1.74	3.52
	Family ($n = 66$)	50% (39)	1.68	1.89	3.58
	Friends ($n = 26$)	24% (19)	2.08	1.62	3.69
	P (ANOVA)	0.46	0.58	0.70	0.97

* $p < 0.05$.

The neem tree (*Azadirachta indica*) and sacred peepal tree (*Ficus religiosa*) were the most frequently named plants, and peacocks (*Pavocristatus*) and dogs (*Canis lupus familiaris*) were the frequently identified animal species (Table 4).

Table 4. Most frequently identified plant and animal species in the parks by the visitors.

Flora		Fauna	
Neem (*Azadirachta indica*)	44 (33%)	Peacock (*Pavo cristatus*)	31 (17%)
Peepal (*Ficus religiosa*)	17 (13%)	Dogs (*Canis lupus familiaris*)	25 (14%)
Keekar (*Prosopis juliflora*)	16 (12%)	Swan (*Cygnus atratus*)	24 (13%)
Jamun (*Syzygium cumini*)	15 (11%)	Crow (*Corvus splendens*)	20 (11%)
Palm (*Arecaceae sp.*)	11 (8%)	Parrot (*Psittacula krameri*)	20 (11%)
Ashoka (*Saraca asoca*)	7 (5%)	Squirrel (*Funambulus pal*)	18 (10%)
Mango (*Mangifera indica*)	6 (5%)	Pigeons (*Columba livia domestica*)	14 (8%)
Amla (*Phyllanthus emblica*)	6 (5%)	Butterflies (*Rhopalocera*)	12 (7%)

3.3. Main Uses of Green Space as Ascertained by Different Population Groups.

The analysis of peoples' perception of uses of green spaces reflects their demands and needs/expectations from green spaces. The respondents were asked about the uses of green space. The visitors' responses were varied. They are categorized into five groups to give us a better understanding of how people value the urban green spaces. The first group, 'social and recreational benefits' comprises of benefits like 'recreation', 'aesthetics', 'social bonding', and 'meeting friends'. 'Mental peace', 'spiritual benefit', 'connect to nature', 'relaxation', 'good for eyes', 'longevity', and 'health benefits' constitute the second category of 'psychological and health benefits'. 'Protection of the environment', 'oxygen generation',/ 'pollution control', 'reduction of global warming', 'fresh air and breeze', 'shade', and 'cooling' have been grouped together to form the third category, 'environmental benefits'. The fourth group includes 'biodiversity benefits'. The fifth category, 'other benefits' comprises of 'economic benefits', 'furniture', 'fruits', and 'environmental education'.

About 87% respondents said that passive use for 'environmental benefits' constitutes the main use of urban green spaces, followed by 'psychological and health benefits' (68%), and about 45% and 13% state that green spaces are important for 'social and recreational' purposes and 'biodiversity benefits', respectively (Figure 4).

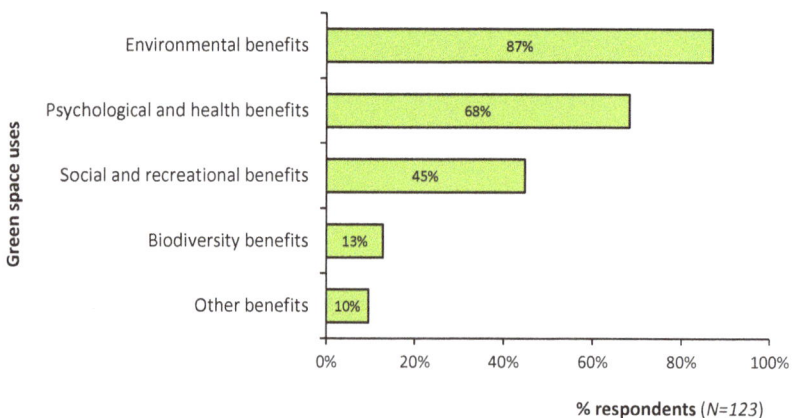

Figure 4. Uses of green space.

There was considerable variation in the perception of green space use between visitors from different age groups (Table 5). Older visitors (>55 years) valued green spaces more for 'environmental benefits' ($p < 0.05$) than younger visitors. Younger (<25 years) visitors (33%) were less aware of the 'psychological and health benefits' ($p = 0.0005$) than other age groups. However, younger (<25 years) visitors were more aware of the 'other benefits'of green space use than the two older age groups. Perception of uses of green space also varied significantly amongst different education groups ($p = 0.01$), occupation groups ($p = 0.003$), and companion groups ($p = 0.01$). Postgraduates (92%) appreciated the 'psychological and health benefits' of urban green space more than graduates and undergraduates. Students (25%) and visitors with friends (42%) did not appreciate the 'psychological and health benefits' of urban nature as much as other groups.

Table 5. Uses of green space by various socio-demographic groups.

Socio-Demographic Groups		Social and Recreational Benefits	Environmental Benefits	Psychological and Health Benefits	Biodiversity Benefits	Other Benefits
Gender	Male (n = 70)	44% (31)	87% (61)	71% (50)	10% (7)	6% (4)
	Female (n = 53)	45% (24)	87% (46)	64% (34)	17% (9)	15% (8)
	P (T-test)	0.91	0.95	0.39	0.27	0.1
Age groups	<25 years (n = 21)	43% (9)	86% (18)	33% (7)	24% (5)	24% (5)
	25 to 55 years (n = 71)	45% (32)	83% (59)	76% (54)	13% (9)	8% (6)
	>55 years (n = 31)	45% (14)	97% (30)	74% (23)	6% (2)	3% (1)
	P (ANOVA)	0.98	0.02 *	0.0005 **	0.19	0.04 *
Education	Under graduation (n = 28)	46% (13)	79% (22)	61% (17)	11% (3)	21%(6)
	Graduation (n = 69)	45% (31)	93% (64)	62% (43)	14% (10)	7% (5)
	Post-graduation (n = 26)	42% (11)	81% (21)	92% (24)	12% (3)	4% (1)
	P (ANOVA)	0.95	0.1	0.01 *	0.86	0.05
Occupation	Business/Self-employed (n = 27)	59% (16)	93% (25)	78% (21)	11% (3)	11% (3)
	Service (n = 54)	41% (22)	81% (44)	67% (36)	11% (6)	7% (4)
	Students (n = 12)	50% (6)	83% (10)	25% (3)	25% (3)	25% (3)
	At home (n = 30)	33% (10)	93% (28)	80% (25)	13% (4)	7% (2)
	P (ANOVA)	0.231	0.339	0.003 *	0.629	0.281
Companion	Alone (n = 31)	52% (16)	77% (24)	74% (23)	13% (4)	13% (4)
	Family (n = 66)	42% (28)	91% (60)	76% (50)	12% (8)	8% (5)
	Friends (n = 26)	42% (11)	88% (23)	42% (11)	15% (4)	12% (3)
	P (ANOVA)	0.68	0.18	0.01 *	0.91	0.54

$* p < 0.05; ** p < 0.01.$

3.4. Quality of Nature

The respondents were asked the question 'What is your assessment of the quality of this park?'. About 49% male and 38% female respondents found the quality of the park to be good (Figure 5). The perception of quality of nature varied significantly in terms of gender and occupational groups (Table 6). Only the females (100%) considered the parks to be 'bad' ($p < 0.0001$). However very few (3%) students considered the parks to be 'very good' ($p = 0.004$). The perception of quality of nature also varied according to the frequency of visits. A significant proportion (65%) of first time visitors considered the parks to be good ($p = 0.02$). The respondents were then asked 'How do you think this park can be improved so that more people come here?'. About 33% of respondents wanted biodiversity improvements in the park, including an increase in the number of trees and flowering plants and more birds. A majority (66%) of the respondents would like infrastructural improvements, ranging from better security and accessibility to the provision of separate play areas for children and increased advertisement of the park. Twenty-six percent of the visitors who consider the park to be very good do not suggest any further improvements ($p < 0.001$). Visitors' expectations for park improvement did not vary across different socio-demographic groups (Table 6).

Table 6. Quality of nature and expected improvements by different socio-demographic groups.

Socio-Demographic Groups		Quality of Nature				Expected Improvements		
		Very Good	Good	Satisfactory	Bad	Not Required	Biodiversity	Infrastructural
Gender	Male (*n* = 70)	59% (24)	63%(34)	46% (12)	0% (0)	62% (10)	54% (20)	60% (51)
	Female (*n* = 53)	41% (17)	37% (20)	54% (14)	100% (2)	38% (6)	46% (17)	40% (34)
	P (T-test)	0.46	0.42	0.11	<0.0001 *	0.23	0.39	0.28
Age groups	<25 years (*n* = 21)	10% (4)	18% (10)	23% (6)	50% (1)	14% (2)	16% (6)	15% (13)
	25 to 55 years (*n* = 71)	61% (25)	56% (30)	58% (15)	50% (1)	43% (7)	68% (25)	59% (50)
	>55 years (*n* = 31)	29% (12)	26% (14)	19% (5)	0% (0)	43% (7)	16% (6)	26% (22)
	P (ANOVA)	0.30	0.90	0.57	0.41	0.19	0.28	0.74
Education	Under graduation (*n* = 28)	10% (4)	30% (16)	31% (8)	0% (0)	13% (2)	30% (11)	22% (19)
	Graduation (*n* = 69)	63% (26)	57% (31)	38% (10)	100% (2)	62% (10)	54% (20)	57% (48)
	Post-graduation (*n* = 26)	27% (11)	13% (7)	31% (8)	0% (0)	25% (4)	16% (6)	21% (18)
	P (ANOVA)	0.05	0.08	0.12	0.46	0.58	0.42	0.99
Occupation	Business/Self-employed (*n* = 27)	39% (16)	15% (8)	12% (3)	0% (0)	25% (4)	16% (6)	22% (19)
	Service (*n* = 54)	34% (14)	56% (30)	35% (9)	50% (1)	44% (7)	54% (20)	44% (37)
	Students (*n* = 12)	3% (1)	11% (6)	15% (4)	50% (1)	0% (0)	6% (2)	12% (10)
	At home (*n* = 30)	24% (10)	18% (10)	38% (10)	0% (0)	31% (5)	24% (9)	22% (19)
	P (ANOVA)	0.004 *	0.08	0.11	0.23	0.54	0.39	0.66
Companion	Alone (*n* = 31)	27% (11)	28% (15)	19% (5)	0% (0)	31% (5)	27% (10)	25% (21)
	Family (*n* = 66)	56% (23)	50% (27)	58% (15)	50% (1)	56% (9)	57% (21)	54% (46)
	Friends (*n* = 26)	17% (7)	22% (12)	23% (6)	50% (1)	13% (2)	16% (6)	21% (18)
	P (ANOVA)	0.74	0.76	0.73	0.52	0.63	0.68	0.98

* $p < 0.01$.

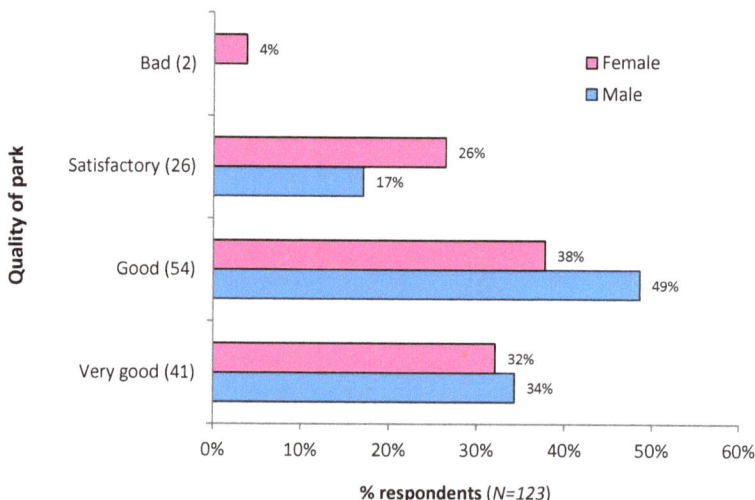

Figure 5. Quality of park according to the visitors.

3.5. Distance to Green Space and Frequency of Use

For about 36% of the respondents, the surveyed parks were closest to their place of residence. The number of total visitors to the surveyed park tends to increase with the distance of place of residence from the parks. Only 7% of the visitors to the surveyed parks come from a distance of <0.5 km, while the majority of the visitors travelled more than one km from their place of residence to visit the surveyed parks. Especially, 33% of the visitors travelled more than 10 km to from their place of residence to visit the parks (Figure 6). There was not much variation in the distance travelled by the respondents to visit the parks from their place of residence in terms of gender, education, and companion (Table 7). However there was significant variation amongst different age and occupation groups. About 13% of older visitors traveled more than 10 km from their place of residence to visit the park ($p = 0.008$), whereas 58% of younger adults (25–55 years) visited parks located at a distance beyond 10 km from their place of residence. In comparison to student visitors

(8%), a greater proportion of business and self-employed (37%) people traveled 5 to 10 km from their place of residence to visit parks ($p = 0.034$).

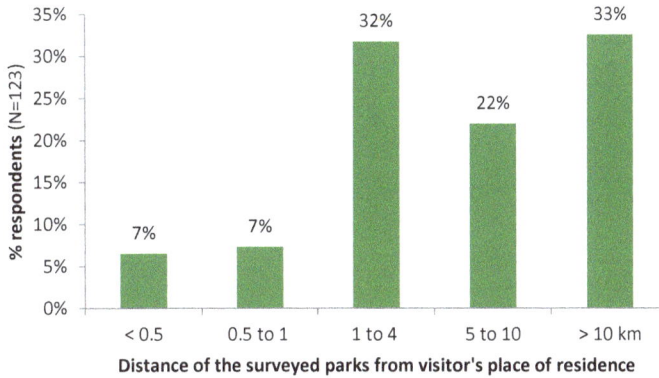

Figure 6. Distance of the surveyed parks from visitor's place of residence.

Table 7. Distance travelled by different socio-demographic groups to visit surveyed parks.

Socio-Demographic Groups		Distance to the Surveyed Park from Place of Residence				
		<0.5 km	0.5 to 1 km	1 to 4 km	5 to 10 km	>10 km
Gender	Male (n = 70)	62% (5)	67% (6)	62% (24)	59% (16)	48% (19)
	Female (n = 53)	38% (3)	33% (3)	38% (15)	41% (11)	52% (21)
	P (T-test)	0.21	0.07	0.35	0.40	0.23
Age groups	<25 years (n = 21)	25% (2)	11% (1)	10% (4)	8% (2)	30% (12)
	25 to 55 years (n = 71)	63% (5)	56% (5)	57% (22)	59% (16)	58% (23)
	>55 years (n = 31)	12% (1)	33% (3)	33% (13)	33% (9)	12% (5)
	P (ANOVA)	0.65	0.79	0.22	0.25	0.008 *
Education	Under graduation (n = 28)	25% (2)	11% (1)	18% (7)	26% (7)	27% (11)
	Graduation (n = 69)	38% (3)	45% (4)	56% (22)	59% (16)	60% (24)
	Post-graduation (n = 26)	37% (3)	44% (4)	26% (10)	15% (4)	13% (5)
	P (ANOVA)	0.45	0.19	0.57	0.65	0.25
Occupation	Business/Self-employed (n = 27)	25% (2)	0% (0)	28% (11)	37% (10)	10% (4)
	Service (n = 54)	50% (4)	67% (6)	49% (19)	22% (6)	48% (19)
	Students (n = 12)	0% (0)	0% (0)	8% (3)	8% (2)	17% (7)
	At home (n = 30)	25% (2)	33% (3)	15% (6)	33% (9)	25% (10)
	P (ANOVA)	0.82	0.21	0.33	0.034 *	0.05
Companion	Alone (n = 31)	13% (1)	33% (3)	33% (13)	22% (6)	20% (8)
	Family (n = 66)	62% (5)	45% (4)	46% (18)	70% (19)	50% (20)
	Friends (n = 26)	25% (2)	22% (2)	21% (8)	8% (2)	30% (12)
	P (ANOVA)	0.70	0.82	0.35	0.08	0.23

*$p < 0.05$.

About 24% of the respondents visited these parks daily, while 28% were first time visitors. The majority (63%) of the daily users resided within 0.5 km of the surveyed parks, and only 10% lived beyond a distance of 10 km from the parks. There was significant variation in the frequency of green space use in terms of gender, education and companion (Table 8). A greater proportion of male respondents (71%) visited the parks several times a week than the females ($p = 0.03$). In comparison to the older age groups, a very small proportion (7%) of younger visitors visited the parks everyday ($p = 0.012$). Fifty-two percent of the visitors preferred to visit the park alone every day ($p = 0.001$). About 76% of the 'first time' visitors preferred to visit the park with family ($p = 0.004$), a greater proportion of these being graduates (47%) ($p = 0.037$).

Table 8. Frequency of visits to the surveyed parks by different socio-demographic groups.

Socio-Demographic Groups		Frequency of Visits						
		Everyday	Several Times a Week	Weekly	Monthly	Half Yearly	Yearly	First Time
Gender	Male (*n* = 70)	62% (18)	71% (12)	68% (15)	45% (5)	50% (4)	50% (1)	44% (15)
	Female (*n* = 53)	38% (11)	29% (5)	32% (7)	55% (6)	50% (4)	50% (1)	56% (19)
	P (T-test)	0.29	0.03 *	0.07	0.05	0.15	0.15	0.11
Age groups	<25 years (*n* = 21)	7% (2)	0% (0)	23% (5)	36% (4)	38% (3)	0% (0)	20% (7)
	25 to 55 years (*n* = 71)	48% (14)	76% (13)	45% (10)	55% (6)	62% (5)	100% (2)	62% (21)
	>55 years (*n* = 31)	45% (13)	24% (4)	32% (7)	9% (1)	0% (0)	0% (0)	18% (6)
	P (ANOVA)	0.012 *	0.10	0.44	0.15	0.12	0.48	0.47
Education	Under graduation (*n* = 28)	21% (6)	12% (2)	4% (1)	36% (4)	25% (2)	0% (0)	38% (13)
	Graduation (*n* = 69)	48% (14)	53% (9)	73% (16)	55% (6)	75% (6)	100% (2)	47% (16)
	Post-graduation (*n* = 26)	31% (9)	35% (6)	23% (5)	9% (1)	0% (0)	0% (0)	15% (5)
	P(ANOVA)	0.33	0.23	0.07	0.41	0.31	0.46	0.04 *
Occupation	Business/Self-employed (*n* = 27)	28% (8)	29% (5)	27% (6)	18% (2)	25% (2)	0% (0)	12% (4)
	Service (*n* = 54)	45% (13)	53% (9)	28% (6)	46% (5)	25% (2)	50% (1)	53% (18)
	Students (*n* = 12)	3% (1)	0% (0)	18% (4)	9% (1)	38% (3)	0% (0)	9% (3)
	At home (*n* = 30)	24% (7)	18% (3)	27% (6)	27% (6)	12% (1)	50% (1)	26% (9)
	P(ANOVA)	0.56	0.37	0.26	0.99	0.05	0.76	0.36
Companion	Alone (*n* = 31)	52% (15)	23% (4)	27% (6)	0% (0)	25% (2)	50% (1)	9% (3)
	Family (*n* = 66)	31% (9)	65% (11)	46% (10)	64% (7)	25% (2)	50% (1)	76% (26)
	Friends (*n* = 26)	17% (5)	12% (2)	27% (6)	36% (4)	50% (4)	0% (0)	15% (5)
	P (ANOVA)	0.001 **	0.53	0.66	0.10	0.10	0.63	0.005 **

* $p < 0.05$; ** $p < 0.01$.

Only 16% of the respondents had a neighbourhood park within 0.5 km of their residence, while more than 44% respondents had parks located at a distance greater than 2 km from their residence (Figure 7).

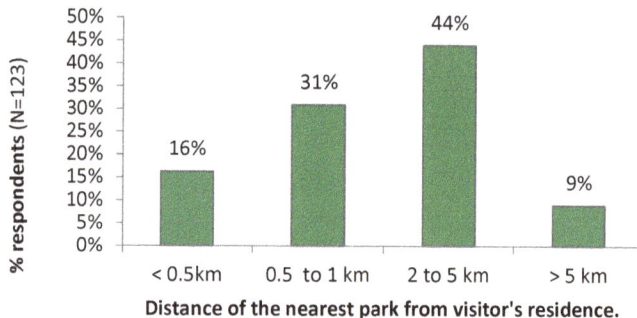

Figure 7. Distance of neighbourhood parks from the respondent's place of residence.

Even though only 16% of respondents had other smaller parks in their closest vicinity (<0.5 km), 50% of respondents were daily visitors to the surveyed parks (Figure 8). Thus we find that even though the nearest park was within easy reach (<0.5 km) to most visitors, their preference was for visiting the surveyed parks, possibly because these represent large, well-maintained, and attractive green spaces.

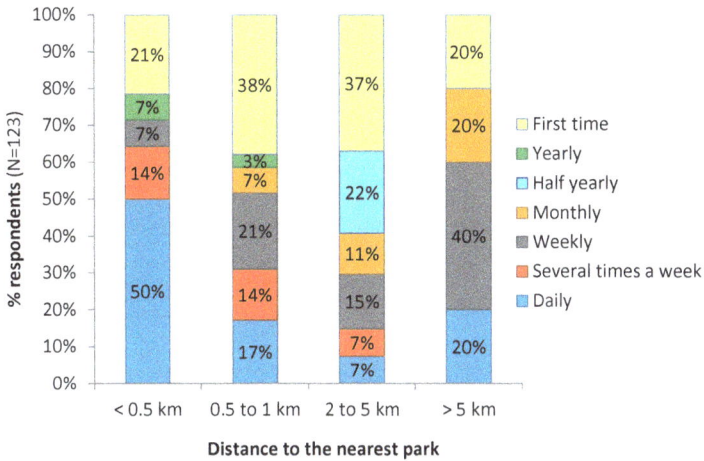

Figure 8. Relation between the frequency of visits and distance to the nearest park.

Visitors to the parks have been categorized into seven categories based on their frequency of visit to the park (Figure 9). The proportion of the daily visitors to the surveyed parks tends to decrease as the distance to the park increases. A majority (63%) of the respondents staying within 0.5 km of the park visited the park daily ($p < 0.01$), while this proportion decreases to 10%, for those who covered more than 10 km daily. There was also significant variation in the frequency of visits and distance to the surveyed parks. The large proportion (50%) of first time visitors ($p < 0.001$) traveled longer distances (>10 km) to visit the park than the rest of the visitors. The visitors tend to travel 5 to 10 km on holidays ($p < 0.05$) to visit surveyed parks. This was again probably because the surveyed parks are larger than the neighbourhood parks and more attractive.

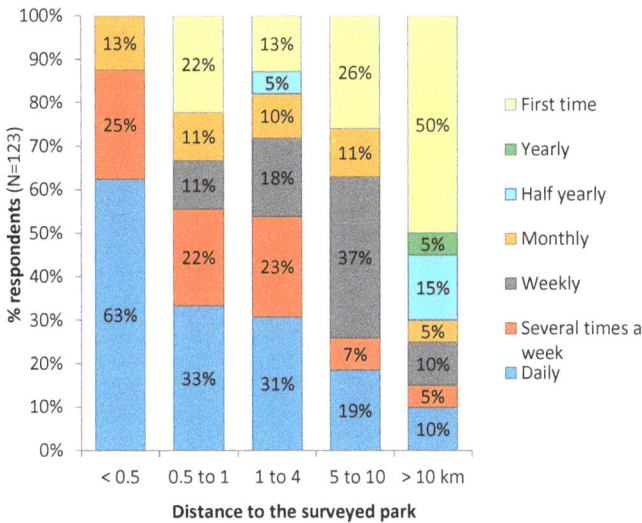

Figure 9. Relation between the frequency of visits and distance to the surveyed park.

4. Discussion

4.1. Visitor Characteristics

It is very important to understand the socio-demographic characteristics of visitors to the park because personal characteristics, companion, work and living situations tend to determine recreation activity response [72]. The population of the New Delhi district is thinly spread in comparison to the National Capital Territory of Delhi. All four surveyed parks are known to be visited by people from a range of socio-economic backgrounds, Lodhi garden having higher visitation by the bureaucrats, politicians, wealthier and more influential visitors. Amongst the visitors, it is seen that a lesser proportion of women visitors visit the park than men, which is probably because women perceive more constraints to outdoor recreation participation like personal safety concerns, inadequate facilities and preoccupation with office and household work [73] than men do. A majority of the respondents had attained university level education, which is in contrary with the study by Jim and Chen [36] in Guangzhou, China. Most of the visitors were employed, are from the age group of 25 years to 55 years, and probably visit the park to relax and escape the stressful and polluted Delhi city life. A female, aged 65 said that *'park ke swach bwatabaran mein humko bahut shanti milti hai aur Dilli ke dhul aur pradushan se mukti milti hai'*, i.e. 'my mind gets lot of peace in the clean environment of the park and it also gives respite from the dust and pollution of Delhi'. A lesser proportion of visitors were from the age group of >55 years. It has been observed in a study of 20 countries, including India [74], that physical activity declines with age. The visitors were mostly accompanied by family members as parks are attractive places to do recreational activities with families and contribute to enhanced social interaction [75]. Single people are reasonably well represented in the study, conforming to the study by Jim and Chen [36].

4.2. Environmental Awareness of the Users

Environmental awareness is a dynamic process aimed at augmenting our knowledge and understanding of the environment [76]. The emotional involvement of individuals tends to shape environmental awareness and attitudes [77]. The association of attitudes and behavior has led to interest in environmental attitudes as predictors of environmentally based actions and participation decisions [78]. A pro-environmental attitude is a powerful predictor of ecological behavior [79]. Pro-environmental attitudes tend to be related to resilient faith on honest intentions for species protection [78]. It was observed that many of the visitors took part in protection of nature, but a lesser proportion of young adults took part in conservation of nature. On the contrary, a study in Cleveland, Ohio has shown that older age groups preferred recreation rather than conservation [80]. It has been observed that young people have a lesser attraction to and interest in nature due to their growing up in highly urbanized areas and in over protected environments [81]. Most of the respondents who took park in nature conservation indicate that they were motivated to do so because of environmental education components in their school educational curriculum. In India, environmental education was introduced as a compulsory school subject in 2003. Following this, various programs are held in the schools to impart environmental education to students; for example, trees are planted and nurtured in the school and within the premises of houses [82]. Hence we find that a significant number of young adults attribute their participation in nature conservation to the school curriculum. A study by Kudryavtsev et al. [83] in the Bronx, New York shows that environmental education enhances environmental stewardship in urban communities. Lack of time is the main hindrance for not taking part in the conservation of nature in cities, as corroborated in studies by Mowen and Confer [84]. A study by Qiu et al. [85] shows that ecological knowledge tends to have a positive influence on a preference for biodiversity. The respondents were able to identify more animals than plant species in the parks, as also observed in Singapore [69], where the interviewees' knowledge of plants was less than their knowledge of animal species. A study by Rupprecht et al. [47] in Brisbane and Sapporo reveals that the respondents didn't consider themselves very conversant about local nature when it comes to knowledge of wild plants, wild animals, and birds. The people taking part in the protection of nature tend to identify more species, as observed in Singapore [69], with members of nature societies tending to identify a greater number of species. The poor knowledge of the plants and animals of most visitors can be attributed to their limited encounters with nature and learning opportunities in the urban concrete built environment, which has eventually led to detachment from and insensitiveness to their natural surroundings [36]. The neem tree and the peacock are more widely recognized by the visitors than the other plant and animal species in the parks. It is probably because the neem tree is deeply interwoven with the fabric of Indian culture [86] and widely used for its natural therapeutic properties in India [87], whereas the peacock, with its distinctive plumage and being the national bird of India, is easily recognized by the visitors to the parks.

4.3. Main Uses of Green Spaces as Ascertained by Different Population Groups

Studies carried out by Swamy and Devy [88] in Bengaluru show that large urban heritage parks are much valued for their environmental benefits like 'regulation of noise and temperature' and 'fresh air and breeze'. A similar pattern of responses can also be observed in our study. 'Recreation', 'aesthetics' and 'socializing' together appear more valued by the respondents in Bengaluru, while in Delhi it is observed that a much greater proportion of visitors appreciate green space for 'Psychological and health benefits'. This can be related to the severe levels of air pollution in Delhi, which has been described as one of the world's most polluted cities [89].

The uses of green spaces are not perceived similarly by all individuals in a population. The older visitors and visitors with higher education tend to appreciate the 'psychological and health benefits' of urban green space more than the rest of the population in Delhi. Perceptions tend to vary with socio-economic and demographic factors, as shown by several different studies [68,90]. In Guangzhou, China, women are more reluctant to participate in outdoor recreation than men [91]. In Delhi, given the overall insecurity of women's safety in public places, women do not access green spaces as much as men. Women tend to value green spaces for passive recreational activities such as socializing [92]. It is evident that social interaction contributes to better social cohesion [93]. Parks also serve as refuges for visitors wanting to escape the stress of the city [20,94].

Priego et al.'s [67] study of three countries reveals that people with different social and cultural backgrounds use and perceive the urban landscape in different ways. Age, gender, education level, income, retirement status, residential neighbourhood and length of stay tend to influence peoples' perception [16,91,95–97]. Varied responses to nature have also been noted between communities varying in race or class [90,98]. However, Qureshi et al. [97] in study in Karachi found no impact of age group or gender on the behavioural pattern of respondents.

4.4. Quality of Nature

The majority of the visitors are happy with the quality of the park, but only females consider the parks to be 'bad'. Thus there is prevalence of negative emotions amongst the female park visitors in Delhi. The 'bad' assessment of the parks may be the outcome of perceived social dangers by the female visitors [99]. The first time visitors rated the parks to be 'good'. It is seen that the different degrees of visitors' familiarity with the parks also leads to differences in their evaluative appraisal [100]. There is a clear preference for safety, cleanliness and maintenance in the parks, as found in other studies like one in Los Angeles [101]. Concerns about safety have been voiced by urban residents in Los Angeles [101] and Hong Kong [96]. Increased maintenance tends to increase the preference and sense of safety [102]. Visitors prefer promotion and advertisement for parks in order to enhance their use by the public. A study by Scott and Jackson [103] in Cleveland, Ohio confirmed that improved promotion encourages park use.

4.5. Distance to Green Space and Frequency of Use

Smaller distance to green spaces in urban areas is important because it contributes to improved human health and wellbeing [104] and is also a key planning and management issue [105]. The surveyed parks are located in the New Delhi district, and visitors come from both the New Delhi district and places far from the New Delhi district. Only 19% of the total visitors have access to green spaces within a distance of 500 m. This confirms that the accessibility in terms of the distance to green space is very low in Delhi in comparison to other countries, where the distance to green spaces is relatively lower [68,106]. We find that the proportion of visitors increases with distance. It is likely to be as a result of the attractiveness of large parks. This confirms studies in Perth [40] and Bengaluru [88] that found that the access to large attractive green spaces is associated with higher level of walking, and distance alone is not a deterrent. This study finds that younger visitors travel longer distances than older visitors, and male visitors tend to visit the parks several times a week. Women face greater constraints to outdoor recreation participation than men [107]. With age and reduced mobility, older people are reluctant to travel longer distances. This is in congruence with the study by Lo and Jim [96], in which less mobile retired people were not willing to travel longer distances to visit large parks. At the same time, students are not very enthusiastic about traveling longer distances to visit the parks. Studies have shown that the lively social environments of the parks tend to attract students, and their park use is strongly guided by their friends' park use [108]. The number of daily visitors tends to decrease with the increase in distance to the park, which has been corroborated by studies in Denmark [68] and Belgium [72]. Mowen and Confer [84] stated that shorter distances are critical for establishing a stronger user base. At the same time, a large proportion of first time and weekly visitors travel longer distances to visit the surveyed parks because they are larger and more attractive than their neighbourhood parks [109]. A study by Swamy [88] also established the fact that large attractive parks tend to attract visitors from far and wide. New Delhi park visitors who visit the parks every day do so without family and friends. Similarly in Guangzhou [110], single visitors visit parks regularly, yet a large number of 'first time' visitors visit the parks with their family. The surveyed parks are attractive tourist destinations, and people are motivated to travel to attractive tourist places as a family as it gives them the opportunity to be together [111].

5. Conclusions

This survey of park visitors in the megalopolis of Delhi aims to understand perceptions of and expectations from nature in urban communities living in a crowded Indian city. This study adds to information that can help in the better design of parks in response to peoples' expectations in Delhi. It also contributes to the rather limited information on peoples' perceptions and needs from urban green spaces in cities in the Global South, in contrast to the relatively greater information available from cities in Europe and North America [16,67,68], and adds to a small but steadily growing database of studies from Asia [88,91,97]. Some findings, such as the increased attractiveness of large parks with more visitors, specifically women and families, due to considerations of safety, may be specific to cities such as Delhi, with its relatively high rate of crime and insecurity for women. Other findings, such as the tendency of daily visitors to visit alone but for one-time visitors to visit with family, are trends that can be seen across other parks. Further research in parks across Delhi and other cities in India is required to understand the Delhi-specific and India-specific characteristics of the case study and, consequently, the implications for planning policies in other urban contexts.

In the stressful and polluted city life of Delhi, visitors value urban green spaces most for the 'environmental benefits'. Older adults and postgraduate educated visitors especially tend to appreciate the 'psychological and health. We also find that the accessibility in terms of distance to green space is very low in Delhi compared to other cities. Large parks tend to attract more visitors from further distances, despite their having small neighbourhood parks in the vicinity of their homes. The visitors want better quality parks with proper maintenance and better infrastructural facilities like separate play areas for children and better security. There are various barriers to park use like lack of time, transport problem, poverty, lack of companion, and safety. Safety represents a constraint on the use of parks, particularly for women and families and especially in places like Delhi, where the interior areas of these large parks are often 'safe havens' for criminals after sunset [112]. An increased focus on security will help residents to access the parks in greater numbers without compromising their personal safety. Such measures will be likely to increase the uses of green space by a diverse mix of gender and age groups. This will ensure socially favourable urban parks for the wellbeing of the urban community. Public perceptions of green spaces in Delhi throw up an interesting mix of requirements, with an increased focus on the aesthetic and the environmental benefits in preference to biodiversity. This study has policy implications for planners and urban designers, as well as for environmental organizations. The pro-environmental attitude of the respondents is expressed by the presence of 'plants at home', taking part in 'conservation of nature', and their awareness for 'need for more parks/green spaces'. Due to India's age-old tradition and culture of using neem for various purposes, this tree is easily identified by park visitors, but still there is a gap in the public understanding of biodiversity in Delhi's parks, with the result that one out of three visitors cannot name a single plant species, while one out of four visitors were unable to name a single common animal or insect that they had observed in the park. Further research is required to understand and take into account the opinion of visitors, capture the views of diverse age, gender, and ethnic-cultural groups in the planning and designing of urban parks and green spaces. It will be interesting to look at the differences of perception of the visitors residing in the New Delhi district and visitors to the New Delhi district. Further analysis combining survey data on socio-economic information with socio-economic data from the census, based on visitors' places of residence, is necessary to find out whether our results hold up for a larger sample of the population and in more recent years. This will ensure effective plans and designs capable of satisfying the needs of the urban community. The methodological approach of the study has exportability, and the empirical information obtained in study has comparability to other Indian cities and cities of the Global South. Given the rapid urbanization currently underway in India, concomitant with the disappearance of trees, wetlands, and green spaces, access to parks provides an increasingly rare opportunity for exposure to nature for many Indian urbanites. This study aims to contribute to an increased focus on the importance of green spaces in the urban context in India, where the current focus on smart cities [113] threatens to obscure the importance of low-technology options for improving urban resilience and wellbeing through a renewed focus on urban green spaces.

Acknowledgments: We thank the Royal Norwegian Embassy (RNE: IND-3025 1210050) and the Tata Social Welfare Trust (TSWT:TSWT/IG/SP/BM/sm/CM/24/MNRL/2011-12) for doctoral research support to Somajita Paul, and we thank Azim Premji University for research support to Harini Nagendra.

Author Contributions: Harini Nagendra and Somajita Paul jointly conceived the study and wrote the paper; Somajita Paul surveyed and analyzed the data; and Harini Nagendra supervised the study.

Conflicts of Interest: The authors declare no conflict of interest. The founding sponsors had no role in the design of the study; in the collection, analyses, or interpretation of data; in the writing of the manuscript; or in the decision to publish the results.

References

1. Rose, S.; Peters, N.E. Effects of urbanization on streamflow in the Atlanta area (Georgia, USA): A comparative hydrological approach. *Hydrol. Process.* **2001**, *15*, 1441–1457. [CrossRef]
2. Booth, D.B.; Jackson, C.R. Urbanization of aquatic systems: Degradation thresholds, storm water detection, and the limits of mitigation. *JAWRA J. Am. Water Resour. Assoc.* **1997**, *33*, 1077–1090. [CrossRef]
3. Khera, N.; Mehta, V.; Sabata, B. Interrelationship of birds and habitat features in urban greenspaces in Delhi, India. *Urban For. Urban Green.* **2009**, *8*, 187–196. [CrossRef]
4. Riley, S.P.; Sauvajot, R.M.; Fuller, T.K.; York, E.C.; Kamradt, D.A.; Bromley, C.; Wayne, R.K. Effects of urbanization and habitat fragmentation on bobcats and coyotes in southern California. *Conserv. Biol.* **2003**, *17*, 566–576. [CrossRef]

5.   McKinney, M.L. Urbanization, biodiversity, and conservation: The impacts of urbanization on native species are poorly studied, but educating a highly urbanized human population about these impacts can greatly improve species conservation in all ecosystems. *Bioscience* **2002**, *52*, 883–890. [CrossRef]

6.   Wang, L.; Lyons, J.; Kanehl, P.; Bannerman, R. Impacts of urbanization on stream habitat and fish across multiple spatial scales. *Environ. Manag.* **2001**, *28*, 255–266. [CrossRef]

7.   Vailshery, L.S.; Jaganmohan, M.; Nagendra, H. Effect of street trees on microclimate and air pollution in a tropical city. *Urban For. Urban Green.* **2013**, *12*, 408–415. [CrossRef]

8.   Shepherd, J.M.; Pierce, H.; Negri, A.J. Rainfall modification by major urban areas: Observations from spaceborne rain radar on the TRMM satellite. *J. Appl. Meteorol.* **2002**, *41*, 689–701. [CrossRef]

9.   Huff, F.; Changnon, S.A., Jr. Climatological assessment of urban effects on precipitation at St. Louis. *J. Appl. Meteorol.* **1972**, *11*, 823–842. [CrossRef]

10. Kaufmann, R.K.; Seto, K.C.; Schneider, A.; Liu, Z.; Zhou, L.; Wang, W. Climate response to rapid urban growth: Evidence of a human-induced precipitation deficit. *J. Clim.* **2007**, *20*, 2299–2306. [CrossRef]

11. Li, J.; Song, C.; Cao, L.; Zhu, F.; Meng, X.; Wu, J. Impacts of landscape structure on surface urban heat islands: A case study of Shanghai, China. *Remote Sens. Environ.* **2011**, *115*, 3249–3263. [CrossRef]

12. Grimmond, S. Urbanization and global environmental change: Local effects of urban warming. *Geogr. J.* **2007**, *173*, 83–88. [CrossRef]

13. Baker, L.A.; Brazel, A.J.; Selover, N.; Martin, C.; McIntyre, N.; Steiner, F.R.; Nelson, A.; Musacchio, L. Urbanization and warming of Phoenix (Arizona, USA): Impacts, feedbacks and mitigation. *Urban Ecosyst.* **2002**, *6*, 183–203. [CrossRef]

14. Saitoh, T.; Shimada, T.; Hoshi, H. Modeling and simulation of the Tokyo urban heat island. *Atmos. Environ.* **1996**, *30*, 3431–3442. [CrossRef]

15. Bolund, P.; Hunhammar, S. Ecosystem services in urban areas. *Ecol. Econ.* **1999**, *29*, 293–301. [CrossRef]

16. Chiesura, A. The role of urban parks for the sustainable city. *Landsc. Urban Plan.* **2004**, *68*, 129–138. [CrossRef]

17. Gobster, P.H. Perception and use of a metropolitan greenway system for recreation. *Landsc. Urban Plan.* **1995**, *33*, 401–413. [CrossRef]

18. Mactavish, J.B.; Schleien, S.J. Playing together growing together: Parent's perspectives on the benefits of family recreation in families that include children with a developmental disability. *Ther. Recreat. J.* **1998**, *32*, 207.

19. Pretty, J.; Peacock, J.; Hine, R.; Sellens, M.; South, N.; Griffin, M. Green exercise in the UK countryside: Effects on health and psychological well-being, and implications for policy and planning. *J. Environ. Plan. Man.* **2007**, *50*, 211–231. [CrossRef]

20. De Vries, S.; Verheij, R.A.; Groenewegen, P.P.; Spreeuwenberg, P. Natural environments—healthy environments? An exploratory analysis of the relationship between greenspace and health. *Environ. Plan. A* **2003**, *35*, 1717–1731. [CrossRef]

21. Lee, A.C.; Maheswaran, R. The health benefits of urban green spaces: A review of the evidence. *J. Public Health* **2011**, *33*, 212–222. [CrossRef] [PubMed]

22. Nielsen, T.S.; Hansen, K.B. Do green areas affect health? Results from a Danish survey on the use of green areas and health indicators. *Health Place* **2007**, *13*, 839–850. [CrossRef] [PubMed]

23. Fuller, R.A.; Irvine, K.N.; Devine-Wright, P.; Warren, P.H.; Gaston, K.J. Psychological benefits of greenspace increase with biodiversity. *Biol. Lett.* **2007**, *3*, 390–394. [CrossRef] [PubMed]

24. Van den Berg, A.E.; Maas, J.; Verheij, R.A.; Groenewegen, P.P. Green space as a buffer between stressful life events and health. *Soc. Sci. Med.* **2010**, *70*, 1203–1210. [CrossRef] [PubMed]

25. Cattell, V.; Dines, N.; Gesler, W.; Curtis, S. Mingling, observing, and lingering: Everyday public spaces and their implications for well-being and social relations. *Health Place* **2008**, *14*, 544–561. [CrossRef] [PubMed]

26. Maas, J.; Van Dillen, S.M.; Verheij, R.A.; Groenewegen, P.P. Social contacts as a possible mechanism behind the relation between green space and health. *Health Place* **2009**, *15*, 586–595. [CrossRef] [PubMed]

27. Peters, K.; Elands, B.; Buijs, A. Social interactions in urban parks: Stimulating social cohesion? *Urban For. Urban Green.* **2010**, *9*, 93–100. [CrossRef]

28. Bell, S.; Tyrväinen, L.; Sievänen, T.; Pröbstl, U.; Simpson, M. Outdoor recreation and nature tourism: A European perspective. *Living Rev. Landsc. Res.* **2007**, *1*, 1–46. [CrossRef]

29. Kaczynski, A.T.; Henderson, K.A. Environmental correlates of physical activity: A review of evidence about parks and recreation. *Leis. Sci.* **2007**, *29*, 315–354. [CrossRef]

30. Tyrväinen, L.; Väänänen, H. The economic value of urban forest amenities: An application of the contingent valuation method. *Landsc. Urban Plan.* **1998**, *43*, 105–118. [CrossRef]
31. Van Herzele, A.; Wiedemann, T. A monitoring tool for the provision of accessible and attractive urban green spaces. *Landsc. Urban Plan.* **2003**, *63*, 109–126. [CrossRef]
32. Tajima, K. New estimates of the demand for urban green space: Implications for valuing the environmental benefits of Boston's big dig project. *J. Urban Aff.* **2003**, *25*, 641–655. [CrossRef]
33. Conway, D.; Li, C.Q.; Wolch, J.; Kahle, C.; Jerrett, M. A spatial autocorrelation approach for examining the effects of urban greenspace on residential property values. *J. Real Estate Financ. Econom.* **2010**, *41*, 150–169. [CrossRef]
34. Morancho, A.B. A hedonic valuation of urban green areas. *Landsc. Urban Plan.* **2003**, *66*, 35–41. [CrossRef]
35. Tyrväinen, L. Economic valuation of urban forest benefits in Finland. *J. Environ. Manag.* **2001**, *62*, 75–92. [CrossRef] [PubMed]
36. Jim, C.; Chen, W.Y. Perception and attitude of residents toward urban green spaces in Guangzhou (China). *Environ. Manag.* **2006**, *38*, 338–349. [CrossRef] [PubMed]
37. Martín-López, B.; Montes, C.; Benayas, J. The non-economic motives behind the willingness to pay for biodiversity conservation. *Biol. Conserv.* **2007**, *139*, 67–82. [CrossRef]
38. Lamarque, P.; Tappeiner, U.; Turner, C.; Steinbacher, M.; Bardgett, R.D.; Szukics, U.; Schermer, M.; Lavorel, S. Stakeholder perceptions of grassland ecosystem services in relation to knowledge on soil fertility and biodiversity. *Reg. Environ. Chang.* **2011**, *11*, 791–804. [CrossRef]
39. Matsuoka, R.H.; Kaplan, R. People needs in the urban Landscape Urban Plann: Analysis of landscape and urban planning contributions. *Landsc. Urban Plan.* **2008**, *84*, 7–19. [CrossRef]
40. Giles-Corti, B.; Broomhall, M.H.; Knuiman, M.; Collins, C.; Douglas, K.; Ng, K.; Lange, A.; Donovan, R.J. Increasing walking: How important is distance to, attractiveness, and size of public open space? *Am. J. Prev. Med.* **2005**, *28*, 169–176. [CrossRef] [PubMed]
41. Nutsford, D.; Pearson, A.L.; Kingham, S. An ecological study investigating the association between access to urban green space and mental health. *Public Health* **2013**, *127*, 1005–1011. [CrossRef] [PubMed]
42. Boone, C.G.; Buckley, G.L.; Grove, J.M.; Sister, C. Parks and People: An Environmental Justice Inquiry in Baltimore, Maryland. *Assoc. Am. Geogr.* **2009**, *99*, 767–787. [CrossRef]
43. Lin, B.; Fuller, R.A.; Bush, R.; Gaston, K.J.; Shanahan, D.F. Opportunity or Orientation? Who Uses Urban Parks and Why. *PLoS ONE* **2014**, *9*, e87422. [CrossRef] [PubMed]
44. Larson, L.R.; Jennings, V.; Cloutier, S.A. Public Parks and Wellbeing in Urban Areas of the United States. *PLoS ONE* **2016**, *11*, e0153211. [CrossRef] [PubMed]
45. Thaiutsa, B.; Puangchit, L.; Kjelgren, R.; Arunpraparut, W. Urban green space, street tree and heritage large tree assessment in Bangkok, Thailand. *Urban For. Urban Green.* **2008**, *7*, 219–229. [CrossRef]
46. Yuen, B.; Hien, W.N. Resident perceptions and expectations of rooftop gardens in Singapore. *Landsc. Urban Plan.* **2005**, *73*, 263–276. [CrossRef]
47. Rupprecht, C.D.; Byrne, J.A.; Ueda, H.; Lo, A.Y. 'It's real, not fake like a park': Residents' perception and use of informal urban green-space in Brisbane, Australia and Sapporo, Japan. *Landsc. Urban Plan.* **2015**, *143*, 205–218. [CrossRef]
48. Chang, H.-S.; Chen, T.-L. Decision making on allocating urban green spaces based upon spatially-varying relationships between urban green spaces and urban compaction degree. *Sustainability* **2015**, *7*, 13399–13415. [CrossRef]
49. Yang, J.; Li, C.; Li, Y.; Xi, J.; Ge, Q.; Li, X. Urban green space, uneven development and accessibility: A case of Dalian's Xigang District. *Chin. Geogr. Sci.* **2015**, *25*, 644–656. [CrossRef]
50. Tan, P.Y.; Ismail, M.R.B. The effects of urban forms on photosynthetically active radiation and urban greenery in a compact city. *Urban Ecosyst.* **2015**, *18*, 937–961. [CrossRef]
51. Lin, B.; Meyers, J.; Barnett, G. Understanding the potential loss and inequities of green space distribution with urban densification. *Urban For. Urban Green.* **2015**, *14*, 952–958. [CrossRef]
52. Paul, S.; Nagendra, H. Vegetation change and fragmentation in the mega city of Delhi: Mapping 25 years of change. *Appl. Geogr.* **2015**, *58*, 153–166. [CrossRef]
53. Government of NCT of Delhi. Available online: http://www.delhi.gov.in/wps/wcm/connect/doit_dpg/DoIT_DPG/Home (accessed on 20 November 2014).

54. Chandramouli, C.; General, R. Census of India 2011. In *Provisional Population Totals*; Government of India: New Delhi, India, 2011.

55. Dupont, V. Conflicting stakes and governance in the peripheries of large Indian metropolises–an introduction. *Cities* **2007**, *24*, 89–94. [CrossRef]

56. Buch, M. Lutyens' New Delhi—yesterday, today and tomorrow. *India Int. Cent. Q.* **2003**, *30*, 29–40.

57. Prasad, G.; Kumar, D.; Nain, G. New Delhi as a Tourism Region. *Glob. J. Res. Anal.* **2015**, *4*, 8.

58. Mohan, M. Climate change: Evaluation of ecological restoration of Delhi ridge using remote sensing and GIS technologies. *Int. Arch. Photogramm. Remote Sens.* **2000**, *33*, 886–894.

59. Lodi, B. Lodi dynasty. Available online: https://www.revolvy.com/main/index.php?s=Lodi%20Dynasty& item_type=topic (accessed on 5 March 2017).

60. Sharma, J.P. The British treatment of historic gardens in the Indian subcontinent: The transformation of Delhi's nawab Safdarjung's tomb complex from a funerary garden into a public park. *Gard. Hist.* **2007**, *35*, 210–228.

61. Feilden, B.M. Bhuli Bhatiyari ka Mahal, Delhi. *Archit. Plus Des.* **1992**, *9*, 63.

62. Rosenthal, R.; Rosnow, R.L. *Essentials of Behavioral Research: Methods and Data Analysis*; McGraw-Hill: New York, NY, USA, 1991.

63. Tucker, P.; Gilliland, J.; Irwin, J.D. Splashpads, swings, and shade: Parents' preferences for neighbourhood parks. *Can. J. Public Health.* **2007**, *98*, 198–202. [PubMed]

64. D'Souza, R.; Nagendra, H. Changes in public commons as a consequence of urbanization: The Agara lake in Bangalore, India. *Environ. Manag.* **2011**, *47*, 840. [CrossRef] [PubMed]

65. Krenichyn, K. 'The only place to go and be in the city': Women talk about exercise, being outdoors, and the meanings of a large urban park. *Health Place* **2006**, *12*, 631–643. [CrossRef] [PubMed]

66. Nordh, H.; Alalouch, C.; Hartig, T. Assessing restorative components of small urban parks using conjoint methodology. *Urban For. Urban Green.* **2011**, *10*, 95–103. [CrossRef]

67. Priego, C.; Breuste, J.; Rojas, J. Perception and value of nature in urban landscapes: A comparative analysis of cities in Germany, Chile and Spain. *Landsc. Online* **2008**, *7*, 22. [CrossRef]

68. Schipperijn, J.; Ekholm, O.; Stigsdotter, U.K.; Toftager, M.; Bentsen, P.; Kamper-Jørgensen, F.; Randrup, T.B. Factors influencing the use of green space: Results from a Danish national representative survey. *Landsc. Urban Plan.* **2010**, *95*, 130–137. [CrossRef]

69. Briffett, C.; Sodhi, N.; Yuen, B.; Kong, L. Green corridors and the quality of urban life in Singapore. In Proceedings the 4th International Urban Wildlife Symposium, Arizona, AR, USA, 1–5 May 2004; pp. 56–63.

70. Randler, C.; Höllwarth, A.; Schaal, S. Urban park visitors and their knowledge of animal species. *Anthrozoos* **2007**, *20*, 65–74. [CrossRef]

71. Cressie, N. *Statistics for Spatial Data*; John Wiley Sons: New York, NY, USA, 2015.

72. Roovers, P.; Hermy, M.; Gulinck, H. Visitor profile, perceptions and expectations in forests from a gradient of increasing urbanisation in central Belgium. *Landsc. Urban Plan.* **2002**, *59*, 129–145. [CrossRef]

73. Huda, S.S.M.S.; Akhtar, A. Leisure behaviour of working women of Dhaka, Bangladesh. *Int. J. Urban Labour Leis.* **2005**, *7*, 1–30.

74. Bauman, A.; Bull, F.; Chey, T.; Craig, C.L.; Ainsworth, B.E.; Sallis, J.F.; Pratt, M. The international prevalence study on physical activity: Results from 20 countries. *Int. J. Behav. Nutr. Phys. Act.* **2009**, *6*, 21. [CrossRef] [PubMed]

75. Ujang, N.; Moulay, A.; Zakariya, K. Sense of well-being indicators: Attachment to public parks in Putrajaya, Malaysia. *Procedia-Soc. Behav. Sci.* **2015**, *202*, 487–494. [CrossRef]

76. Rego, A.B.; Muthoka, M.G. Education for environmental awareness. *Soc. Relig. Concerns East Afr. A Wajibu Anthol.* **2005**, *10*, 197.

77. Kollmuss, A.; Agyeman, J. Mind the gap: Why do people act environmentally and what are the barriers to pro-environmental behavior? *Environ. Educ. Res.* **2002**, *8*, 239–260. [CrossRef]

78. Kotchen, M.J.; Reiling, S.D. Environmental attitudes, motivations, and contingent valuation of nonuse values: A case study involving endangered species. *Ecol. Econ.* **2000**, *32*, 93–107. [CrossRef]

79. Kaiser, F.G.; Ranney, M.; Hartig, T.; Bowler, P.A. Ecological behavior, environmental attitude, and feelings of responsibility for the environment. *Eur. Psychol.* **1999**, *4*, 59. [CrossRef]

80. Payne, L.L.; Mowen, A.J.; Orsega-Smith, E. An examination of park preferences and behaviors among urban residents: The role of residential location, race, and age. *Leis. Sci.* **2002**, *24*, 181–198. [CrossRef]

81. Kong, L.; Yuen, B.; Sodhi, N.S.; Briffett, C. The construction and experience of nature: Perspectives of urban youths. *Tijdschr. Econ. Soc. Geogr.* **1999**, *90*, 3–16. [CrossRef]
82. Sonowal, C.J. Environmental education in schools: The Indian scenario. *J. Hum. Ecol.* **2009**, *28*, 15–36.
83. Kudryavtsev, A.; Krasny, M.E.; Stedman, R.C. The impact of environmental education on sense of place among urban youth. *Ecosphere* **2012**, *3*, 1–15. [CrossRef]
84. Mowen, A.J.; Confer, J.J. The relationship between perceptions, distance, and socio-demographic characteristics upon public use of an urban park "in-fill". *J. Park Recreat. Adm.* **2003**, *21*, 58–74.
85. Qiu, L.; Lindberg, S.; Nielsen, A.B. Is biodiversity attractive?—On-site perception of recreational and biodiversity values in urban green space. *Landsc. Urban Plan.* **2013**, *119*, 136–146. [CrossRef]
86. Marden, E. The neem tree patent: International conflict over the commodification of life. *Boston Coll. Int. Comp. Law Rev.* **1999**, *22*, 279.
87. Verma, V.C.; Gond, S.K.; Kumar, A.; Kharwar, R.N.; Strobel, G. The endophytic mycoflora of bark, leaf, and stem tissues of Azadirachta indica A. Juss (Neem) from Varanasi (India). *Microb. Ecol.* **2007**, *54*, 119–125. [CrossRef] [PubMed]
88. Swamy, S.; Devy, S. Forests, heritage green spaces, and neighbourhood parks: Citizen's attitude and perception towards ecosystem services in Bengaluru. *J. Resour. Energy Dev.* **2010**, *7*, 117–122.
89. Guttikunda, S.K.; Calori, G. A gis based emissions inventory at 1 km × 1 km spatial resolution for air pollution analysis in Delhi, India. *Atmos. Environ.* **2013**, *67*, 101–111. [CrossRef]
90. Crow, T.; Brown, T.; De Young, R. The riverside and Berwyn experience: Contrasts in landscape structure, perceptions of the urban landscape, and their effects on people. *Lands. Urban Plan.* **2006**, *75*, 282–299. [CrossRef]
91. Jim, C.; Shan, X. Socioeconomic effect on perception of urban green spaces in Guangzhou, China. *Cities* **2013**, *31*, 123–131. [CrossRef]
92. Björk, J.; Albin, M.; Grahn, P.; Jacobsson, H.; Ardö, J.; Wadbro, J.; Östergren, P.-O.; Skärbäck, E. Recreational values of the natural environment in relation to neighbourhood satisfaction, physical activity, obesity and wellbeing. *J. Epidemiol. Community Health* **2008**, *62*, e2. [CrossRef] [PubMed]
93. Maloutas, T. Editorial: Urban segregation and the European context. *Greek Rev. Soc. Res.* **2004**, *113*, 3–24. [CrossRef]
94. Bishop, I.; Ye, W.-S.; Karadaglis, C. Experiential approaches to perception response in virtual worlds. *Landsc. Urban Plan.* **2001**, *54*, 117–125. [CrossRef]
95. Tyrväinen, L.; Mäkinen, K. *Tools for Mapping Social Values and Meaning of Urban Woodlands and Other Open Space*; COST Action C11 Green Structures and Urban Planning–Final report; COST: Brussels, Belgium, 2004.
96. Lo, A.Y.; Jim, C.Y. Citizen attitude and expectation towards greenspace provision in compact urban milieu. *Land Use Policy* **2012**, *29*, 577–586. [CrossRef]
97. Qureshi, S.; Breuste, J.H.; Jim, C. Differential community and the perception of urban green spaces and their contents in the megacity of Karachi, Pakistan. *Urban Ecosyst.* **2013**, *16*, 853–870. [CrossRef]
98. Pincetl, S.; Gearin, E. The reinvention of public green space. *Urban Geogr.* **2005**, *26*, 365–384. [CrossRef]
99. Sreetheran, M.; Van Den Bosch, C.C.K. A socio-ecological exploration of fear of crime in urban green spaces–A systematic review. *Urban For. Urban Green.* **2014**, *13*, 1–18. [CrossRef]
100. Wong, K.K.; Domroes, M. The visual quality of urban park scenes of Kowloon Park, Hong Kong: Likeability, affective appraisal, and cross-cultural perspectives. *Environ. Plan. B Plan Des.* **2005**, *32*, 617–632. [CrossRef]
101. Gearin, E.; Kahle, C. Teen and adult perceptions of urban green space Los Angeles. *Child. Youth Environ.* **2006**, *16*, 25–48.
102. Kuo, F.E.; Bacaicoa, M.; Sullivan, W.C. Transforming inner-city landscapes trees, sense of safety, and preference. *Environ. Behav.* **1998**, *30*, 28–59. [CrossRef]
103. Scott, D.; Jackson, E.L. Factors that limit and strategies that might encourage people's use of public parks. *J. Park Recreat. Adm.* **1996**, *14*, 1–17.
104. Sotoudehnia, F.; Comber, A. Poverty and Environmental Justice: A GIS Analysis of Urban Greenspace Accessibility for Different Economic Groups. In Proceedings of the 13th AGILE Int Conf on Geographic Information Science, Guimaraes, Portugal, 11–14 May 2010; pp. 10–14.
105. Allison, M.T. *Culture, Conflict, and Communication in the Wildland-Urban Interface*; Access and boundary maintenance: Serving culturally diverse populations; Eastview Press: Boulder, CO, USA, 1992; pp. 99–108.

106. Barbosa, O.; Tratalos, J.A.; Armsworth, P.R.; Davies, R.G.; Fuller, R.A.; Johnson, P.; Gaston, K.J. Who benefits from access to green space? A case study from Sheffield, UK. *Landsc. Urban Plan.* **2007**, *83*, 187–195. [CrossRef]

107. Johnson, C.Y.; Bowker, J.M.; Cordell, H.K. Outdoor recreation constraints: An examination of race, gender, and rural dwelling. *South. Rural Sociol.* **2001**, *7*, 111–133.

108. Ries, A.V.; Voorhees, C.C.; Roche, K.M.; Gittelsohn, J.; Yan, A.F.; Astone, N.M. A quantitative examination of park characteristics related to park use and physical activity among urban youth. *J. Adolesc. Health* **2009**, *45*, S64–S70. [CrossRef] [PubMed]

109. Chaudhry, P.; Bagra, K.; Singh, B. Urban greenery status of some Indian cities: A short communication. *Int. J. Environ. Sci. Dev.* **2011**, *2*, 98. [CrossRef]

110. Jim, C.Y.; Chen, W.Y. Recreation–amenity use and contingent valuation of urban greenspaces in Guangzhou, China. *Landsc. Urban Plan.* **2006**, *75*, 81–96. [CrossRef]

111. Yoon, Y.; Uysal, M. An examination of the effects of motivation and satisfaction on destination loyalty: A structural model. *Tour. Manag.* **2005**, *26*, 45–56. [CrossRef]

112. Pankhuri, Y. Times of India. Available online: http://timesofindia.indiatimes.com/city/delhi/parks-safe-haven-for-criminals-at-night/articleshow/57239080.cms (accessed on 4 April 2017).

113. Government of India. Ministry of Urban Development. Available online: http://moud.gov.in/cms/schemes-or-programmes.php (accessed on 1 February 2017).

land

MDPI

Article

Informal Urban Green Space: Residents' Perception, Use, and Management Preferences across Four Major Japanese Shrinking Cities

Christoph D. D. Rupprecht [ORCID]

FEAST Project, Research Institute for Humanity and Nature, Kyoto 6038047, Japan; crupprecht@chikyu.ac.jp; Tel.: +81-75-707-2499

Received: 3 August 2017; Accepted: 22 August 2017; Published: 25 August 2017

Abstract: Urban residents' health depends on green infrastructure to cope with climate change. Shrinking cities could utilize vacant land to provide more green space, but declining tax revenues preclude new park development—a situation pronounced in Japan, where some cities are projected to shrink by over ten percent, but lack green space. Could informal urban green spaces (IGS; vacant lots, street verges, brownfields etc.) supplement parks in shrinking cities? This study analyzes residents' perception, use, and management preferences (management goals, approaches to participatory management, willingness to participate) for IGS using a large, representative online survey ($n = 1000$) across four major shrinking Japanese cities: Sapporo, Nagano, Kyoto and Kitakyushu. Results show that residents saw IGS as a common element of the urban landscape and their daily lives, but their evaluation was mixed. Recreation and urban agriculture were preferred to redevelopment and non-management. For participative management, residents saw a need for the city administration to mediate usage and liability, and expected an improved appearance, but emphasized the need for financial and non-financial support. A small but significant minority (~10%) were willing to participate in management activities. On this basis, eight principles for participatory informal green space planning are proposed.

Keywords: vacant land; land use; urban planning; Japan; wasteland; green infrastructure; recreation; landscape; participatory management; depopulation

---

1. Introduction

Urban green spaces as an essential element of green infrastructure are increasingly linked to human wellbeing [1–5]. However, the benefits they provide come at a price, from maintenance of facilities and vegetation to day-to-day management [6]. This financial burden is particularly pronounced in countries such as Japan, where municipalities face a shrinking tax base due to demographic trends of aging and depopulation [7]. Furthermore, maintenance costs for urban green spaces in Japan are rising, as parks created during the 1960s and 1970s period of economic and population growth are increasingly in need of refurbishment [8]. On the other hand, Japanese cities provide residents only with a comparatively low amount of green space (ca. 10 m^2 per person in major cities [9]). This raises the question of how local governments can procure additional green space areas, not least to adapt to climate change, rising temperatures and associated heat waves. Cities in North America and Europe have been looking toward participative vacant land management to meet residents' needs [10–14]. However, despite introducing participative management for formal green spaces since the 1990s [15], local governments in Japan have been slow to explore this direction for non-traditional green spaces. Yet such schemes might create more recreational green space, provide benefits associated with green infrastructure, and could simultaneously alleviate the costs of maintenance. One reason for the lack of similar initiatives in Japan may lie in the scarcity of related research—we still know little about what residents in Japan think of informal urban green spaces such as vacant lots, street verges or brownfields, let alone participative management approaches. To fill this gap, this paper seeks to provide some insight by analyzing how residents in four major shrinking Japanese cities perceive and use informal urban green spaces, what management goals and approaches they prefer, and how willing they are to participate in managing such spaces.

Japanese cities are facing a major demographic challenge with consequences for urban land and green space management. This makes the country a useful object of study, because similar demographic trends are expected in other countries such as South Korea or China as the population peaks and the economy enters the post-growth stage. Until 2040, many are projected to experience both rapid aging and population decline, with some likely to lose over 10% of their total population [16]. The major effects of this trend are fourfold. First, the cumulative effects of aging and depopulation are eroding cities' tax base, which in turn forces them to balance expenses for maintaining green spaces with competing demands such as aging water infrastructure. This puts budgets for green space under scrutiny and leaves them at risk of being cut, even though many Japanese cities already fail to provide the 10 m^2 of park area per person set in governmental standards [17]. Second, the number of both vacant houses and vacant lots are increasing [18], a trend that will likely accelerate as population decline intensifies. While this process could be seen as an opportunity for municipalities to buy land at a moderate price to increase public green space, their financial situation makes this difficult. In fact, strategic park planning in Japan has recently focused on protecting current levels of green space rather than expanding them, as cities are struggling to cover maintenance costs for parks built during the decades of high population and economic growth. Third, as the population demographics change, so do people's green space and nature needs [19]. Recent years have seen a strong demand for recreational urban agriculture, with long waiting lists for community garden parcels being a common occurrence. While some cities such as Yokohama have started retrofitting parks with areas for growing vegetables, such initiatives are still rare. Finally, Japanese cities will need to invest heavily into green infrastructure to adapt to and mitigate the effects of rising global temperatures due to climate change [20–22]. The cities' aging population is particular at risk from heat-related health problems, making this also an urgent issue of public health. These demographic transition-related major effects have led researchers to investigate the potential of non-traditional green spaces [23].

Recent research has shown that both in Japan and abroad, informal urban green spaces (IGS) such as vacant lots, street verges, brownfields, power line corridors and waterside spaces can make up about 5% of urban land [24]. Moreover, residents are already using these spaces for recreation, both as adults and children [23,25]. However, not all residents evaluate these spaces favorably as they are largely unmanaged. In particular, the aesthetics of wild nature do not necessarily directly translate from a Western to a Japanese cultural context [26], where many residents prefer to see a human touch as evidence of human care and attention for a space. Against an international background of residents using and re-using—often spontaneously—land considered derelict or unwanted by conventional urban planning [27–29], researchers have thus called for exploring ways informal green spaces in shrinking Japanese cities could be managed by the local community using participatory approaches [23]. However, a number of questions about participatory IGS management in Japan remain unanswered. Addressing the following gaps in the literature will support municipalities in planning and implementing such management approaches, with the intention of contributing to residents' wellbeing.

Under what circumstances and in what form might participatory IGS management be feasible in shrinking Japanese cities? To answer these questions, two major gaps in the existing literature need to be addressed. First, existing exploratory studies [23,25] on how residents in Japan perceive, use, and evaluate IGS have been limited in scope (single city) and sample size (<200), were not representative and, as mail-back surveys, may have suffered from bias where only residents interested in the topic responded. Second, we know little about how residents think of participatory IGS management, from their preferred management goals and approaches to acceptable levels and forms of participation.

This paper thus focuses on the following questions: (1) how do residents perceive, use and evaluate IGS across major shrinking Japanese cities; (2) what management goals do residents prefer for IGS; (3) how do residents think about different ways, approaches and circumstances in participatory IGS management; and (4) how willing are residents to engage in participatory IGS management? Addressing these questions is important, because the resulting findings and insights may inform urban planning, climate adaptation and participatory management in post-growth industrialized countries, using Japan as a case where depopulation is most advanced. The study thus contributes to the local (site-specific), national and international discourse.

2. Materials and Methods

2.1. Study Sites

This study focused on the Japanese cities Sapporo, Kyoto, Kitakyushu and Nagano, all four projected to lose over 10% of their current population until 2040 [16]. They were selected to represent the regional Japanese cities expected to shrink most severely as a result of Japan's projected national population decline. Differences in city age, population size, population density and green space per capita (Table 1) and geographic location (Figure 1) provide a range variety of geographic context. Additionally, including Sapporo allows comparison of results with previous research on IGS in the city, in particular addressing potential nonresponse bias in the previous mail-back survey.

Figure 1. Location of the four targeted shrinking cities in Japan (green).

Table 1. Comparison of study sites.

Characteristics	Sapporo	Kyoto	Kitakyushu	Nagano
City status	1902 (founded 1824)	1889 (founded ~600 AD)	1963 (formerly Mojigaseki, founded ~645 AD)	1897 (founded ~642 AD)
Population (2015)	1.95 million	1.48 million	961,000	378,000
Projected population (2040)	1.71 million	1.28 million	784,000	302,000
Area (km^2)	1121	828	492	835
Population density (inhabitants/km^2)	1741	1782	1954	452
Climate (Köppen-Geiger)	Dfa	Cfa	Cfa	Cfa
Green space per capita (2015)	12.5	4.4	12.0	9.4

Sources: Wikipedia Japan [16], IPSS 2013, Statistics Japan 2015 (https://www.e-stat.go.jp/SG1/estat/eStatTopPortalE.do)

2.2. Data Collection, Survey Instrument and Data Analysis

The survey was conducted across the four study sites through an online survey coordinated through Rakuten Research, a major Japanese online polling service. Respondents were recruited from the service's panel, consented to and received compensation for their participation in accordance to the service's policies (the exact compensation is not disclosed by the company, but is less than ¥1000/US$10/10 €). Responses were collected over a period of two weeks in late 2016, and sampled to be representative for the population demographic of the respective cities. The study was approved by the home institution's research ethics committee (RIHN2017-1).

The survey instrument (see File S1) consisted of 24 questions in four parts: (1) respondents' perception, recreational use and evaluation of IGS; (2) respondents' preferences for IGS management goals, management approaches, and willingness to participate in management; (3) respondents' opinion about the value of urban nature and general preferences for urban land use directions in shrinking cities; and (4) respondents' socio-demographic data. Question types included multiple choice, Likert-scale, modified Likert-scale (agree with option A or B) and open comment questions. The cover page of the survey instrument contained a brief explanation of what IGS are alongside a typology with color photographs (Figure 2; File S1). Native Japanese speakers helped to ensure the survey instrument was linguistically correct and easy to read. Questions in part one drew on previous research about residents' perception, use and evaluation of IGS in Sapporo and Brisbane [23] and used a modified version of this study's survey instrument [30]. Multiple-choice questions were replaced by Likert scales to improve measurement precision. Questions in part two address calls in the literature to explore participative IGS and urban green space management approaches [23,25].

Choices for management goals ranged from urban greening, recreation and conservation to redevelopment, parking space (a common use for vacant land in Japan) and an option to forgo management altogether. Regarding management approaches, questions sought to clarify who respondents believe is responsible for managing IGS, whether existing green space management approaches are appropriate for IGS, what participatory IGS management would mean for its aesthetics and issues such as liability, whether use could be temporary, and what degree of support (financial, training) residents-as-managers would require from authorities. Willingness to participate in IGS management was tested using questions adopted from recent Japanese research on volunteers' willingness to work in participatory green space conservation [31]. Questions covered willingness to contribute time or money, participation frequency, willingness to assume a leadership role, and a variety of participatory activities. Questions in part three consisted of a modified NEP scale developed for prior research on IGS [23], while also asking respondents what strategic planning goals they preferred for shrinking cities (growth, larger housing, more recreational green space, more urban agriculture, or returning space to nature). These questions were intended to create a larger normative background against which more concrete, local scale IGS management goals and approaches can be discussed. Finally, questions in part four asked for respondents' socio-demographic data, including type of housing and residential green space, length of local residency, level of educational attainment, yearly income and post code.

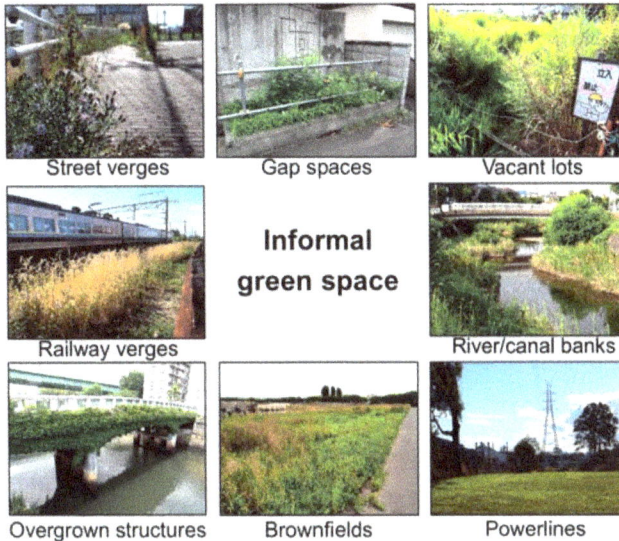

Figure 2. Typology of IGS (translated to English) with example photographs of different IGS types, as shown to respondents in the online survey. From top left, clockwise: street verge, gap, vacant lot, waterside, power-line corridor, brownfield, structural IGS, railway verge.

Data was analyzed using descriptive and inferential statistics following procedures described by Field and colleagues [32] using R [33] and its packages *likert* [34] and FactoMineR [35] as well as JASP [36] and jamovi [37]. The map of study locations (Figure 1) was created with QGIS [38], but no spatial analysis was performed for this study. The analyzed data set is available as supplementary material. Multiple correspondence analysis was used to explore the multivariate relationship of key IGS and demographic variables. Non-parametric tests (chi-square, Wilcoxon rank-sum, Kruskal-Wallis, Spearman) were used as analysis indicated data did not fulfill the assumptions of parametric tests (normal distribution, homogeneity). Post-hoc corrections for pairwise comparisons with the Kruskal-Wallis test were performed using the Dwass-Steel-Critchlow-Fligner test to identify significant differences between individual pairs, which are reported in the text using abbreviated study location pairs (e.g., Sapporo-Nagano (SP-NA), Kyoto-Kitakyushu (KY-KK)). The reliability of core scales (IGS benefits, IGS problems, management goals, management styles, willingness to participate) was adequate or better (Cronbach's alpha > 0.7). The analyzed data set is available as supplementary material (File S2).

3. Results

3.1. Sample Characteristics

A representative sample for the population demographic of the respective cities with a total of 1000 valid responses was collected (see Table 2). Income and length of residency did not differ significantly between cities. Level of educational attainment differed significantly ($X^2(12) = 23.93$, $p < 0.05$), with Kyoto (71%) having the most and Kitakyushu (57%) the least respondents with a university or post-graduate degree. Housing also differed significantly ($X^2(12) = 149.0$, $p < 0.001$).

In Nagano, most respondents were living in a house with garden (67%), while in Sapporo most were living in an apartment without shared green space (46%). In Kyoto, a quarter of respondents (24%) lived in a house without garden, whereas in all other cities less than 10% of respondents did. Respondents had been living in their respective city for 20 years on average (range 0–69), with no significant difference between cities.

Table 2. Demographic characteristics of respondents (*n* = 1000).

Age	Mean/SD/Lowest/Highest	46/13/20/69
	Respondents in their 20s	16%
	Respondents in their 30s	19%
	Respondents in their 40s	23%
	Respondents in their 50s	20%
	Respondents in their 60s	22%
Sex	Female	50%
	Male	50%
Education (highest attained)	Junior high school	2%
	High school	31%
	University	57%
	Postgraduate	7%
	Other	4%
Housing	Detached with garden	43%
	Detached without garden	10%
	Apartment, shared green space	12%
	Apartment, no shared green space	33%
	Other	1%
Income	Under ¥2 million	10%
	¥2–4 million	24%
	¥400–600 million	19%
	¥600–800 million	13%
	¥800–1000 million	9%
	¥1000–1250 million	4%
	Over ¥1250 million	3%
	Don't know/Don't want to answer	19%

Asked about their attitude towards urban nature, respondents highly valued urban nature (Figure 3), with no significant differences between cities or respondent sex. With increasing age, respondents valued green space in their neighborhood more (Figure 3, Q3; $p < 0.05$, $r_s = -0.07$; the negative effect size results from the coding of agreement (low) to disagreement (high)), were more likely to agree nature has intrinsic value (Figure 3, Q2; $p < 0.05$, $r_s = -0.06$), and were more willing to donate (Figure 3, Q1, $p < 0.05$, $r_s = -0.06$). Asked about their opinion on how to proceed when cities shrink, respondents preferred converting the land to green space (for recreation or agriculture) over increased housing size or growth at all cost (Figure 4). The only significant difference between the four cities was respondents' support for urban agriculture (Figure 4, Q5; H(3) = 8.30, $p < 0.05$; significant pairwise differences: Sapporo-Nagano (SP-NA), $p = 0.005$). Sex had no effect on opinion about shrinking cities. In contrast to their attitude towards urban nature, with rising age respondents were less likely to support using space in shrinking cities to increase green space (Figure 4, Q4; $p < 0.001$, $r_s = 0.11$).

Please tell us how you think about urban nature.

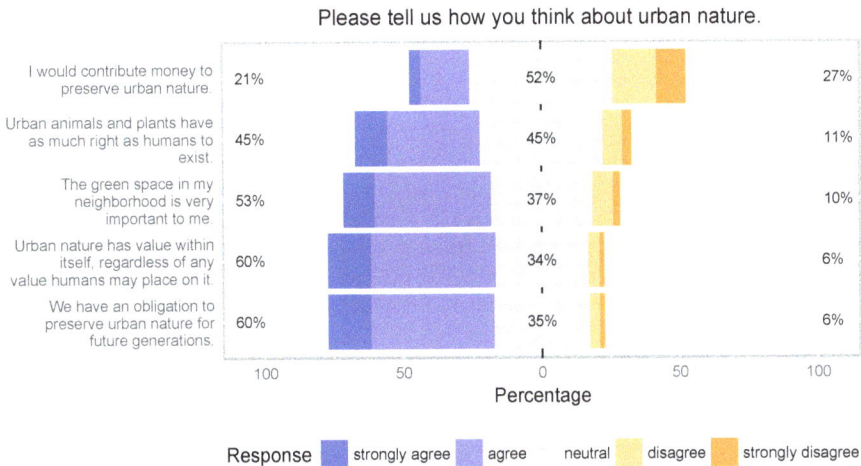

Response: strongly agree | agree | neutral | disagree | strongly disagree

Figure 3. Respondents' attitude toward urban nature.

When cities shrink...

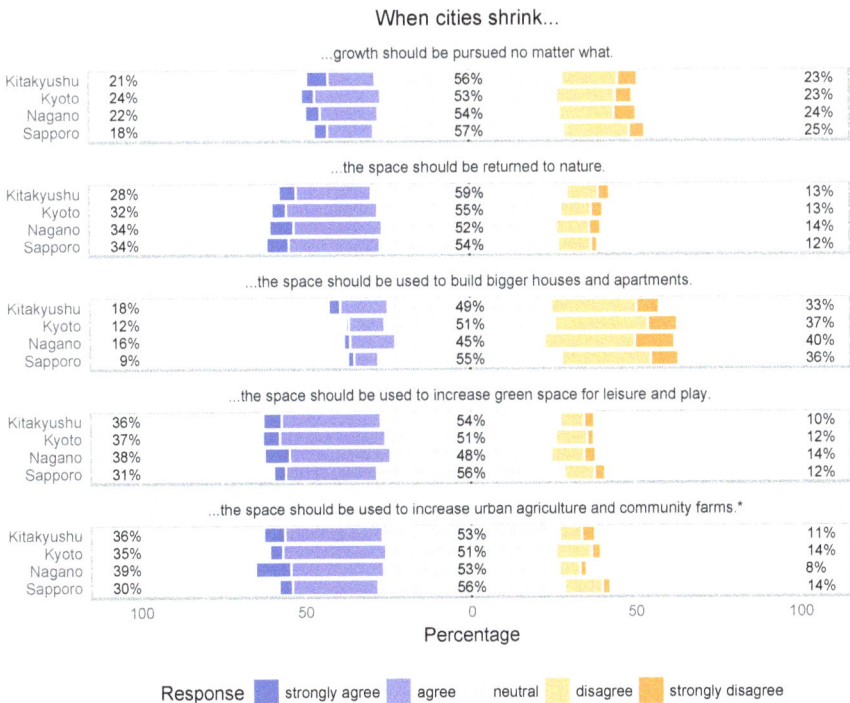

Response: strongly agree | agree | neutral | disagree | strongly disagree

Figure 4. Respondents' opinion on how to proceed when cities are shrinking.

3.2. Respondents' Perception, Use, and Evaluation of IGS

Most respondents (70%) knew IGS in their neighborhood (Table 3). Among those, vacant lots, street verges, waterside and gap spaces were reported most often. Almost half of the respondents who knew IGS in their neighborhood perceived it as relatively biodiverse. Familiarity with IGS in the neighborhood did not differ significantly between cities, but perceived biodiversity did ($X^2(12) = 29.37$, $p < 0.01$)). Nagano had the highest perceived biodiversity, Sapporo the lowest. Sex had no effect on familiarity or perceived biodiversity.

Most respondents who knew of IGS in their neighborhood did not use it for recreational purposes (79%, Table 4), but some reported using it weekly or daily. Use frequency did not differ between the four cities. Respondents used IGS for wide variety of activities, including going for a stroll, exercise-oriented walking, playing with children, walking the dog, BBQs, general exercise, taking a rest, collecting insects, taking photos, observing plants and animals, fishing, doing nothing, and using it as a shortcut on the way to somewhere else. The most common reasons IGS users gave for preferring it over parks or gardens were that IGS was close to their home, was not crowded, had no use restrictions, and was wild and exciting with many and/or different species of plants and animals around. Most IGS users reported no problems with using IGS for recreational purposes, while littering was the most commonly mentioned problem encountered. Respondents (including non-users) had a wide variety of ideas to make using IGS easier. Main ideas included a minimum level of maintenance and management (such as picking up trash, removing dog droppings, planting flowers, mowing), adding facilities such as seating, improving accessibility by removing barriers such as fences, adding signs to show the space may be used, improve safety and/or safe appearance, and converting it to formal green space (e.g., park, community garden). However, some also emphasized there was no need to do anything, voicing concerns that increased use could cause IGS to lose its special characteristics (see also Section 3.3).

Table 3. Respondents' perception of IGS.

Question Asked	Response Options [1]	Answers (%)
How many informal green spaces (as introduced above) exist in your neighborhood? ($n = 1000$)	None	30.4
	A few (1–5)	44.5
	Some (5–10)	14.8
	Many (over 10)	10.3
What types of informal green spaces do you know of in your neighborhood? Please only select spaces with vegetation other than parks, gardens or plazas. (multiple answers possible; $n = 696$)	Vacant lots	73.6
	Structures (overgrown walls, fences, roofs etc.)	34.8
	Railway verges	29.2
	Street verges	68.0
	Brownfields	10.8
	Waterside (river banks, river beds etc.)	54.7
	Power-line corridors	15.8
	Gaps (between walls or fences etc.)	44.8
How many species of animals and plants do you think live in informal green spaces? ($n = 696$)	Very few	4.5
	Few	31.0
	Many	38.8
	Very many	9.9
	I don't know	15.8

[1] Response options are listed in order of appearance in the survey instrument.

Table 4. Respondents' use of IGS for recreational purposes.

Question Asked	Response Options [1]	Answers (%)
How often do you use informal green space for recreation, exercise or play etc.? ($n = 696$)	Never	78.7
	Daily	0.9
	Once a week	6.2
	Once a month	5.0
	A few times per year	9.2

Table 4. *Cont.*

Question Asked	Response Options [1]	Answers (%)
Why do you use informal green space and not a park or garden? (*n* = 148)	It's near my home	66.2
	It's wild and exciting	19.6
	It's not crowded	39.2
	There are more or different animals or plants	19.6
	It has better privacy (nobody watching)	9.5
	There are no use restrictions (e.g., no dogs, no ball play)	23.6
	It can be used for many things (e.g., gardening)	11.5
	There are no nice parks near my home	12.2
	I don't have a garden or similar green space	14.9
	Other	6.8
Did you experience any problems when using informal green space? (*n* = 148)	No	62.8
	Hard to access (fence, signs etc.)	8.8
	I was scared to use it	8.8
	Dangerous animals	5.4
	Dangerous plants	4.1
	Danger of injury	10.1
	Lots of litter	19.6
	Conflict with the owner	3.4
	Conflict with police	0.0
	Conflict with other users	2.7
	Criminals were present	0.7
	Drug users were present	0.0
	Prostitutes were present	0.0
	Other	2.0

[1] Response options are listed in order of appearance in the survey instrument.

Benefits respondents most commonly associated with IGS were related to city greening, ecosystem services such as air filtration and cooling, wildlife habitat, and providing an opportunity for nature contact (Figure 5). While more respondents saw IGS as a space children can use to play, this was not true for leisure activities in general. Food-related benefits received the lowest agreement. Residents' opinion on IGS benefits differed between the four cities only regarding the ability of IGS to sequester carbon (Figure 5, Q15; H(3) = 10.06, $p < 0.05$; SP-NA, $p = 0.002$), with agreement strongest in Nagano (54%) and weakest in Sapporo (39%). Women felt more strongly than men (W = 134,895, $p < 0.05$) that every bit of green in the city is good (Figure 5, Q17), while other IGS benefit perceptions were not affected by sex. With increasing age respondents agreed significantly less with a number of IGS benefits related to recreation, child play, a more interesting neighborhood and the general benefit of green in the city (Figure 5, Q4/8/3/17; $p < 0.01$ to $p < 0.001$), but effect sizes were limited ($r_s = 0.07$ to $r_s = 0.13$).

In contrast, respondents perceived littering, weeds and pest animals, as well as a disorderly look to be the main potential problems associated with IGS (Figure 6). However, IGS was mostly not perceived as a waste of space. Likewise, the common Japanese social issue of conflicts around noise associated with children's play was not seen as a problem related to IGS by most respondents. Five problems were perceived differently between the study locations: pest animals (Figure 6, Q2; H(3) = 28.46, $p < 0.001$; SP-NA, $p < 0.001$; SP-KY, $p < 0.001$; SP-KK, $p = 0.006$), weeds (Figure 6, Q6; H(3) = 24.06, $p < 0.001$; all study location pairs $p < 0.03$ except KY-KK), fire hazard (Figure 6, Q5; H(3) = 14.15, $p < 0.01$; SP-NA, $p = 0.033$; SP-KY, $p < 0.001$; SP-KK, $p = 0.002$), littering (Figure 6, Q9; H(3) = 14.07, $p < 0.01$; SP-NA, $p < 0.001$; NA-KY, $p = 0.014$), and disorderly appearance (Figure 6, Q10; H(3) = 14.61, $p < 0.01$; SP-NA, $p < 0.001$; SP-KY, $p = 0.006$). Women felt more strongly about liability, vandalism, criminals, danger to children, unwanted individuals and fire hazard than men (Figure 6, Q8/12/4/11/1/5; $p < 0.05$ to $p < 0.001$), similar to how with increasing age respondents felt more strongly about vandalism, graffiti, disorderly appearance, danger to children and fire hazard (Figure 6, Q12/7/10/11/5; $p < 0.05$ to $p < 0.001$; $r_s = 0.07$ to $r_s = 0.14$).

Overall, respondents felt that IGS affected their daily lives to different degrees. Of all respondents, 40.1% felt IGS did not affect their daily lives for the better or the worse. On the other hand, 15.5% answered IGS made their daily life better, and 10.6% disagreed, stating it made their daily life worse. Finally, 33.8% perceived IGS as both a positive and negative influence on their daily lives. Respondents' IGS evaluation did not differ significantly between the four cities. While a detailed analysis of qualitative answers about the reasons how respondents felt IGS affected their daily lives goes beyond the scope of this paper, overall themes brought up were as follows. Overall, the ambiguity of IGS in terms of management, responsibility, ownership, and usability was perceived as a source of problems. Many respondents noted that some IGS influenced their daily lives positively, others negatively, leading to their overall evaluation of "both" in the previous question. Others commented such evaluation would likely depend not only on the IGS in question, but also on individual aesthetic and recreational preferences as well as individual attitudes in general and towards nature in particular, and on how IGS would be used.

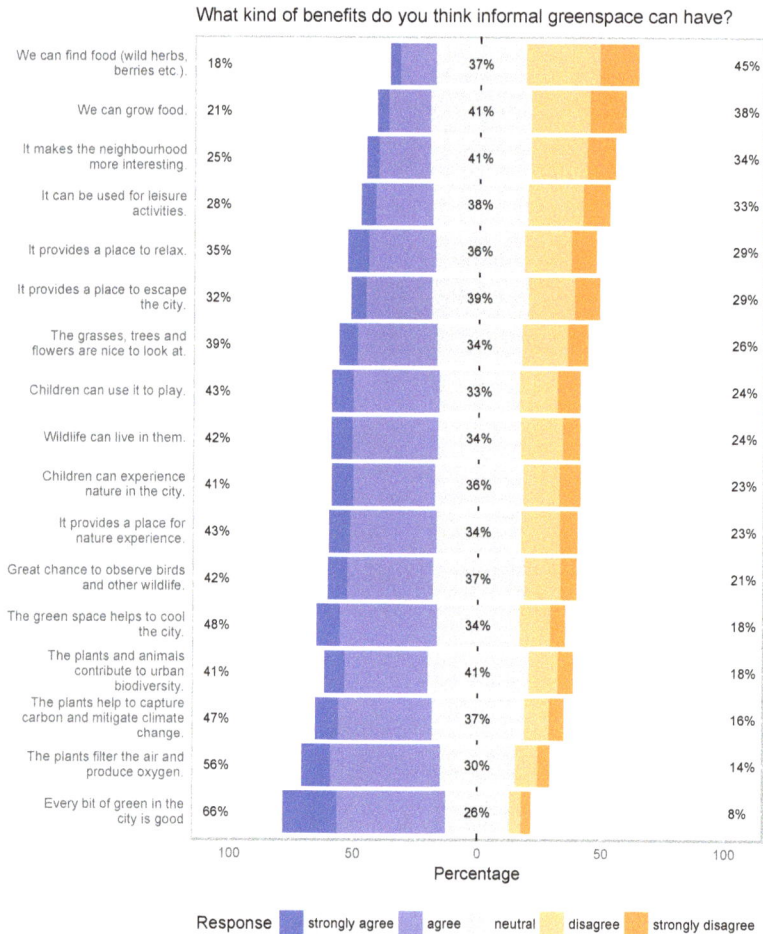

Figure 5. Respondents' level of agreement to potential IGS benefits.

What kind of problems do you think informal greenspace can pose?

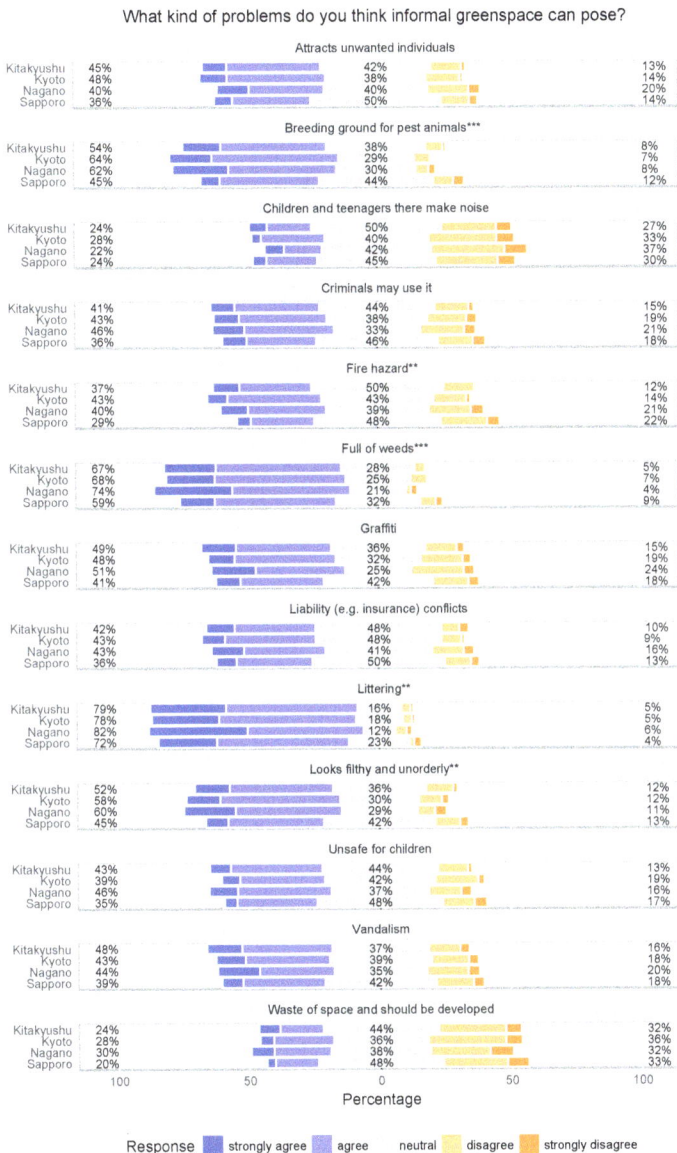

Figure 6. Respondents' level of agreement to potential IGS problems. Significant differences between the four cities are noted as * ($p < 0.05$), ** ($p < 0.01$), and *** ($p < 0.001$).

3.3. Respondents' Preferences for IGS Management: Goals, Approaches, Willingness to Participate

Regarding management goals for IGS, respondents clearly favored preserving IGS as green space over redevelopment (Figure 7). On the other hand, not managing IGS was clearly disfavored. Opinions differed signficiantly between the four cities regarding the management goals of beautification (Figure 7, Q2; H(3) = 11.78, $p < 0.01$; SP-NA, $p < 0.001$; SP-KY, $p = 0.009$; SP-KK, $p = 0.036$), urban agriculture (Figure 7, Q10; H(3) = 17.59, $p < 0.01$; SP-NA, $p < 0.001$; SP-KK, $p = 0.002$; NA-KY, $p = 0.006$; KY-KK,

$p = 0.018$), and parking space (Figure 7, Q7; H(3) = 9.98, $p < 0.05$; SP-KK, $p = 0.004$; KY-KK, $p = 0.011$). Women were more strongly opposed to not managing IGS (Figure 7, Q6; W=115669, $p < 0.05$). With rising age, respondents were more strongly opposed to child recreation, animal conservation but also redevelopment as IGS management goals (Figure 7, Q5/1/9; $p < 0.05$ to $p < 0.001$, $r_s = 0.07$ to 0.11). Respondents suggested a number of additional or alternative management goals, including using IGS for temporary events such as live music by indie bands or flea markets, off-leash areas for dogs, power generation (e.g., using solar panels), and building child-care or aged-care centers.

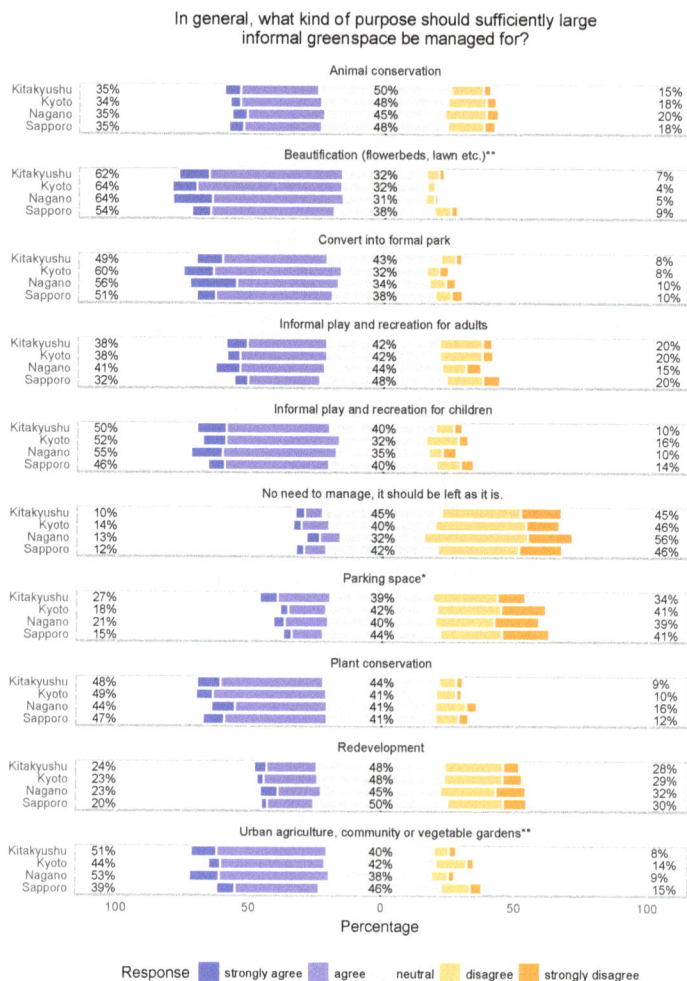

In general, what kind of purpose should sufficiently large informal greenspace be managed for?

Animal conservation

City	Agree	Neutral	Disagree
Kitakyushu	35%	50%	15%
Kyoto	34%	48%	18%
Nagano	35%	45%	20%
Sapporo	35%	48%	18%

Beautification (flowerbeds, lawn etc.)**

City	Agree	Neutral	Disagree
Kitakyushu	62%	32%	7%
Kyoto	64%	32%	4%
Nagano	64%	31%	5%
Sapporo	54%	38%	9%

Convert into formal park

City	Agree	Neutral	Disagree
Kitakyushu	49%	43%	8%
Kyoto	60%	32%	8%
Nagano	56%	34%	10%
Sapporo	51%	38%	10%

Informal play and recreation for adults

City	Agree	Neutral	Disagree
Kitakyushu	38%	42%	20%
Kyoto	38%	42%	20%
Nagano	41%	44%	15%
Sapporo	32%	48%	20%

Informal play and recreation for children

City	Agree	Neutral	Disagree
Kitakyushu	50%	40%	10%
Kyoto	52%	32%	16%
Nagano	55%	35%	10%
Sapporo	46%	40%	14%

No need to manage, it should be left as it is.

City	Agree	Neutral	Disagree
Kitakyushu	10%	45%	45%
Kyoto	14%	40%	46%
Nagano	13%	32%	56%
Sapporo	12%	42%	46%

Parking space*

City	Agree	Neutral	Disagree
Kitakyushu	27%	39%	34%
Kyoto	18%	42%	41%
Nagano	21%	40%	39%
Sapporo	15%	44%	41%

Plant conservation

City	Agree	Neutral	Disagree
Kitakyushu	48%	44%	9%
Kyoto	49%	41%	10%
Nagano	44%	41%	16%
Sapporo	47%	41%	12%

Redevelopment

City	Agree	Neutral	Disagree
Kitakyushu	24%	48%	28%
Kyoto	23%	48%	29%
Nagano	23%	45%	32%
Sapporo	20%	50%	30%

Urban agriculture, community or vegetable gardens**

City	Agree	Neutral	Disagree
Kitakyushu	51%	40%	8%
Kyoto	44%	42%	14%
Nagano	53%	38%	9%
Sapporo	39%	46%	15%

Percentage (100 — 50 — 0 — 50 — 100)

Response: strongly agree / agree / neutral / disagree / strongly disagree

Figure 7. Respondents' preferred management goals for IGS. Significant differences between the four cities are noted as * ($p < 0.05$) and ** ($p < 0.01$).

In regard to management approaches (Figure 8), respondents thought IGS management was an issue of concern for residents in the neighborhood. Yet while residents were perceived to know better how to manage IGS than the city administration, respondents still thought the administration to be responsible for management. Qualitative responses clarified that this seemingly contradictory result was partly based on the perception that the city administration was shirking responsibilities, leaving residents to deal with problems without sufficient support. Overall, respondents were slightly in favor of stricter management and IGS use only with permission, while placing the liability with users. Respondents largely agreed participatory IGS management would improve its appearance, but felt both financial and other support by the city administration to be necessary. They also favored using IGS long-term over temporary use.

Significant differences in opinions on management approaches were found between the cities for some questions, and the extremes in opinion were as follows. Respondents rejected IGS use without permission in Kyoto (35% vs. 17%) more strongly than in Kitakyushu (24% vs. 24%; Figure 8, Q5; H(3) = 10.01, $p < 0.05$; NA-KY, $p = 0.041$; KY-KK, $p = 0.003$). Respondents were most optimistic that resident IGS management would improve its appearance in Nagano (38% vs. 8%), but least optimistic in Kyoto (27% vs. 14%; Figure 8, Q7; H(3) = 11.00, $p < 0.05$; SP-NA, $p = 0.008$; NA-KY, $p = 0.003$; NA-KK, $p = 0.049$). Similarly, respondents were more convinced financial support for management would be necessary in Nagano (43% vs. 9%) than in Kyoto (36% vs. 14%; Figure 8, Q8; H(3) = 8.95, $p < 0.05$; SP-NA, $p = 0.035$; NA-KY, $p = 0.015$; KY-KK, $p = 0.042$). Finally, long-term IGS use had the highest support over temporary use in Nagano (41% vs. 8%) and the lowest in Kyoto (32% vs. 11%; Figure 8, Q10; H(3) = 10.13, $p < 0.05$; NA-KY, $p = 0.012$; KY-KK, $p = 0.005$). Women were less strongly in favor of placing the liability with users than men (Figure 8, Q6; W = 115,950, $p < 0.05$), but sex did not affect preference for other management approaches. With rising age, respondents favored IGS management following strict rules (Figure 8, Q4; $p < 0.05$, $r_s = -0.07$) and requiring permission before using IGS (Figure 8, Q5; $p < 0.01$, $r_s = -0.10$).

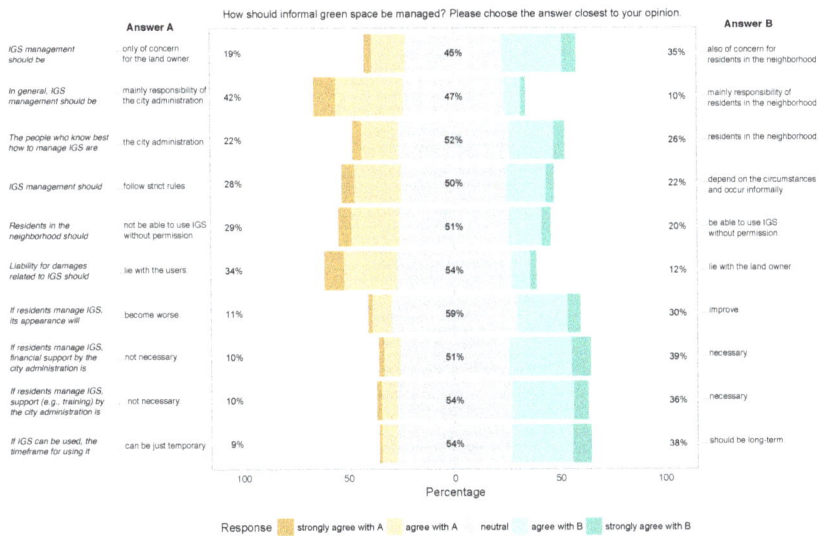

Figure 8. Respondents' opinion on approaches to IGS management.

Willingness to participate in participatory IGS management showed that less than 20% of respondents wanted to contribute, and less than 10% wanted to try organizing or leading such activities (Figure 9). On the other hand, when asked about concrete activities, respondents were more positive towards participation. However, a large number of respondents strongly disagreed across all types of participation. The only significant differences in responses between the four cities was in willingness to pick up trash (Figure 9, Q6; H(3) = 12.36, $p < 0.01$; SP-NA, $p = 0.004$; NA-KY, $p = 0.004$) and willingness to plant flowers or trees (Figure 9, Q8; H(3) = 12.25, $p < 0.01$; SP-NA, $p < 0.001$; NA-KY, $p = 0.016$). Men were more willing to contribute time, skills, and money to IGS management, were more willing to participate monthly or weekly, try themselves at organizing or leading, and were more willing to mow, build equipment or collaborate in research (Figure 9, Q1–5/7/9–10; $p < 0.05$ for money, $p < 0.01$ for all others). With rising age, respondents were less willing to participate in planting activities (Figure 9, Q8; $p < 0.05$, $r_s = 0.08$) or in building play equipment (Figure 9, Q9; $p < 0.001$, $r_s = 0.12$). Respondents commented in qualitative responses that participative IGS management could be integrated into traditional local structures of community management (e.g., neighborhood associations), but also stressed that participation should not become a community duty (such as cleaning street gutters still is in many urban neighborhoods).

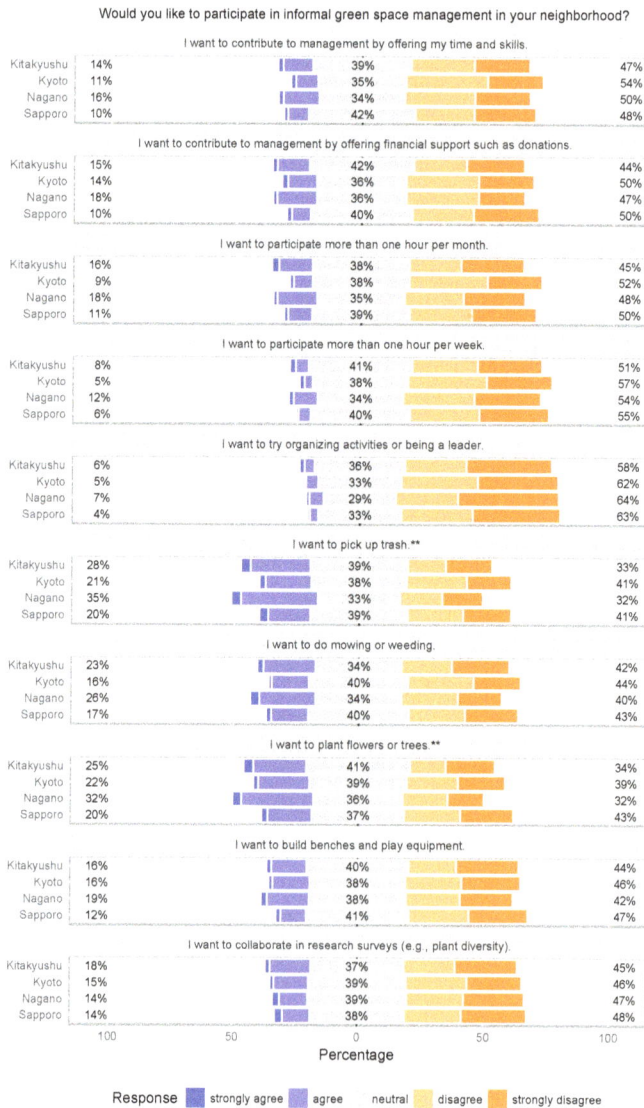

Would you like to participate in informal green space management in your neighborhood?

I want to contribute to management by offering my time and skills.

City	Disagree	Neutral	Agree
Kitakyushu	14%	39%	47%
Kyoto	11%	35%	54%
Nagano	16%	34%	50%
Sapporo	10%	42%	48%

I want to contribute to management by offering financial support such as donations.

City	Disagree	Neutral	Agree
Kitakyushu	15%	42%	44%
Kyoto	14%	36%	50%
Nagano	18%	36%	47%
Sapporo	10%	40%	50%

I want to participate more than one hour per month.

City	Disagree	Neutral	Agree
Kitakyushu	16%	38%	45%
Kyoto	9%	38%	52%
Nagano	18%	35%	48%
Sapporo	11%	39%	50%

I want to participate more than one hour per week.

City	Disagree	Neutral	Agree
Kitakyushu	8%	41%	51%
Kyoto	5%	38%	57%
Nagano	12%	34%	54%
Sapporo	6%	40%	55%

I want to try organizing activities or being a leader.

City	Disagree	Neutral	Agree
Kitakyushu	6%	36%	58%
Kyoto	5%	33%	62%
Nagano	7%	29%	64%
Sapporo	4%	33%	63%

*I want to pick up trash.***

City	Disagree	Neutral	Agree
Kitakyushu	28%	39%	33%
Kyoto	21%	38%	41%
Nagano	35%	33%	32%
Sapporo	20%	39%	41%

I want to do mowing or weeding.

City	Disagree	Neutral	Agree
Kitakyushu	23%	34%	42%
Kyoto	16%	40%	44%
Nagano	26%	34%	40%
Sapporo	17%	40%	43%

*I want to plant flowers or trees.***

City	Disagree	Neutral	Agree
Kitakyushu	25%	41%	34%
Kyoto	22%	39%	39%
Nagano	32%	36%	32%
Sapporo	20%	37%	43%

I want to build benches and play equipment.

City	Disagree	Neutral	Agree
Kitakyushu	16%	40%	44%
Kyoto	16%	38%	46%
Nagano	19%	38%	42%
Sapporo	12%	41%	47%

I want to collaborate in research surveys (e.g., plant diversity).

City	Disagree	Neutral	Agree
Kitakyushu	18%	37%	45%
Kyoto	15%	39%	46%
Nagano	14%	39%	47%
Sapporo	14%	38%	48%

100   50   0   50   100
Percentage

Response ■ strongly agree ■ agree ■ neutral ■ disagree ■ strongly disagree

Figure 9. Respondents' willingness to participate IGS management. Significant differences between the four cities are noted as ** ($p < 0.01$).

3.4. Multiple Correspondence Analysis of Key IGS and Demographic Variables

A multiple correspondence analysis of key IGS variables (number of IGS known in the neighborhood, perceived number of species living in IGS, perceived influence of IGS on everyday life, frequency of recreational IGS use) and demographic variables (level of educational attainment, household income, housing type, study location) was performed to explore relationships in the data through multivariate analysis. Likert-scale questions (perception of benefits and problems of IGS, attitude toward management goals and style) were excluded, as their addition did not substantially increase the variation explained and their inter-scale relationships were already analyzed using Cronbach's alpha. The results (Figure 10) show a clear distinction between respondents who knew and did not know IGS in their neighborhood. Furthermore, number of IGS known in the neighborhood seemed to be associated with perceived number of species in IGS, frequency of recreational use, and demographic characteristics such as detached housing with garden, higher educational attainment level, higher household income and the study location Nagano. However, the displayed first two dimensions only explain a cumulative variation of 14% (unadjusted).

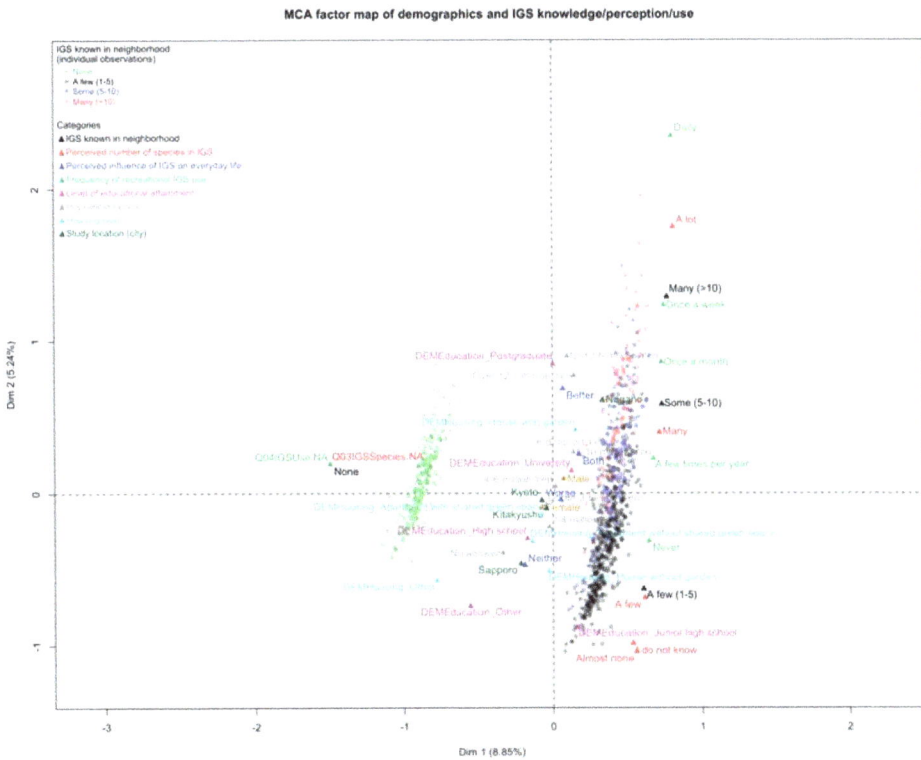

Figure 10. Multiple correspondence analysis of key IGS and demographic variables.

4. Discussion

4.1. Perception, Evaluation, and Use of IGS

The results of this survey suggest IGS is a part of everyday life and a common feature of the urban landscape, not only in Sapporo, where IGS in Japan was first studied, but also across shrinking Japanese cities. Furthermore, the findings provide strong evidence supporting the following conclusions from prior IGS research in Japan [23,25]. Many respondents perceive IGS to be biodiverse and to possess a range of potential benefits, from ecosystem services such as air filtration and cooling to wildlife habitat and opportunities for nature contact. Despite the near-absent discussion of recreational IGS use in the Japanese literature, every fifth respondent across all cities reported already engaging in such use—a majority of them without encountering problems. Proximity was again the most common reason why respondents used IGS, a topic that merits renewed attention as Japan's population ages and its mobility declines. On the other hand, as in prior research [23], many respondents felt IGS affected their daily lives in both positive and negative ways, with littering, weeds and pest animals, and the aesthetic appearance of IGS identified as major potential problems. These findings underline how important it is both to realize the potential of IGS and develop adequate approaches to manage it, especially in shrinking Japanese cities.

Differences between earlier results and this study were mostly in degree, not direction. Overall, fewer respondents knew IGS in their neighborhood or used it for recreational purposes in Sapporo in this study than in the postal survey conducted in 2012 [23]. These differences could be the result of a slight bias in responses of the original postal survey towards residents interested in the topic. In contrast, differences were less pronounced in regard to evaluation as well as perceived benefits and problems. Following the recommendations outlined in the earlier study, it is thus important to take a closer look at residents' preferences for management goals and approaches.

4.2. Preferred Management Goals

Respondents preferred active IGS management to a hands-off approach. Furthermore, they strongly favored management as green space over conversion to parking space or other urban land use. One such management goal is the creation of new parks using IGS where size and characteristics are suitable. These results are in line with earlier research that found Japanese respondents are overall hesitant to embrace the concept of urban wilderness, a concept that has figured prominently in work on IGS from Europe, North America and Australia [27,28,39–50]. Yet this does not imply that residents do not perceive the value of IGS as a different kind of urban green space. However, it suggests cultural factors could be influencing residents' perception and evaluation of IGS, something that has been suggested before in more general discussions of nature perception and culture [26,51]. This, again, may be a matter of degree rather than direction, as some respondents critical of IGS in Australia also mentioned impressions of neglect and abandonment as reasons for their negative perception of IGS. What follows is a dilemma: on one hand, vacancy often has negative cultural associations [52], while on the other hand a freedom of purpose can be a freedom from purpose, opening up space and possibilities that would otherwise not exist. Such notions have been explored in detail in prior work on IGS [53–55]. This issue then brings into focus more general opinions on using space in shrinking cities.

In their opinions about strategic directions for shrinking cities in general, respondents not only favored using space that becomes available for recreational green space and urban agriculture, they also supported giving up land use for human purposes to return it to nature. In contrast, respondents rejected using space opened up through population decline to increase housing size, even though Japanese houses and apartments are on average much smaller than housing in Western cities. These opinions align well with calls by researchers to focus on contact with nature and green infrastructure as a source of improved human wellbeing [4,56]. In a larger context, the results may reflect a shift in focus from material wealth to non-material wellbeing in the Japanese public that has occurred since the early 1970s (Figure 11, [57]). Overall, respondents' opinions signal support for Japanese urban green planning to expand its ambitions beyond the unambitious current policies, which often target only preserving existing green space rather than creating new ones. As budget constraints are partly to blame, the question then is to what degree participatory management approaches are supported by residents.

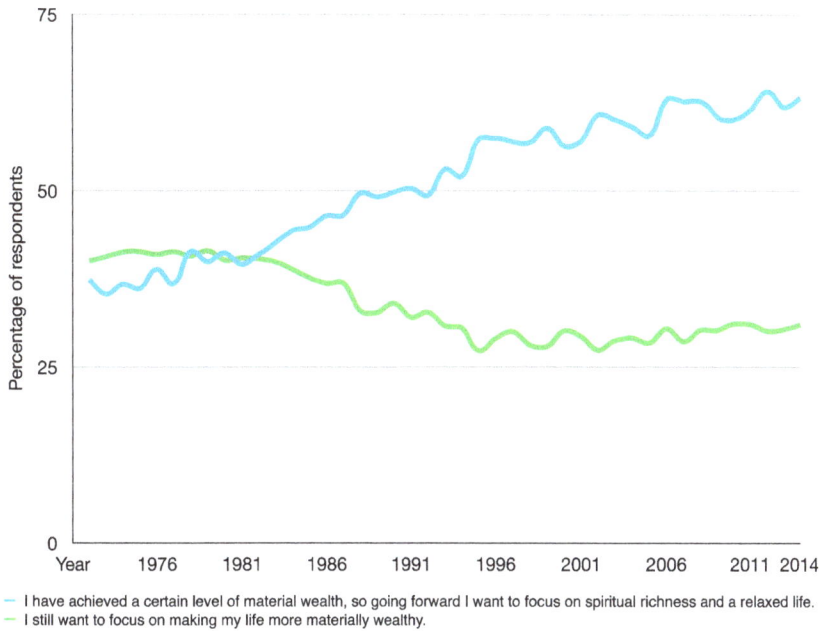

Figure 11. Long-term shift in life goals from material to non-material wellbeing in the Japanese public [57].

4.3. Preferred Management Approaches

Despite numerous benefits participatory IGS management could provide (e.g., realizing recreational potential, reducing financial burden to public funding etc.), the results of this survey suggest that residents will only accept such arrangements if they think it will improve IGS appearance—a primary concern reflected throughout perceived IGS benefits and problems as well as preferred IGS management goals. However, respondents did expect IGS appearance would improve through participatory management. Having residents manage IGS then stops being simply a strategy proposed by scholars to solve surrounding issues, and emerges as a new management approach supported by both professionals and stakeholders. This expected positive outcome also provides the basis on which the details of participatory IGS management can be discussed.

Principal issues of participatory IGS management are the questions of whom it should concern, and who should be responsible for it. It is not surprising that most respondents identified IGS management as a topic of concern for those living in the neighborhood, not just the land owner, as many knew of IGS in their neighborhood and felt it affected their daily lives. Yet at the same time, most saw the main responsibility for managing IGS lying with the city administration—even though overall respondents thought residents in the neighborhood would know better than the administration how such management should happen. This shows IGS management is situated in a triangle of concerned parties—owners (whether present/known or absent), neighbors and the city administration.

The opinion that taking care of land is an issue of concern to neighbors is not unique to IGS. Regulations across different countries affect everything from how houses may be built to how private green space should look. However, with IGS a number of complicating factors are introduced. First, depending on the type of IGS, the degree to which appearance and management is considered an issue likely varies, both among land owners and neighbors—a gap between walls or fences likely draws less attention than an overgrown vacant lot or fence. The issue becomes even more problematic if the owner is absent (e.g., living in a different area, corporate or public owner) or unknown, a problem that has lead the Japanese Ministry of Land, Infrastructure and Transport to produce over 300 pages of guidelines for dealing with unclear land ownership [58]. Not only does this mean the owner may be less invested in managing the space, neighbors may also have a harder time communicating their concerns about the space. This trend is to some degree exacerbated by land ownership fragmentation, which can lead to the generation of informal green spaces even in growing cities such as Tokyo [59]. Furthermore, eminent domain (or resumption/expropriation) is an exceedingly complicated issue in Japan [60,61]. On the other hand, social pressure can be perceived by owners as meddling in someone else's affairs. In the context of rapid aging, property maintenance can also be a task older people may no longer be physically able to carry out, creating a potential source of neighborhood friction. Even if relevant regulations exist, it may be difficult for the city administration to enforce them, and providing assistance in managing privately owned IGS would require using already strained public funding. A participatory IGS management framework would thus need to mediate between neighbors and space owners without the city administration abdicating from the responsibility respondents perceive it to have.

Three core issues are likely to play an important role in the outcome of participatory IGS management. First, the variety of IGS characteristics (type, size, ownership etc.), the inherent subjectivity of IGS evaluation, and the social nature of the relationships involved would favor a flexible approach to management. However, survey results show a plurality of respondents favored strict rules over case-by-case and informal management. This further supports a role of the city administration in providing a framework for IGS management, even though such a framework would need enough inherent flexibility to process diverse cases. Second, a principal goal of IGS management should be to realize its potential for recreation [29] and conservation [62,63]—a goal this survey shows is supported by respondents. However, a plurality of respondents rejected allowing residents to use IGS without permission. This means participatory IGS management would require a mechanism to establish consent to IGS use, whether through communal opt-in or opt-out approaches or incentive/disincentive-based arrangements. For temporary uses, agents involved in the event planning can facilitate this process [64], while in other cases non-governmental organizations fill this role (e.g., 596 Acres in New York). Third, because the issue of liability has been identified as an important barrier to IGS use, participatory IGS management should attempt to at least ameliorate this problem. Liability has a long history alongside a record of court decisions seeking to solve this problem [65]. However, current arrangements often require land owners to make IGS inaccessible to avoid liability, thus reducing its use value for recreation (for conservation, the outcome likely depends on the particular species and types of barriers to movement involved). In this case, respondents in general saw IGS users liable for damages, suggesting that this issue could be ameliorated, possibly by offering reduced owner liability in exchange for consenting to IGS use by residents. Once the three core issues above have been addressed, what remains to be discussed are necessary support and timeframes of IGS use.

The history of participatory green space management shows that simply handing over such tasks to residents can easily fail without necessary support. While participatory approaches are often intended to reduce financial pressure on strained public finances, respondents were clear in their assessment that financial support by the city administration would be necessary if residents were to manage IGS. This has been pointed out by Arnstein in her "ladder of citizen participation" as early as 1969 [66], but actual implementation has been slow. On the other hand, financial support is likely not sufficient. Respondents acknowledged this, and while the need for non-financial support through training opportunities, knowledge sharing or provision of facilities may seem obvious, it merits confirmation to ensure such support is made part of participatory management arrangements.

4.4. Willingness to Volunteer

Participatory IGS management in principle draws on resources provided by residents to achieve its goals, meaning its viability depends on how willing residents are to engage in such activities. Only a minority of respondents was willing to participate in IGS management either through contributing time or money. Yet over 10% of respondents were willing to participate one hour or more per month. In comparison, Takase and colleagues [67–70], who have done extensive work on participatory green space in Japan, found in a survey of residents interested in participatory green space activities that their average willingness to work was 11.6 days per year [31]. On this basis, they recommended a frequency of one activity per month. This could prove sufficient, as managing IGS likely requires less effort than ordinary green space maintenance. Prior research even warns of the danger of applying traditional planning tools or standards to IGS, which can lead to diminished recreational potential as attractive features of IGS are lost [71]. Even in a Japanese context, where wild urban nature seems to be less valued in comparison to Europe, respondents emphasized in this study that even a minimal level of IGS management could improve both perception and recreational value of IGS. Some participants also voiced concerns that over-management might risk losing the characteristics making IGS special.

In regard to possible ways of organizing participation, respondents proposed drawing upon traditional institutions of local self-management such as neighborhood associations (chōnaikai). These associations seem to be an obvious choice, but on further examination a number of problems emerge. First, the very process of depopulation that drives an increase in vacant lots and other IGS is, together with other changes to Japan's social fabric, leading to a decline in organization level among neighborhood associations. Furthermore, neighborhood association membership is often limited by social class or housing type, where apartment dwellers and/or renters are not invited to participate as they are seen as transitory residents with no stake in the neighborhood. Second, neighborhood associations also have a history of resisting changes to the status quo, which leads back to their conservative background as a state co-opted tool of maintaining the public order during the second world war [72]. During the 1970s environmental pollution protests, neighborhood associations often acted to suppress protest and thus constituted an important barrier to institutionalization. As a result, it might be prudent to avoid framing participatory IGS management as a civic duty, but rather organize it based on interest and by emphasizing benefits to participating. Indeed, Takase and colleagues show that participation in green space conservation efforts is associated with benefits similar to those attributed to IGS [31]. Research on community gardens has also found that participatory management can help to strengthen community ties [73–75], a process that might alleviate some of the issues resulting from social change and depopulation. Existing examples from Japan could serve as references for implementing participatory arrangements, such as the conversion of power-line corridor IGS into a community garden in Nagoya or the informal urban agriculture practiced along a riverbank in Sapporo [76]. Such examples also propose a potential solution to the problem of environmental gentrification associated with increasing traditional types of urban green spaces [56,77].

4.5. Main Contributions of This Study to Our Understanding of IGS

This study contributes to our understanding of IGS in three major ways. First, unlike existing exploratory work [23,25], this study is based on a representative sample, so the results can (with some limitations, see below) be generalized for the four cities. The results thus make a compelling argument for the importance of IGS in general and IGS planning specifically. Second, this study is the first comprehensive examination of residents' preferences for IGS management goals and approaches using a large sample rather than exploratory or qualitative (e.g., case studies, interviews etc.) methods. The study thus provides insight into residents' attitudes that goes beyond common methods in participatory planning (such as public consultations, which usually employ an opt-in approach). Third, the study probes the feasibility of participatory IGS management by examining residents' willingness to participate, applying prior groundbreaking work on participatory green space management in Japan to informal green spaces for the first time. These major contributions provide a solid basis on which more general principles for IGS planning can be proposed. This task is taken up in the conclusion.

4.6. Limitations

This study has a number of limitations. While the sample is representative for the four study locations, all of these cities are projected to be shrinking, so results could differ for urban areas still experiencing growth (e.g., Nagoya, Tokyo). As the survey was conducted online, the opinion of residents without Internet access may not be adequately represented. However, research has shown online surveys to be in general robust and reliable. Another limitation is the age range of respondents (20–69 years). Past research, as well as response in this survey, has shown that IGS plays an important role for children's recreation, but due to constraints by the survey provider, this study was not able to collect data underage respondents. What role IGS plays for recreation of adults over 70 is unclear, but this age group was also not included in this study. The sample also showed a large number of respondents to have a neutral opinion about many questions. While this may indicate a lack of interest in the topic, including these responses avoids the potential non-response bias found in postal surveys. Due to budget constraints, this survey was limited to shrinking cities in Japan, even though participatory management is likely also of interest to shrinking municipalities in other countries. The scarcity of prior research on both IGS in Japan and participatory IGS management in general also limits comparisons and possible discussion of similarities and differences, an issue that should be addressed by additional future research.

5. Conclusions

This study has analyzed perception, use and management preferences of residents in four major shrinking Japanese cities on the basis of a large-scale (n = 1000) representative online survey. The results have important implications for planning and management of IGS in shrinking Japanese cities, and provide clear directions for managing non-traditional green spaces to urban planners. Drawing upon the reported findings, the following eight major planning principles derived from the findings are proposed as a potential basis of future strategic IGS planning policy in Japan:

1. IGS is an integral part of the everyday urban landscape and residents' daily lives. As such, planners should consider ways to integrate it into existing green plans.
2. IGS has a wide variety of potential benefits and problems, but whether IGS affects residents' positively or negatively depends on how it is managed. Current (non-)management produces positive outcomes for some residents, but remains suboptimal.
3. Residents strongly support recreation, urban agriculture and conservation as three central functions and goals for IGS management. These preferences should form the basis of IGS management planning.
4. Residents strongly support recreation, urban agriculture and returning space to nature as overarching uses for space opened up through urban shrinkage. These preferences question current efforts of national, regional and local governments to attract new residents and halt or even reverse urban shrinkage. Instead, residents' preferences should form the basis for larger strategic urban planning.
5. The city administration is expected to play a role in managing IGS, even in participatory planning arrangements. Participatory management should not lead to a complete retreat of government involvement in managing IGS, but actively draw on the local knowledge of residents.
6. Realizing the potential of IGS for recreation likely requires a clear framework of rules around IGS use, mechanisms of consent to improve accessibility, and strategies to ameliorate liability issues. These three requirements indicate an important role to play for planners and green space managers.
7. Participatory IGS management is expected to improve the urban landscape aesthetic. To achieve this outcome, residents require both financial and non-financial support from the city administration.

8. A small but significant minority (~10%) of residents is willing to participate through offering time, skills and donations. As IGS only requires limited management to be evaluated considerably higher by residents, the basis of participatory IGS management likely exists. Participation should be voluntary, not forced.

The important role IGS plays for residents and as a common type of space in the urban landscape suggests further research is necessary to address questions that could not be covered in this study. First, it remains unclear whether IGS perception, use, and management preferences of residents in shrinking cities are identical or similar to those of residents in Japanese cities with growing population (e.g., Nagoya, Tokyo). Future research should explore what role this and other factors play, both in Japan and internationally. Second, we still know little about perception, use, and management preference of underage residents and residents over 70 years of age. Such research would likely need to draw on non-internet-based research methods. Third, despite recent work on the topic, examples of participatory IGS management are understudied, making it difficult for planners to learn from real-world cases. A detailed case comparison study could provide valuable insights on benefits and drawbacks of different levels of informality or regulation in IGS management, ideally identifying best-practice examples. Fourth, the recent attention on green infrastructure has raised the question of how IGS can function as one type of green infrastructure, including in comparison to private green spaces and associated participatory management approaches. Finally, the thorough understanding of residents' attitude toward participatory IGS management on which the proposed planning principles are based is currently limited to Japan. Further research is necessary to understand to what degree the principles hold true in other geographic and cultural contexts.

Supplementary Materials: The following are available online at www.mdpi.com/2073-445X/6/3/59/s1, File S1: Survey instrument (In Japanese), File S2: Raw survey dataset (In Japanese).

Acknowledgments: I would like to thank all respondents for participating in this study, and the members of the FEAST Project for their assistance with this study. This research was supported by the FEAST Project (No. 14200116), Research Institute for Humanity and Nature (RIHN), an Early Career Researcher Support Grant by RIHN, and by JSPS KAKENHI Grant Numbers JP17K08179, JP17K15407.

Conflicts of Interest: The author declares no conflict of interest. The funding sponsors had no role in the design of the study; in the collection, analyses, or interpretation of data; in the writing of the manuscript, and in the decision to publish the results.

References

1. Keniger, L.; Gaston, K.; Irvine, K.; Fuller, R. What are the Benefits of Interacting with Nature? *Int. J. Environ. Res. Public Health* **2013**, *10*, 913–935. [CrossRef] [PubMed]
2. Lee, A.C.K.; Maheswaran, R. The health benefits of urban green spaces: A review of the evidence. *J. Public Health* **2011**, *33*, 212–222. [CrossRef] [PubMed]
3. Bratman, G.N.; Daily, G.C.; Levy, B.J.; Gross, J.J. The benefits of nature experience: Improved affect and cognition. *Landsc. Urban Plan.* **2015**, *138*, 41–50. [CrossRef]
4. Tzoulas, K.; Korpela, K.; Venn, S.; Yli-Pelkonen, V.; Kaźmierczak, A.; Niemela, J.; James, P. Promoting ecosystem and human health in urban areas using Green Infrastructure: A literature review. *Landsc. Urban Plan.* **2007**, *81*, 167–178. [CrossRef]
5. Van den Berg, M.; Wendel-Vos, W.; van Poppel, M.; Kemper, H.; van Mechelen, W.; Maas, J. Health Benefits of Green Spaces in the Living Environment: A Systematic Review of Epidemiological Studies. *Urban For. Urban Green.* **2015**, *14*, 806–816. [CrossRef]
6. Naumann, S.; Davis, M.; Kaphengst, T.; Pieterse, M.; Rayment, M. *Design, Implementation and Cost Elements of Green Infrastructure Projects*; Ecologic Institute and GHK Consulting: Berlin, Germany, 2010.
7. Yokohari, M.; Bolthouse, J. Planning for the slow lane: The need to restore working greenspaces in maturing contexts. *Landsc. Urban Plan.* **2011**, *100*, 421–424. [CrossRef]
8. Kobayashi, R.; Sakai, A. Study on transition of regenerating for city park in Sapporo city. *J. City Plan. Inst. Jpn.* **2008**, *43.3*, 583–588. [CrossRef]

9.   2014 Status of Urban Green Space per Capita on Prefectural Basis.   Available online: http://www.webcitation.org/query?url=https%3A%2F%2Fwww.mlit.go.jp%2Fcrd%2Fpark%2Fjoho%2Fdatabase%2Ft_kouen%2Fpdf%2F04_h26.pdf&date=2017-08-24 (accessed on 24 August 2017).

10. Eanes, F.; Ventura, S.J. Inventorying Land Availability and Suitability for Community Gardens in Madison, Wisconsin. *Cities Environ. CATE* **2015**, *8*, 2.

11. Savino, S.D.M. Facilitating Social-Ecological Transformation of a Vacant Lot on an Urban Campus: The Houston-Congolese Connection. *Cities Environ. CATE* **2015**, *8*, 4.

12. Delgado, C. Answer to the Portuguese Crisis: Turning Vacant Land into Urban Agriculture. *Cities Environ. CATE* **2015**, *8*, 5.

13. Dennis, M.; James, P. User participation in urban green commons: Exploring the links between access, voluntarism, biodiversity and well being. *Urban For. Urban Green.* **2016**, *15*, 22–31. [CrossRef]

14. Gasperi, D.; Pennisi, G.; Rizzati, N.; Magrefi, F.; Bazzocchi, G.; Mezzacapo, U.; Centrone Stefani, M.; Sanyé-Mengual, E.; Orsini, F.; Gianquinto, G. Towards Regenerated and Productive Vacant Areas through Urban Horticulture: Lessons from Bologna, Italy. *Sustainability* **2016**, *8*, 1347. [CrossRef]

15. Sugita, S.; Doi, Y. A Fundamental Study on Stewardship of Public Space by Local Community Groups. *J. City Plan. Inst. Jpn.* **2012**, *47*, 469–474. [CrossRef]

16. IPSS Regional Population Projections for Japan: 2010–2040 (March 2013). Available online: http://www.ipss.go.jp/pp-shicyoson/e/shicyoson13/t-page.asp (accessed on 19 April 2016).

17. Japanese Ministry of Land, Infrastructure and Transport. Toshi Kōenhō Unyō Shishin (dai 2 ban) (In Japanese; Urban Park Law Operation Guidelines, 2nd Edition). 2012. Available online: https://www.mlit.go.jp/crd/townscape/pdf/koen-shishin01.pdf (accessed on 24 August 2017).

18. Yamada, C.; Terada, T.; Tanaka, T.; Yokohari, M. Directions for Vacant Lot Management in the Outer Suburbs of the Tokyo Metropolitan Region. *Urban Reg. Plan. Rev.* **2016**, *3*, 66–84. [CrossRef]

19. Takano, T.; Nakamura, K.; Watanabe, M. Urban residential environments and senior citizens' longevity in megacity areas: The importance of walkable green spaces. *J. Epidemiol. Community Health* **2002**, *56*, 913–918. [CrossRef] [PubMed]

20. Gill, S.E.; Handley, J.F.; Ennos, A.R.; Pauleit, S. Adapting cities for climate change: The role of the green infrastructure. *Built Environ.* **2007**, *33*, 115–133. [CrossRef]

21. Byrne, J.; Ambrey, C.; Portanger, C.; Lo, A.; Matthews, T.; Baker, D.; Davison, A. Could urban greening mitigate suburban thermal inequity?: The role of residents' dispositions and household practices. *Environ. Res. Lett.* **2016**, *11*, 95014. [CrossRef]

22. Kusaka, H.; Hara, M.; Takane, Y. Urban Climate Projection by the WRF Model at 3-km Horizontal Grid Increment: Dynamical Downscaling and Predicting Heat Stress in the 2070's August for Tokyo, Osaka, and Nagoya Metropolises. *J. Meteorol. Soc. Jpn. Ser. II* **2012**, *90*, 47–63. [CrossRef]

23. Rupprecht, C.D.D.; Byrne, J.A.; Ueda, H.; Lo, A.Y.H. "It"s real, not fake like a park': Residents' perception and use of informal urban green-space in Brisbane, Australia and Sapporo, Japan. *Landsc. Urban Plan.* **2015**, *143*, 205–218. [CrossRef]

24. Rupprecht, C.D.D.; Byrne, J.A. Informal urban green-space: Comparison of quantity and characteristics in Brisbane, Australia and Sapporo, Japan. *PLoS ONE* **2014**, *9*, e99784. [CrossRef] [PubMed]

25. Rupprecht, C.D.D.; Byrne, J.A.; Lo, A.Y.H. Memories of vacant lots: How and why residents used informal urban greenspace as children and teenagers in Brisbane, Australia and Sapporo, Japan. *Child. Geogr.* **2016**, *14*, 340–355. [CrossRef]

26. Flint, C.G.; Kunze, I.; Muhar, A.; Yoshida, Y.; Penker, M. Exploring empirical typologies of human–nature relationships and linkages to the ecosystem services concept. *Landsc. Urban Plan.* **2013**, *120*, 208–217. [CrossRef]

27. Jorgensen, A.; Keenan, R. (Eds.) *Urban Wildscapes*; Routledge: Abingdon, UK, 2012; ISBN 978-0-415-58106-6.

28. Campo, D. *The Accidental Playground*; Fordham University Press: New York, NY, USA, 2013.

29. Rupprecht, C.D.D.; Byrne, J.A. Informal urban greenspace: A typology and trilingual systematic review of its role for urban residents and trends in the literature. *Urban For. Urban Green.* **2014**, *13*, 597–611. [CrossRef]

30. Informal Urban Greenspace Perception and Use: Survey Instrument. Available online: https://doi.org/10.13140/RG.2.1.3830.7448 (accessed on 24 August 2017).

31. Takase, Y.; Furuya, K.; Sakuraba, S. The Relationship between Citizens' Willingness to Work for Participation in Green Space Conservation Activities and Attributes, Citizens' Willingness to Work and Attitude toward Participation. *J. Jpn. Inst. Landsc. Archit.* **2015**, *78*, 619–624. [CrossRef]

32. Field, A.; Miles, J.; Field, Z. *Discovering Statistics Using R*; SAGE: Newcastle upon Tyne, UK, 2012; ISBN 978-1-4462-5846-0.

33. R Development Core Team R: A Language and Environment for Statistical Computing. 2017. Available online: http://www.r-project.org (accessed on 24 August 2017).

34. Bryer, J.; Speerschneider, K.; Bryer, M.J. Package "Likert". 2016. Available online: http://cran.pau.edu.tr/web/packages/likert/likert.pdf (accessed on 24 August 2017).

35. Husson, F. Package "FactoMineR". 2017. Available online: https://cran.r-project.org/web/packages/FactoMineR/FactoMineR.pdf (accessed on 24 August 2017).

36. JASP Team. T. JASP (Version 0.8.1.1). 2017. Available online: https://jasp-stats.org/ (accessed on 24 August 2017).

37. Jamovi Team. T. Jamovi (Version 0.7.5.7). 2017. Available online: https://www.jamovi.org (accessed on 24 August 2017).

38. QGIS Development Team QGIS Geographic Information System; Open Source Geospatial Foundation. 2017. Available online: http://qgis.osgeo.org (accessed on 24 August 2017).

39. Jorgensen, A.; Tylecote, M. Ambivalent landscapes—Wilderness in the urban interstices. *Landsc. Res.* **2007**, *32*, 443–462. [CrossRef]

40. Buijs, A.E.; Elands, B.H.M.; Langers, F. No wilderness for immigrants: Cultural differences in images of nature and landscape preferences. *Landsc. Urban Plan.* **2009**, *91*, 113–123. [CrossRef]

41. Konijnendijk, C. Between fascination and fear—The impacts of urban wilderness on human health and wellbeing. *Socialmed. Tidskr.* **2012**, *89*, 289–295.

42. Mathey, J.; Rink, D. Urban Wastelands–A Chance for Biodiversity in Cities? Ecological Aspects, Social Perceptions and Acceptance of Wilderness by Residents. In *Urban Biodiversity and Design*; Müller, N., Werner, P., Kelcey, J.G., Eds.; Wiley-Blackwell: Oxford, UK, 2010; pp. 406–424.

43. Rink, D.; Herbst, H. From wasteland to wilderness—Aspects of a new form of urban nature. In *Applied Urban Ecology: A Global Framework*; Richter, M., Weiland, U., Eds.; Wiley-Blackwell: Chichester, UK, 2011; pp. 82–92.

44. Vicenzotti, V.; Trepl, L. City as Wilderness: The Wilderness Metaphor from Wilhelm Heinrich Riehl to Contemporary Urban Designers. *Landsc. Res.* **2009**, *34*, 379–396. [CrossRef]

45. Jeans, D.N. Wilderness, Nature and Society: Contributions to the history of an environmental attitude. *Aust. Geogr. Stud.* **1983**, *21*, 170–182. [CrossRef]

46. Head, L.; Muir, P. Suburban life and the boundaries of nature: Resilience and rupture in Australian backyard gardens. *Trans. Inst. Br. Geogr.* **2006**, *31*, 505–524. [CrossRef]

47. Instone, L. Encountering Native Grasslands: Matters of Concern in an Urban Park. *Aust. Humanit. Rev.* **2010**, *49*, 91–117.

48. Foster, J. Restoration of the Don Valley Brick Works: Whose Restoration? Whose Space? *J. Urban Des.* **2005**, *10*, 331–351. [CrossRef]

49. Kowarik, I. Urban wilderness: Supply, demand, and access. *Urban For. Urban Green.* **2017**. [CrossRef]

50. Davis, M. *Dead Cities: And Other Tales*; New Press: New York, NY, USA, 2002.

51. Kellert, S.R. Attitudes, Knowledge, and Behavior Toward Wildlife Among the Industrial Superpowers: United States, Japan, and Germany. *J. Soc. Issues* **1993**, *49*, 53–69. [CrossRef]

52. Corbin, C.I. Vacancy and the Landscape: Cultural Context and Design Response. *Landsc. J.* **2003**, *22*, 12–24. [CrossRef]

53. Franck, K.A.; Stevens, Q. (Eds.) *Loose Space: Possibility and Diversity in Urban Life*; Routledge: Abingdon, UK, 2007.

54. Schneekloth, L. Unruly and Robust: An abandoned industrial river. In *Loose Space: Possibility and Diversity in Urban Life*; Franck, K.A., Stevens, Q., Eds.; Routledge: Abingdon, UK, 2007; pp. 253–270.

55. Unt, A.-L.; Travlou, P.; Bell, S. Blank Space: Exploring the Sublime Qualities of Urban Wilderness at the Former Fishing Harbour in Tallinn, Estonia. *Landsc. Res.* **2013**, *39*, 267–286. [CrossRef]

56. Wolch, J.R.; Byrne, J.; Newell, J.P. Urban green space, public health, and environmental justice: The challenge of making cities "just green enough". *Landsc. Urban Plan.* **2014**, *125*, 234–244. [CrossRef]

57. Cabinet Office Japan. Kokumin Seikatsu ni Kansuru Yoron Chōsa (In Japanese; Public Opinion Poll on the Public's Everyday Life). Available online: http://survey.gov-online.go.jp/h26/h26-life/index.html (accessed on 19 June 2017).

58. Guidelines for Discovery and Use of Land in Cases of Unclear Ownership. Available online: http://www.webcitation.org/query?url=http%3A%2F%2Fwww.mlit.go.jp%2Fcommon%2F001178691. pdf&date=2017-08-24 (accessed on 24 August 2017).

59. Rahmann, H.; Jonas, M. Void Potential: Spatial Dynamics and Cultural Manifestations of Residual Spaces. In *Terrain Vague: Interstices at the Edge of the Pale*; Mariani, M., Barron, P., Eds.; Routledge: Abingdon, UK, 2013; pp. 89–104.

60. Sorensen, A. Conflict, consensus or consent: Implications of Japanese land readjustment practice for developing countries. *Habitat Int.* **2000**, *24*, 51–73. [CrossRef]

61. Funabiki, T. A Study on the Relations between Land Use Regulation and Public Compensation in the Methods of Green-space Acquisition and Enhancement System. *Pap. Environ. Inf. Sci.* **2009**, *23*, 13–18. [CrossRef]

62. Rupprecht, C.D.D.; Byrne, J.A.; Garden, J.G.; Hero, J.-M. Informal urban green space: A trilingual systematic review of its role for biodiversity and trends in the literature. *Urban For. Urban Green.* **2015**, *14*, 883–908. [CrossRef]

63. Bonthoux, S.; Brun, M.; di Pietro, F.; Greulich, S.; Bouché-Pillon, S. How can wastelands promote biodiversity in cities? A review. *Landsc. Urban Plan.* **2014**, *132*, 79–88. [CrossRef]

64. Oswalt, P.; Overmeyer, K.; Misselwitz, P. Patterns of the Unplanned. In *Pop Up City*; Schwarz, T., Rugare, S., Eds.; Cleveland Urban Design Collaborative: Cleveland, OH, USA, 2009; pp. 5–18.

65. Explosives. Dumping Refuse on Vacant Lots. Injury to Children. Travell v. Bannerman, 75 N.Y. Supp. 866. *Yale Law J.* **1902**, *12*, 47. [CrossRef]

66. Arnstein, S.R. A Ladder Of Citizen Participation. *J. Am. Inst. Plann.* **1969**, *35*, 216–224. [CrossRef]

67. Takase, Y.; Furuya, K. Issues in Promoting Participation in Open Space Conservation Activities and Priority Ranking of Solutions from University Students' Perspectives. *J. Jpn. Inst. Landsc. Archit.* **2013**, *76*, 717–722. [CrossRef]

68. Takase, Y.; Furuya, K.; Sakuraba, S. Promotion Process for Green Space Conservation Activity Participation from Residents' Perspectives. *J. Archit. Plan. Trans. AIJ* **2014**, *79*, 2241–2249. [CrossRef]

69. Takase, Y.; Furuya, K.; Sakuraba, S. Challenges to Promote Participation in Conservation Activities Based on differences of attitude between Citizens and Open space Conservation Activity Organizations. *J. Jpn. Inst. Landsc. Archit.* **2014**, *77*, 553–558. [CrossRef]

70. Takase, Y.; Furuya, K. Study on promoting public participation in green space conservation activities run by local governments. *J. City Plan. Inst. Jpn.* **2016**, *51*, 1016–1023. [CrossRef]

71. Qviström, M. Taming the wild: Gyllin's Garden and the urbanization of a wildscape. In *Urban Wildscapes*; Jorgensen, A., Keenan, R., Eds.; Routledge: Abingdon, UK, 2012; pp. 187–200.

72. Sorensen, A. Centralization, urban planning governance, and citizen participation in Japan. In *Cities, Autonomy, and Decentralization in Japan*; Hein, C., Pelletier, P., Eds.; Routledge: Abingdon, UK, 2006; pp. 101–127.

73. DelSesto, M. Cities, Gardening, and Urban Citizenship: Transforming Vacant Acres into Community Resources. *Cities Environ. CATE* **2015**, *8*, 3.

74. Rosol, M. Grassroots Gardening Initiatives: Community Gardens in Berlin. In *Enterprising Communities: Grassroots Sustainability Innovations*; Robertson, D.P., Ed.; Emerald Group Publishing Limited: Bingley, UK, 2012; pp. 123–143.

75. Chan, J.; DuBois, B.; Tidball, K.G. Refuges of local resilience: Community gardens in post-Sandy New York City. *Urban For. Urban Green.* **2015**, *14*, 625–635. [CrossRef]

76. Rupprecht, C.D.D.; Byrne, J.A. Informal urban green space as anti-gentrification strategy? In *Just Green Enough: Urban Development and Environmental Gentrification*; Routledge Equity, Justice and the Sustainable City; Curran, W., Hamilton, T., Eds.; Routledge: London, UK, 2017.

77. Curran, W.; Hamilton, T. Just green enough: Contesting environmental gentrification in Greenpoint, Brooklyn. *Local Environ.* **2012**, *17*, 1027–1042. [CrossRef]

Review

A Review on Remote Sensing of Urban Heat and Cool Islands

Azad Rasul [1,2,*], **Heiko Balzter** [2,3], **Claire Smith** [2], **John Remedios** [3], **Bashir Adamu** [2,4], **José A. Sobrino** [5], **Manat Srivanit** [6] **and Qihao Weng** [7]

1   Department of Geography, Soran University, Kawa Street, Soran 44008, Erbil, Iraq
2   Department of Geography, Centre for Landscape and Climate Research, University of Leicester, University Road, Leicester LE1 7RH, UK; hb91@le.ac.uk (H.B.); cls53@le.ac.uk (C.S.); ba108@alumni.le.ac.uk (B.A.)
3   National Centre for Earth Observation, University of Leicester, University Road, Leicester LE1 7RH, UK; jjr8@leicester.ac.uk
4   Department of Geography, Modibbo Adama University of Technology, Yola P.M.B. 2076, Nigeria
5   Global Change Unit, Department of Thermodynamics, Faculty of Physics, University of Valencia, Valencia E-46071, Spain; sobrino@uv.es
6   Faculty of Architecture and Planning, Thammasat University, Pathumthani 12121, Thailand; s.manat@gmail.com
7   Department of Earth and Environmental Systems, College of Arts and Sciences, Indiana State University, St. Terre Haute, IN 47809, USA; qweng@indstate.edu
*   Correspondence: aor4@alumni.le.ac.uk or azad977@gmail.com; Tel.: +964-(0)750-735-8574

Academic Editors: Andrew Millington, Harini Nagendra and Monika Kopecka
Received: 20 April 2017; Accepted: 6 June 2017; Published: 9 June 2017

Abstract: The variation between land surface temperature (LST) within a city and its surrounding area is a result of variations in surface cover, thermal capacity and three-dimensional geometry. The objective of this research is to review the state of knowledge and current research to quantify surface urban heat islands (SUHI) and surface urban cool islands (SUCI). In order to identify open issues and gaps remaining in this field, we review research on SUHI/SUCI, the models for simulating UHIs/UCIs and techniques used in this field were appraised. The appraisal has revealed some great progress made in surface UHI mapping of cities located in humid and vegetated (temperate) regions, whilst few studies have investigated the spatiotemporal variation of surface SUHI/SUCI and the effect of land use/land cover (LULC) change on LST in arid and semi-arid climates. While some progress has been made, models for simulating UHI/UCI have been advancing only slowly. We conclude and suggest that SUHI/SUCI in arid and semi-arid areas requires more in-depth study.

Keywords: land surface temperature (LST); urban climate; surface urban cool island (SUCI); remote sensing; review

1. Introduction

Surface urban heat islands (SUHI) are one of the crucial topics in urban climatology studies. The comfort of the urban inhabitants is influenced by surface temperature through modified air temperature of the lowest layer of the urban atmosphere [1]. The land surface is a complex feature that can be described as a combination of green vegetation, water surfaces, impervious surface materials and exposed soils. As a result of this complexity, LST varies spatially and temporally. Impervious surface differs considerably between urban and suburban areas and it is the main contributor to the SUHI effect [2–4]. The results by Rasul et al. [5,6] from Landsat and MODIS LST indicate the existence of SUCI in semi-arid cities during different times of the day and not only in the morning as stated in other literature [7–10].

Since the early 1900s, the UHI intensity of hundreds of urban areas around the world have been assessed [11] and this field remains an extensive area of study within urban climatology [12]. The growth and strength of the heat island areas during this time bring challenges for energy, the health of urban residents, water supplies, urban infrastructure and social comfort [13]. In addition, it exacerbates heat waves and creates a negative effect on life expectancy on urban inhabitants [14].

Ignoring atmospheric correction means assuming that atmospheric effects are the same in all places, while in reality, water vapor and pollutant contents vary horizontally in urban areas. If the atmospheric correction is neglected or is incorrectly, estimated surface temperature SUHI intensity may be incorrectly derived [15,16]. Typically, the average surface emissivity in urban areas is about 2% lower than the typical rural areas [17]. Without emissivity correction and neglecting this difference, temperature retrievals of urban-rural environments can show differences of 1.5 °C or more. Therefore, the urban heat island effect can typically be underestimated [15].

Reviews of the retrieval of LST, SUHI, generating, determination and mitigation UHI was carried out by a number of authors [18–20]. However, a review of the SUHI/SUCI in dry climate and methods used for studying the SUHI is still lacking.

The objective of this paper is to review the state of knowledge and current research to quantify the SUHI/SUCI. This paper provided knowledge on the techniques used for SUHI and SUHI/SUCI that were based on different climatic regions, specifically for the arid and semi-arid climates.

The articles reviewed in this paper are based on techniques and methods. Moreover, sampling of research for different remote sensing data and SUHI/SUCI from different climatic regions was reviewed. There extensive research on UHI in humid regions, thus this paper focused on SUHI/SUCI in dry climate.

2. Techniques and Statistics Used in Urban Heat Island Studies

2.1. Methods to Compare Multi-Temporal LST Images

In the literature, various techniques have been applied for analyzing the temporal change of satellite based LST. In the first technique, some researchers directly compare two or more LST images without any modification [21]. This approach lacks scientific rigor because the atmospheric situation is not the same at the time of image acquisition. Having a high temperature in the second image compared to the first one, for example, does not mean the temperature has increased because it is possible that at that time the temperature was high for other reasons. The second technique to account for this and to better establish the SUHI is through the normalization of the temperature based on the mean and standard deviation in high and low temperature areas [22,23]. The third technique is common normalization of temperature based on min and max LST of the same image in the same way as for NDVI [24]. A Normalized Ratio Scale (NRS) technique is proposed by Rasul et al. [25], to normalize the value of each pixel-based ratio to make the LST images from different times comparable and at the same time maintaining the original values.

2.2. Determining the Urban Heat Island

Researchers used various methods to assess UHI; for instance, Tran et al. [26] have used satellite data to assess maximum SUHI. Hafner and Kidder [27] have used a model to assess SUHI. UHI was determined in some studies as a comparison of the mean and maximum temperature between urban and rural areas. Others compared temperature during times such as a season, a month, a year or some days. In some cases, it was selected as temperature changes over time [28]. Moreover, Magee et al. [29] selected UHI as the average changed temperature for both the urban and the rural areas [20].

For determination of SUHI, the comparison of mean urban and rural patterns provides robust results. The use of trends of LST before and after urbanized areas illustrated the significant influence of urbanization on the UHI but many cities have no historic records of LST before urbanization which creates an obstacle for SUHI studies.

2.3. Statistical Analyses of Urban Heat Islands

Weng et al. [30] conducted pixel-by-pixel correlation analysis between surface temperature on the one hand, and NDVI, green vegetation (GV), and impervious surface fractions on the other hand.

Linear regression has been used extensively in UHI studies to show the relationship between LST and NDVI [31–33]. Szymanowski and Kryza [34] conducted Multiple Linear Regression (MLR) to state the land-use situation of the UHI, but inaccurate results have been obtained when the process tended towards non-stationary variables such as the impact of the wind. The common character of meteorological data is non-stationary, hence the application algorithm can be largely limited in case the technique is unable to manipulate it. According to Szymanowski and Kryza [34] and Su et al. [35], geographically weighted regression (GWR) is better suited than MLR and other conventional regression analyses. GWR shows the relationship between temperature and land covers more clearly and it is more successful in the spatial modelling of UHI.

For spatial modelling of the UHI, Szymanowski and Kryza [34] suggested the combined GWR residual kriging (GWRK) method as an alternative to the extensively used MLR model. Florio et al. [36] emphasized that the kriging models estimates temperature better than MLR. RK errors are neutral while regression models are inclined to give biased predicted values. RK and GWR methods have been also been applied to LST [37,38].

3. Surface Urban Heat Island

Surface urban heat island intensity (SUHII) is determined by variations of surface temperature between urban and surrounding rural areas with similar geographic characteristics. Remote sensing sensors, thermal images and field data have all been used to assess the SUHII of urban areas.

3.1. Satellite Measurements of Urban Heat Island

In order to ascertain surface temperature through radiation the traditional technique of aerial surveillance is commonly used [39–41]. Thereafter, Wark et al. [42] and Rao and Winston [43] attempted to utilize satellites to measure surface temperatures. Through data obtained from the Television Infrared Observation Satellite (TIROS II), they found that measuring surface temperature is possible in clear and dry areas [44]. Primarily, LST and SUHI have been derived from the National Oceanic and Atmospheric Administration's (NOAA) AVHRR data [30]. After that, Landsat's Thematic Mapper (TM) and Landsat's Enhanced Thematic Mapper Plus (ETM+) were widely employed to retrieve surface temperatures [30,45,46]. Srivastava et al. [47] estimated surface temperature in the Singhbhum Shear Zone of India. The results indicated that emissivity has a strong relationship with the reflectance of ETM+ band 3. They compared field data with estimated LST from different algorithms. It was found that the use of Valor's emissivity and single channel equations increase the accuracy of the result and is closer to field truth temperature.

Surface UHI can be derived from remote sensing images as a captivating and possibly valuable source [2,15]. Rao [48] reported the first study of SUHI based on imagery data. Through the study of surface temperature patterns of the mid-Atlantic coast of USA, the study utilized thermal Infrared Radiometer (IR) data of the Improved TIROS Operational Satellite (ITOS-1). The research found that the center of the city is the warmest part. Matson et al. [49] and Price [50] detected the UHI by utilizing satellite data. Since then, the SUHI and surface temperature have been observed through utilizing different sensors such as satellites, aircrafts, and ground-based sensors. Later, in 1989, Roth et al. [19] studied the thermal urban climates.

The AVHRR sensor has been used to discern the surface temperature [51–53] and to analyze the regional-scale of UHI effects [22,54,55]. Airborne acquired high-resolution images were also used to assess the thermal determiners of urban surfaces such as sky view factor and surface materials [19]. The ASTER is another sensor of the TIR image that collects both daytime and night-time data and has been used for determining the UHI effect in many cities [56–58].

Landsat images are widely used to investigate the growth of SUHIs and to assess the relationship between LST and land use/land cover (LULC) [59–61]. Unfortunately, calibration problems with Landsat 8 TIRS have restricted its use. Clinton and Gong [62] used MODIS at 1 km special resolution with high temporal resolution to investigate UHIs and Urban Heat Sinks (UHSs) of cities on a global scale. Furthermore, MODIS data has been used to analyze daily differences of LST and UHI in Abu Dhabi. Standard nocturnal UHIs were found in the city, while during the day the city center was cooler than its surroundings [63].

The selection of LST data for SUHI studies varies based on the purpose of the research and the availability of remotely sensed data. Landsat images (with 30 m spatial resolution) are appropriate for investigating the spatial variation of SUCI/SUHI and the effect of LULC change from different samples of classes on LST, whereas MODIS LST (with higher temporal resolution) is effective for studying the temporal variation of SUHI/SUHI (e.g., diurnal, seasonal and decadal) at coarser scales. ASTER LST with high spatial resolution is appropriate for quantifying the variation of SUHI between day and night. In general, aircraft-based LST data have higher resolution, but it is expensive and the areal coverage is irregular and it is a non-standardized product while satellite-based LST has extensive spatial coverage, limited spatiotemporal resolution and is influenced by atmospheric effects on the signal [64].

The result of comparison LST of the urban and rural surroundings may vary based on day/night, location and different climatic patterns of the cities (Table 1). The table illustrates the highest SUHI exists in cities with the "Dwa" and "Csb" Köppen climate types while the highest SUCI is found in cities located in "Bwh" and "BSh" climates.

Table 1. A summary of surface UHII/UCII in different areas of the world.

Type	Study Area	Climate	Reference Study	Approach	UHII/UCII °C
Daytime SUHI	Beijing, China	Dwa: Hot Summer Continental	Tran et al. [26]	Satellite data	10
	Vancouver, Canada	Csb: Warm-summer Mediterranean	Roth et al. [15]	Satellite data	7.5
	Medellin, Colombia	Af: Tropical Rainforest	Peng et al. [65]	Satellite data	7
	Athens, Greece	Csa: Dry-summer Subtropical	Stathopoulou and Cartalis [66]	Satellite data	3.3
Nighttime SUHI	Madrid, Spain	Csa: Dry-summer Subtropical	Sobrino et al. [67]	Airborne	5
	Birmingham, UK	Cfb: Marine West Coast	Tomlinson et al. [68]	Satellite data	5
	Erbil, Iraq	BSh: Subtropical Semiarid (Hot Steppe)	Rasul et al. [6]	Satellite data	4.59
	Manila, Philippines	Aw: Tropical Savanna	Tiangco et al. [57]	Satellite data	2.96
	Atlanta, USA	Cfa: Humid Subtropical	Hafner and Kidder [27]	Modeling	1.2
Daytime SUCI	Abu Dhabi, UAE	Bwh: Subtropical Desert	Lazzarini et al. [63]	Satellite data	−6
	Dubai, UAE	Bwh: Subtropical Desert	Frey et al. [69]	Satellite data	−5
	Erbil, Iraq	BSh: Subtropical Semiarid (Hot Steppe)	Rasul et al. [5]	Satellite data	−3.9
	Cairo, Egypt	Bwh: Subtropical Desert	Shahraiyni et al. [70]	Satellite data	−3.1
	Central India	Cfa: Humid subtropical and (Aw) tropical wet and dry	Shastri et al. [71]	Satellite data	−2.5

3.2. Urban Heat Island in Arid and Semi-Arid Climate

SUHI studies pay more attention to urban areas located in tropical, Mediterranean and cold climatic regions whereas arid regions with extreme high temperatures have been less focused on [72]. Moreover, the effect of LULC change on LST has been widely assessed for cities in the humid climate while in cities located in semi-arid environments requires more focus to be better quantified and understood. Some of the few UHI studies in the literature based in arid regions were carried out in Phoenix and Tucson, Arizona by Tarleton and Katz [73], Kuwait City by Nasrallah et al. [74], Erbil City by Rasul et al. [6] and the Al Ahsa oasis by Al-Ali [72]. The effect of land cover on UHII of the Al Ahsa oasis in Saudi Arabia has been assessed by using both ground data and satellite images. The limitation of approach in such research is in comparing urban area with nearby towns to study UHI and ignoring the bare soil and desert sand surrounding the city that has high LST in arid and semi-arid regions. In semi-arid regions, the importance of changing aridity soil moisture in the rural areas in modifying heat islands has not been studied extensively.

4. Urban Cool Islands

The general conviction that the air temperature in green sites can be cooler than non-green sites was confirmed by many studies on the temperature of parks and forest cover [75]. To explain the effect of parks on the temperature of cities in detail, more research is necessary on the design of urban green area, distribution and type of greening. Studies on many parks indicated that the temperature is cooler in larger parks and those with trees [75]. On average, larger parks are cooler than smaller ones but not always, while the urban cool islands (UCI) of the parks is more related to the characteristics of the parks [76]. The results from the study indicate that 61 parks in Taipei city were confirmed as UCIs whereas around one-fifth of parks with ≥50% paved coverage and little tree and shrub cover, have been warmer than their urban surrounding at midday during the summertime [77]. Several studies have confirmed that this so-called "oasis" exhibits the cold island effect [78,79]. In some environments such as arid, semi-arid, arctic and subarctic, cities have been reported as UCIs (negative UHI) during certain times of the day or during particular seasons [27,80,81].

During the dry season the daytime SUHI intensity in some cities such as Mexico City and Reykjavik is very weak and sometimes exhibits a cold island. As a result of high thermal inertia, urban places in arid areas have the capability of showing both nocturnal SUHIs and diurnal SUCIs [62]. The amount of soil moisture and humidity in urban areas have an effect because evaporation via latent heat reduces LST. As such, the investigation proved the existence of UCIs in Dubai compared to the desert areas [77,82,83].

To date, plenty of research has investigated SUHI and SUCI in green spaces and water bodies within cities whereas only a few studies have investigated Surface Urban Cool Island across a whole urban area so it is requiring greater comprehension. Usually, research of atmospheric UHI uses measured air temperature of some points in and around the city that not represent the study area entirely. However, because SUHI studies usually use remote sensing data it represents the temperature of the whole of the study area with some consistency.

5. Future Research Directions

There is a need to utilize remote sensing data in investigating surface temperatures of cities in dry and semi-dry environments on a large scale. That study is a necessary requirement in the description of surface characteristics in this specific environmental climate class. Furthermore, since urban climate archipelagos produces an aggregate impact on temperature, moisture or precipitation [84], future studies should focus on SUHI archipelagos.

Even higher spatial resolution with more temporal sampling and improved better calibrated data would be very useful. The application of higher resolution remote sensing data facilitates study on UHI characteristics and urban climate. Moreover, a future sensor improving on Landsat and aircraft thermal data are some possible options. On the other hand, in order to determine a temporal variation of LST using satellite data at restricted overpass times, it appears necessary to use field data to investigate diurnal UHI in dry environments. Future research should improve methods to simultaneously derive LST and LSE from hyperspectral TIR, multi spectral-temporal, TIR-microwave data, and methods should consider aerosol and cirrus effects [18].

In addition, another viable angle of future studies should focus on mitigation strategies for night-time SUHI and explore surface materials that can reduce surface temperature in urbanised areas in dry regions. Research should look more closely at different parts of the city. Finally, the area needs the development of more research on techniques to reduce LST in rural areas surrounding the cities in dry regions such as the effect of irrigated vegetation in the dry season and increased soil moisture through artificial streams.

Acknowledgments: The authors would like to thank the HCDP Scholarship Programme and Soran University for their financial support of this research. H. Balzter was supported by the Royal Society Wolfson Research Merit Award, 2011/R3 and the NERC National Centre for Earth Observation.

Author Contributions: Azad Rasul conceived the review research. Azad Rasul, Heiko Balzter and Claire Smith all contributed in designing the review, writing and editing it. All other authors contributed in evaluating, improving and editing the paper. The list of authors from the fifth author to the end ordered alphabetically.

Conflicts of Interest: The authors declare no conflict of interest.

References

1. Srivanit, M.; Hokao, K. Thermal Infrared Remote Sensing for Urban Climate and Environmental Studies: An Application for the City of Bangkok, Thailand. *J. Archit. Plan. Res. Stud.* **2012**, *9*, 83–100.
2. Zhang, Z.; Ji, M.; Shu, J.; Deng, Z.; Wu, Y. Surface Urban Heat Island in Shanghai, China: Examining the Relationship between Land Surface Temperature and Impervious Surface Fractions Derived from Landsat ETM+ imagery. *Int. Arch. Photogramm. Remote Sens. Spat. Inf. Sci.* **2008**, *37*, 601–606.
3. Zhang, Y.; Balzter, H.; Liu, B.; Chen, Y. Analyzing the Impacts of Urbanization and Seasonal Variation on Land Surface Temperature Based on Subpixel Fractional Covers Using Landsat Images. *IEEE J. Sel. Top. Appl. Earth Obs. Remote Sens.* **2017**, *10*, 1344–1356. [CrossRef]
4. Zhang, Y.; Harris, A.; Balzter, H. Characterizing fractional vegetation cover and land surface temperature based on sub-pixel fractional impervious surfaces from Landsat TM/ETM+. *Int. J. Remote Sens.* **2015**, *36*, 4213–4232. [CrossRef]
5. Rasul, A.; Balzter, H.; Smith, C. Spatial variation of the daytime Surface Urban Cool Island during the dry season in Erbil, Iraqi Kurdistan, from Landsat 8. *Urban Clim.* **2015**, *14*, 176–186. [CrossRef]
6. Rasul, A.; Balzter, H.; Smith, C. Diurnal and Seasonal Variation of Surface Urban Cool and Heat Islands in the Semi-Arid City of Erbil, Iraq. *Climate* **2016**, *4*, 42. [CrossRef]
7. Bornstein, R. Observations of the Urban Heat Island Effect in New York City. *Am. Meteorol. Soc.* **1968**, *7*, 575–582. [CrossRef]
8. Oke, T.R. The energetic basis of the urban heat island. *Q. J. R. Meteorol. Soc.* **1982**, *108*, 1–24. [CrossRef]
9. Morris, C.J.G.; Simmonds, I. Associations between varying magnitudes of the urban heat island and the synoptic climatology in Melbourne, Australia. *Int. J. Clim.* **2000**, *20*, 1931–1954. [CrossRef]
10. Miao, S.; Chen, F.; LeMone, M.; Tewari, M.; Li, Q.; Wang, Y. An observational and modeling study of characteristics of urban heat island and boundary layer structures in Beijing. *J. Appl. Meteorol. Clim.* **2009**, *48*, 484–501. [CrossRef]
11. Stewart, I.; Oke, T. Newly developed "thermal climate zones" for defining and measuring urban heat island magnitude in the canopy layer. In Proceedings of the T.R. Oke Symposium: Urban Scales, Urban Systems and the Urban Heat Island (Joint between the Timothy R. Oke Symposium and the Eighth Symposium on the Urban Environment), Garmisch-Partenkirchen, Germany, 12 January 2009.
12. Souch, C.; Grimmond, S. Applied climatology: Urban climate. *Prog. Phys. Geogr.* **2006**, *30*, 270–279. [CrossRef]
13. Ukwattage, N.L.; Dayawansa, N.D.K. *Urban Heat Islands and the Energy Demand: An Analysis for Colombo City of Sri Lanka Using Thermal Remote Sensing Data*; Department of Agricultural Engineering, Faculty of Agriculture, University of Peradeniya: Colombo, Sri Lanka, 2012; pp. 124–131.
14. Tan, J.; Zheng, Y.; Tang, X.; Guo, C.; Li, L.; Song, G.; Zhen, X.; Yuan, D.; Kalkstein, A.; Li, F.; et al. The urban heat island and its impact on heat waves and human health in Shanghai. *Int. J. Biometeorol.* **2010**, *54*, 75–84. [CrossRef] [PubMed]
15. Roth, M.; Oke, T.; Emery, W. Satellite-derived urban heat islands from three coastal cities and the utilization of such data in urban climatology. *Int. J. Remote Sens.* **1989**, *10*, 1699–1720. [CrossRef]

16. Barsi, J.A.; Barker, J.L.; Schott, J.R. An Atmospheric Correction Parameter Calculator for a Single Thermal Band Earth-Sensing Instrument. In Proceedings of the 2003 IEEE International Geoscience and Remote Sensing Symposium, Toulouse, France, 21–25 July 2003.

17. Arnfield, A.J. An approach to the estimation of the surface radiative properties and radiation budgets of cities. *Phys. Geogr.* **1982**, *3*, 97–122.

18. Li, Z.-L.; Tang, B.H.; Wu, H.; Ren, H.; Yan, G.; Wan, Z.; Trigo, I.F.; Sobrino, J.A. Satellite-derived land surface temperature: Current status and perspectives. *Remote Sens. Environ.* **2013**, *131*, 14–37. [CrossRef]

19. Voogt, J.A.; Oke, T.R. Thermal remote sensing of urban climates. *Remote Sens. Environ.* **2003**, *86*, 370–384. [CrossRef]

20. Rizwan, A.M.; Dennis, L.Y.C.; Liu, C. A review on the generation, determination and mitigation of Urban Heat Island. *J. Environ. Sci.* **2008**, *20*, 120–128. [CrossRef]

21. Abdullah, H. The Use of Landsat 5 TM Imagery to Detect Urban Expansion and Its Impact on Land Surface Temperatures in The City of Erbil, Iraqi Kurdistan. Master's Thesis, University of Leicester, Leicester, UK, 2012.

22. Streutker, D.R. A remote sensing study of the urban heat island of Houston, Texas. *Int. J. Remote Sens.* **2002**, *23*, 2595–2608. [CrossRef]

23. Zhang, J.; Wang, Y.; Wang, Z. Change analysis of land surface temperature based on robust statistics in the estuarine area of Pearl River (China) from 1990 to 2000 by Landsat TM/ETM+ data. *Int. J. Remote Sens.* **2007**, *28*, 2383–2390. [CrossRef]

24. Khandelwal, S.; Goyal, R.; Kaul, N.; Singhal, V. Study of Land Surface Temperature Variations with Distance from Hot Spots for Urban Heat Island Analysis. In Proceedings of the Geospatial World Forum: Dimensions and Directions of Geospatial Industry, Hyderabad, India, 18–21 January 2011.

25. Rasul, A.; Balzter, H.; Smith, C. Applying a Normalized Ratio Scale Technique to Assess Influences of Urban Expansion on Land Surface Temperature of the Semi-Arid City of Erbil. *Int. J. Remote Sens.* **2017**, *38*, 3960–3980. [CrossRef]

26. Tran, H.; Uchihama, D.; Ochi, S.; Yasuoka, Y. Assessment with satellite data of the urban heat island effects in Asian mega cities. *Int. J. Appl. Earth Obs. Geoinf.* **2006**, *8*, 34–48. [CrossRef]

27. Hafner, J.; Kidder, S.Q. Urban Heat Island Modeling in Conjunction with Satellite-Derived Surface/Soil Parameters. *J. Appl. Meteorol.* **1999**, *38*, 448–465. [CrossRef]

28. Mochida, A.; Murakami, S.; Ojima, T.; Kim, S.; Ooka, R.; Sugiyama, H. CFD analysis of mesoscale climate in the Greater Tokyo area. *J. Wind Eng. Ind. Aerodyn.* **1997**, *67*, 459–477. [CrossRef]

29. Magee, N.; Curtis, J.; Wendler, G. The urban heat island effect at Fairbanks, Alaska. *Theor. Appl. Clim.* **1999**, *64*, 39–47. [CrossRef]

30. Weng, Q.; Lu, D.; Schubring, J. Estimation of land surface temperature—Vegetation abundance relationship for urban heat island studies. *Remote Sens. Environ.* **2004**, *89*, 467–483. [CrossRef]

31. Sun, D.; Kafatos, M. Note on the NDVI-LST relationship and the use of temperature-related drought indices over North America. *Geophys. Res. Lett.* **2007**, *34*, L24406. [CrossRef]

32. Weng, Q.; Lu, D. A sub-pixel analysis of urbanization effect on land surface temperature and its interplay with impervious surface and vegetation coverage in Indianapolis, United States. *Int. J. Appl. Earth Obs. Geoinf.* **2008**, *10*, 68–83. [CrossRef]

33. Schwarz, N.; Schlink, U.; Franck, U.; Großmann, K. Relationship of land surface and air temperatures and its implications for quantifying urban heat island indicators—An application for the city of Leipzig (Germany). *Ecol. Indic.* **2012**, *18*, 693–704. [CrossRef]

34. Szymanowski, M.; Kryza, M. GIS-based techniques for urban heat island spatialization. *Clim. Res.* **2009**, *38*, 171–187. [CrossRef]

35. Su, Y.F.; Foody, G.M.; Cheng, K.S. Spatial non-stationarity in the relationships between land cover and surface temperature in an urban heat island and its impacts on thermally sensitive populations. *Landsc. Urban Plan.* **2012**, *107*, 172–180. [CrossRef]

36. Florio, E.N.; Lele, S.R.; Chi Chang, Y.; Sterner, R.; Glass, G.E. Integrating AVHRR satellite data and NOAA ground observations to predict surface air temperature: A statistical approach. *Int. J. Remote Sens.* **2004**, *25*, 2979–2994. [CrossRef]

37. Mukherjee, S.; Joshi, P.K.; Garg, R.D. Regression-Kriging technique to downscale satellite-derived land surface temperature in heterogeneous agricultural landscape. *IEEE J. Sel. Top. Appl. Earth Obs. Remote Sens.* **2015**, *8*, 1245–1250. [CrossRef]

38. Kalota, D. Exploring relation of land surface temperature with selected variables using geographically weighted regression and ordinary least square methods in Manipur State, India. *Geocarto Int.* **2016**. [CrossRef]

39. Albrecht, F. *Mikrometeorologische Temperaturmessungen vom Flugzeug aus*; Deutscher Wetterdienst: Hesse, Germany, 1952.

40. Combs, A.C. *Techniques and Results of Infrared Surface-Temperature Measurements in New Jersey and Greenland*; U.S. Army Signal Research and Development Laboratory: Fort Monmouth, NJ, USA, 1961.

41. Lorenz, D. *Messungen der Bodenoberflächentemperatur vom Hubschrauber aus:(mit 9 Tabellen im Text)*; Selbstverlag des Deutschen Wetterdienstes: Hesse, Germany, 1962.

42. Wark, D.Q.; Yamamoto, G.; Lienesch, J. Methods of estimating infrared flux and surface temperature from meteorological satellites. *J. Atmos. Sci.* **1962**, *19*, 369–384. [CrossRef]

43. Rao, P.; Winston, J.S. An Investigation of Some Synoptic Capabilities of Atmospheric "Window" Measurments from Satellite TIROSII. *Appl. Meteorol.* **1963**, *2*, 12–23. [CrossRef]

44. Lenschow, D.H.; Dutton, J.A. Surface temperature variations measured from an airplane over several surface types. *J. Appl. Meteorol.* **1964**, *3*, 65–69. [CrossRef]

45. Chen, X.L.; Zhao, H.M.; Li, P.X.; Yin, Z.Y. Remote sensing image-based analysis of the relationship between urban heat island and land use/cover changes. *Remote Sens. Environ.* **2006**, *104*, 133–146. [CrossRef]

46. Shahmohamadi, P.; Cubasch, U.; Sodoudi, S.; Che-Ani, A.I. *Mitigating Urban Heat Island Effects in Tehran Metropolitan Area*; In Tech Open: Rijeka, Croatia, 2012.

47. Srivastava, P.K.; Majumdar, T.J.; Bhattacharya, A.K. Surface temperature estimation in Singhbhum Shear Zone of India using Landsat-7 ETM+ thermal infrared data. *Adv. Space Res.* **2009**, *43*, 1563–1574. [CrossRef]

48. Rao, P. Remote sensing of urban heat islands from an environmental satellite. *Bull. Am. Meterol. Soc.* **1972**, *53*, 647–648.

49. Matson, M.; Mcclain, E.P.; McGinnis, D.F., Jr.; Pritchard, J.A. Satellite Detection of Urban Heat Islands. *Mon. Weather Rev.* **1978**, *106*, 1725–1734. [CrossRef]

50. Price, J.C. Assessment of the urban heat island effect through the use of satellite data. *Mon. Weather Rev.* **1979**, *107*, 1554–1557. [CrossRef]

51. Ottlé, C.; Vidal-Madjar, D. Estimation of land surface temperature with NOAA9 data. *Remote Sens. Environ.* **1992**, *40*, 27–41. [CrossRef]

52. Gutman, G.G. Multi-annual time series of AVHRR-derived land surface temperature. *Adv. Space Res.* **1994**, *14*, 27–30. [CrossRef]

53. Pinheiro, A.C.T.; Mahoney, R.; Privette, J.L.; Tucker, C.J. Development of a daily long term record of NOAA-14 AVHRR land surface temperature over Africa. *Remote Sens. Environ.* **2006**, *103*, 153–164. [CrossRef]

54. Lopez Garcia, M.J.; CASELLES, V.; Melia, J.; PEREZCUEVA, A. NOAA-AVHRR contribution to the analysis of urban heat islands. In Proceedings of the 5th International Colloquium Physical Measurements and Signatures in Remote Sensing, Courchevel, France, 14–18 Januray 1991; pp. 501–504.

55. Lee, H.Y. An application of NOAA AVHRR thermal data to the study of urban heat islands. *Atmos. Environ. Part B Urban Atmos.* **1993**, *27*, 1–13. [CrossRef]

56. Nichol, J. Remote sensing of urban heat islands by day and night. *Photogramm. Eng. Remote Sens.* **2005**, *71*, 613–621. [CrossRef]

57. Tiangco, M.; Lagmay, A.M.F.; Argete, J. ASTER-based study of the night-time urban heat island effect in Metro Manila. *Int. J. Remote Sens.* **2008**, *29*, 2799–2818. [CrossRef]

58. Liu, L.; Zhang, Y. Urban Heat Island Analysis Using the Landsat TM Data and ASTER Data: A Case Study in Hong Kong. *Remote Sens.* **2011**, *3*, 1535–1552. [CrossRef]

59. Xu, Y.; Qin, Z.; Lv, J. Comparative analysis of urban heat island and associated land cover change based in Suzhou city using landsat data. In Proceedings of the 2008 International Workshop on Education Technology and Training & 2008 International Workshop on Geoscience and Remote Sensing, Shanghai, China, 21–22 December 2008.

60. Bajaj, D.N.; Inamdar, A.B.; Vaibhav, V. Temporal Variation of Urban Heat Island Using Landsat Data: A Case Study of Ahmedabad, India. In Proceedings of the 33rd Asian Conference on Remote Sensing, Pattaya, Thailand, 26–30 November 2012.

61. Li, Y.Y.; Zhang, H.; Kainz, W. Monitoring patterns of urban heat islands of the fast-growing Shanghai metropolis, China: Using time-series of Landsat TM/ETM+ data. *Int. J. Appl. Earth Obs. Geoinf.* **2012**, *19*, 127–138. [CrossRef]

62. Clinton, N.; Gong, P. MODIS detected surface urban heat islands and sinks: Global locations and controls. *Remote Sens. Environ.* **2013**, *134*, 294–304. [CrossRef]

63. Lazzarini, M.; Marpu, P.R.; Ghedira, H. Temperature-land cover interactions: The inversion of urban heat island phenomenon in desert city areas. *Remote Sens. Environ.* **2013**, *130*, 136–152. [CrossRef]

64. Voogt, J. How Researchers Measure Urban Heat Islands. Available online: https://swap.stanford.edu/20120109061918/http://www.epa.gov/heatisland/resources/pdf/EPA_How_to_measure_a_UHI.pdf (accessed on 20 April 2017).

65. Peng, S.; Piao, S.; Ciais, P.; Friedlingstein, P.; Ottle, C.; Breon, F.M.; Nan, H.; Zhou, L.; Myneni, R.B. Surface urban heat island across 419 global big cities. *Environ. Sci. Technol.* **2012**, *46*, 696–703. [CrossRef] [PubMed]

66. Stathopoulou, M.; Cartalis, C. Daytime urban heat islands from Landsat ETM+ and Corine land cover data: An application to major cities in Greece. *Sol. Energy* **2007**, *81*, 358–368. [CrossRef]

67. Sobrino, J.A.; Oltra-Carrió, R.; Sòria, G.; Jiménez-Muñoz, J.C.; Franch, B.; Hidalgo, V.; Mattar, C.; Julien, Y.; Cuenca, J.; Romaguera, M.; et al. Evaluation of the surface urban heat island effect in the city of Madrid by thermal remote sensing. *Int. J. Remote Sens.* **2013**, *34*, 3177–3192. [CrossRef]

68. Tomlinson, C.J.; Chapman, L.; Thornes, J.E.; Baker, C.J. Derivation of Birmingham's summer surface urban heat island from MODIS satellite images. *Int. J. Climatol.* **2012**, *32*, 214–224. [CrossRef]

69. Frey, C.M.; Rigo, G.; Parlow, E. Urban radiation balance of two coastal cities in a hot and dry environment. *Int. J. Remote Sens.* **2007**, *28*, 2695–2712. [CrossRef]

70. Shahraiyni, H.T.; Sodoudi, S.; El-Zafarany, A.; Abou El Seoud, T.; Ashraf, H.; Krone, K. A Comprehensive Statistical Study on Daytime Surface Urban Heat Island during Summer in Urban Areas, Case Study: Cairo and Its New Towns. *Remote Sens.* **2016**, *8*, 643. [CrossRef]

71. Shastri, H.; Barik, B.; Ghosh, S.; Venkataraman, C.; Sadavarte, P. Flip flop of Day-night and Summer-Winter Surface Urban Heat Island Intensity in India. *Sci. Rep.* **2017**, *7*, 40178. [CrossRef] [PubMed]

72. Al-Ali, A.; Mubarak, H. The Effect of Land Cover on the Air and Surface Urban Heat Island of a Desert Oasis. Ph.D. Thesis, Durham University, Durham, UK, 2015.

73. Tarleton, L.F.; Katz, R.W. Statistical explanation for trends in extreme summer temperatures at Phoenix, Arizona. *J. Clim.* **1995**, *8*, 1704–1708. [CrossRef]

74. Nasrallah, H.A.; Brazel, A.J.; Balling, R.C. Analysis of the Kuwait City urban heat island. *Int. J. Clim.* **1990**, *10*, 401–405. [CrossRef]

75. Bowler, D.E.; Buyung-Ali, L.; Knight, T.M.; Pullin, A.S. Urban greening to cool towns and cities: A systematic review of the empirical evidence. *Landsc. Urban Plan.* **2010**, *97*, 147–155. [CrossRef]

76. Chang, C.-R.; Li, M.-H.; Chang, S.-D. A preliminary study on the local cool-island intensity of Taipei city parks. *Landsc. Urban Plan.* **2007**, *80*, 386–395. [CrossRef]

77. Frey, C.M.; Rigo, G.; Parlow, E. Investigation of the Daily Urban Cooling Island (UCI) in Two Coastal Cities in an Arid Environment: Dubai and Abu Dhabi (UAE). Available online: https://www.researchgate.net/profile/E_Parlow/publication/242551628_INVESTIGATION_OF_THE_DAILY_URBAN_COOLING_ISLAND_UCI_IN_TWO_COASTAL_CITIES_IN_AN_ARID_ENVIRONMENT_DUBAI_AND_ABU_DHABI_UAE/links/00b7d5298b3c511cac000000/INVESTIGATION-OF-THE-DAILY-URBAN-COOLING-ISLAND-UCI-IN-TWO-COASTAL-CITIES-IN-AN-ARID-ENVIRONMENT-DUBAI-AND-ABU-DHABI-UAE.pdf (accessed on 20 April 2017).

78. Wen, L.J.; Lü, S.H.; Chen, S.Q.; Meng, X.H.; Bao, Y. Numerical simulation of cold island effect in Jinta Oasis summer. *Plateau Meteorol.* **2005**, *24*, 865–871.

79. Li, S.; Mo, H.; Dai, Y. Spatio-temporal pattern of urban cool island intensity and its eco-environmental response in Chang-Zhu-Tan urban agglomeration. *Commun. Inf. Sci. Manag. Eng.* **2011**, *1*, 1–6.

80. Steinecke, K. Urban climatological studies in the ReykjavmHk subarctic environment. *Atmos. Environ.* **1999**, *33*, 4157–4162. [CrossRef]

81. Keramitsoglou, I.; Kiranoudis, C.T.; Ceriola, G.; Weng, Q.; Rajasekar, U. Identification and analysis of urban surface temperature patterns in Greater Athens, Greece, using MODIS imagery. *Remote Sens. Environ.* **2011**, *115*, 3080–3090. [CrossRef]

82. Frey, C.M.; Rigo, G.; Parlow, E.; Marçal, A. The cooling effect of cities in a hot and dry environment. In Global developments in environmental earth observation from space, Proceedings of the 25th Annual Symposium of the European Association of Remote Sensing Laboratories, Porto, Portugal, 6–11 June 2005.

83. Pielke, R.A.; Davey, C.; Morgan, J. Assessing "global warming" with surface heat content. *EOS Trans. Am. Geophys. Union* **2004**, *85*, 210–211. [CrossRef]

84. Shepherd, J.M.; Andersen, T.; Strother, C.; Horst, A.; Bounoua, L.; Mitra, C. Urban Climate Archipelagos: A New Framework for Urban Impacts on Climate. Available online: https://earthzine.org/2013/11/29/urban-climate-archipelagos-a-new-framework-for-urban-impacts-on-climate/ (accessed on 20 April 2017).

land

Article

Modeling Future Land Cover Changes and Their Effects on the Land Surface Temperatures in the Saudi Arabian Eastern Coastal City of Dammam

Muhammad Tauhidur Rahman [1,*], Adel S. Aldosary [1] and Md. Golam Mortoja [2]

[1] Department of City and Regional Planning, King Fahd University of Petroleum and Minerals,
    KFUPM Box 5053, Dhahran 31261, Saudi Arabia; asdosary@kfupm.edu.sa
[2] ICT-GIS Division, Institute of Water Modelling, House #496, Road # 32, New DOHS, Mohakhali, Dhaka 1206,
    Bangladesh; himu3107@gmail.com
*   Correspondence: mtr@kfupm.edu.sa; Tel.: +966-13-860-7364

Academic Editor: Andrew Millington
Received: 30 April 2017; Accepted: 25 May 2017; Published: 29 May 2017

Abstract: Over the past several decades, Saudi cities have experienced rapid urban developments and land use and land cover (LULC) changes. These developments will have numerous short- and long-term consequences including increasing the land surface temperature (LST) of these cities. This study investigated the effects of LULC changes on the LST for the eastern coastal city of Dammam. Using Landsat imagery, the study first detected the LULC using the maximum likelihood classification method and derived the LSTs for the years 1990, 2002, and 2014. Using the classified results, it then modeled the future LULC for 2026 using the Cellular Automata Markov (CAM) model. Finally, using three thematic indices and linear regression analysis, it then modeled the LST for 2026 as well. The built-up area in Dammam increased by 28.9% between 1990 and 2014. During this period, the average LSTs for the LULC classes increased as well, with bare soil and built-up area having the highest LST. By 2026, the urban area is expected to encompass 55% of the city and 98% of the land cover is envisioned to have average LSTs over 41 °C. Such high temperatures will make it difficult for the residents to live in the area.

Keywords: land use and land cover change; urban growth modeling; Cellular Automata Markov (CAM) model; land surface temperature; Saudi Arabia; urban heat island

1. Introduction

Every year, human migrations to cities are causing urban areas to grow and bringing rapid changes to their ecosystem, biodiversity, natural landscapes, and the environment [1,2]. While such growth is a sign of the region's employment growth and economic prosperity, it has numerous short- and long-term consequences. Among the long-term consequences, increases in the city's land surface temperature (LST) from growing urban built-up areas have received wide attention from geographers, urban planners, and climatologists over the past decade [3,4]. Studies show that urban expansion tends to increase urban areas' LSTs by an average of 2–4 °C when compared to their outskirt rural areas [5]. Rising LSTs and urban heat island (UHI) formations have been linked to high energy consumption, air pollution, and human health problems including asthma and heat-stroke related deaths of children and elders [6,7].

Over the past several decades, cities in the Kingdom of Saudi Arabia have been rapidly expanding from economic prosperity and increasing migrating population from villages and working expatriates from neighboring Asian countries [8]. In 1950, only 21% of the Kingdom's residents lived in a major city [9]. That number increased to 58% in 1975, and today, urban areas house 82% of the country's population, with many of the urban residents having migrated to the cities in an effort to seek a modern lifestyle, better employment, and/or educational opportunities [9,10]. To fulfill the needs of the settling residents, multi-story apartment buildings and private villas along with commercial and industrial facilities were built, and areas that were small villages in the 1940s transformed into present day major cities and metropolitan areas. The eastern coastal city of Dammam is one such city; it was a small fishing village in the 1940s that rapidly expanded in the last three decades due to the booming of oil industries and became a major metropolitan city (with its neighboring cities of Dhahran and Khobar) and the capitol of the Eastern Province of Saudi Arabia [11]. Such expansions significantly changed the land use and land cover (LULC) and are expected to affect and increase the city's LSTs.

Examining LULC changes has become an increasing concern throughout the recent decades because of their roles in reducing biodiversity, modifying the ecosystem, and altering the pattern and composition of the regional and global climate [12]. Detecting and monitoring LULC changes through direct field visits can be difficult, time consuming, and are prone to producing inaccurate results. Improvements and integration of remote sensing and Geographic Information Systems (GIS) in the past several decades have resolved some of these limitations and today are powerful tools for assessing, monitoring, and modelling LULC changes [13–16]. They have been utilized to examine LULC changes in Saudi cities as well. Using Landsat TM data, Alwashe and Bokhari [17] studied the changes in vegetation in the city of Al Madinah. Also using Landsat TM data, Al-Rowili et al. [18] monitored and identified decadal urban areal changes in Jeddah between 1988 and 1998. Aljoufie et al. [19] tried to assess the relationships between transportation systems and urban expansion for Jeddah by aerial photographs, SPOT imagery, and the city's Master plan. For the city of Tabuk, Al-Harbi [20] measured agricultural land use changes based on Landsat TM and Spot 5 imagery collected from 1988 to 2008. Al-Gaadi et al. [21] monitored changes from 1990 to 2006 of Dirab region from Landsat data. Finally, Rahman [8] also used Landsat TM, ETM+, and OLI data to examine growth in Al-Khobar from 1990 and 2013.

Since the early 1970s, the application of remote sensing technologies for measuring LSTs and examining the formation and spatial distribution of UHIs has also been quite promising [22]. Using the various available thermal infrared sensors that can collect data at various spatial resolutions, researchers have studied the LST characteristics (per LULC categories) in different urban settings. Using Landsat TM data collected over Twin Cities, Minnesota, Yuan and Bauer [23] studied the relationships among LSTs, normalized difference vegetation index (NDVI), and percentage of impervious surface area (ISA). Xiao et al. [24] measured LSTs from Landsat TM sensor and determined their quantitative relationships with several biophysical and demographic variables for Beijing city. Also in Beijing, Li et al. [25] observed the correlations between the spatio-temporal trends in LSTs obtained from Landsat TM sensor and the configuration of greenspaces (classified from SPOT imagery). By collecting LSTs retrieved from the AVHRR sensor and combining it with land coverage classified from SPOT-HRV data, Dousset and Gourmelon [26] studied the relationships between different land covers and LSTs in the metropolises of Los Angeles and Paris. Recently, Chaudhuri and Mishra [27] compared the LSTs (from Landsat data) among the various types of land coverage in the border cities of Calcutta (India) and Khulna (Bangladesh). Also for Bangladesh, Ahmed et al. [28] first measured the LSTs and the decadal LULC changes in Dhaka metropolitan area from Landsat sensors. They then modelled the growth of the city and simulated the LSTs of the built-up areas for 2029. Among the Middle Eastern cities, El Abidine et al. [29] modelled the heat waves by examining the relationships between LST and the variations among the LULC categories in the entire state of Qatar. A similar study was also conducted by Rasul et al. [30] using Landsat 8 data to compare LSTs in different LULC categories in the northern Kurdistan Iraqi city of Erbil. Lazzarini et al. [31] combined MODIS, ASTER, and Landsat ETM+ data and found the association between LST, NDVI, and surface UHI at the city and district level for the city of Abu Dhabi, United Arab Emirates. For the Kingdom of Bahrain, Radhi and Sharples [32] used a combination of Landsat imagery, statistics, and weather station data to examine urban developments and their impacts in forming UHIs during the last few decades.

Based on a review of the literature, it was found that examining the LULC changes and their impacts on the LSTs is lacking in Saudi Arabia's Eastern Province in general and particularly in its capital city of Dammam. This study aims to fill that gap. Specifically, we first examined the LULC changes within Dammam over the past two decades (1990–2014) using recent and historical archived Landsat satellite data. Using the Landsat's thermal infrared sensor, we then investigated the changes in the LSTs for each LULC category during these two decades. Finally, based on the city's historical development, we modelled future growth of Dammam and its corresponding changes in LSTs for the year 2026. In Sections 2 and 3, the study area's description along with the detailed methodology will be presented. Section 4 will highlight the results and their discussions will be provided in Section 5. Finally, the concluding remarks and paths for further research are presented in Section 6.

2. Study Area

The entire city of Dammam (26.32° N and 49.50° E) with an area of 653 km^2 was the area under consideration for this study (Figure 1). Dammam currently has an estimated population of 1,106,630. Almost 58% of the residents are Saudis with the rest migrating from the neighboring Middle Eastern, Asian, and Western countries [33]. Having a desert climate, the summers are hot and humid (temperatures varying from 30 °C–45 °C) while the winters are cool and dry (10 °C–21 °C) [34].

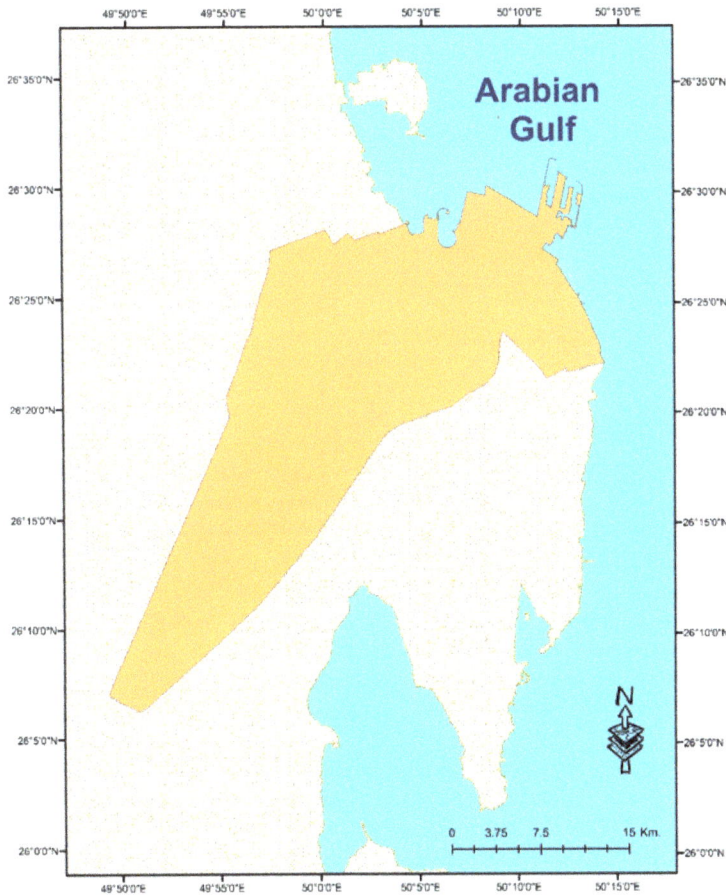

Figure 1. Study area in the eastern coast of Saudi Arabia.

3. Methodology

3.1. Data Collection, Classification, Accuracy Assessments

Three Landsat images processed at level-one terrain-corrected (L1T) level collected over the city of Dammam from the years 1990, 2002, and 2014 were gathered from the United States Geological Survey's (USGS) EarthExplorer website to estimate the LULC changes and calculate the LSTs for the respective years (Table 1). Since L1T data are delivered with corrected radiometric and geometric distortions, no additional geo-rectification or image-to-image registrations were performed. Also, USGS resampled and provided all the Landsat thermal bands to 30 m to align with the multispectral bands of the sensors [35]. Finally, a recent high resolution GeoEye image was acquired for classification and accuracy assessments.

Table 1. Characteristics of Landsat data sets for the study.

Date of Acquisition	Path and Row	Landsat Sensor	Spatial Resolution of Multi-Spectral Bands	Spatial Resolution of Thermal Bands
16 August 1990	164/42	TM	30 m	120 m (resampled to 30 m)
24 July 2002	164/42	ETM+	30 m	60 m (resampled to 30 m)
2 August 2014	164/42	OLI	30 m	100 m (resampled to 30 m)

The imagery was classified into four separate LULC classes based on Level 1 of the Anderson Classification scheme. It is the preferred classification system for classifying Landsat data [36]. Table 2 highlights the LULC features included for each class. To classify the LULC from the three pre-processed Landsat images, the Maximum Likelihood Classification (MLC) technique was used with signatures collected among the four classes identified through field surveys, GeoEye image, and Google Earth imagery.

Table 2. Scheme for land cover classifications.

Land Cover Class	Description
Built-up	Residential, industrial, transportation networks, and commercial infrastructures.
Bare soil	Sand, vacant lands, bare soils.
Vegetation	Trees, parks, playgrounds, grasslands.
Water body	Lakes and coastal water.

To examine the classification accuracies, 295 ground truth reference points were sampled (through stratified random sampling method) across the study area. Historical LULC paper maps (with scales of 1:32,258) were also collected from the Ad-Dammam municipality's office. These historical maps, ground truth reference points, and high resolution GeoEye and Google Earth images were all utilized in combination to measure the overall classification accuracies. For comparison of accuracy results, error matrices were created. The classified results had overall accuracies of 86% (1990), 89% (2002), and 93% (2014) with Kappa coefficients of 0.87, 0.90, and 0.94. Finally, the changes between the classified data sets were examined using a post-classification change detection method.

3.2. Derivation of LSTs

The LSTs for this study were obtained from Landsat's thermal band(s). Landsat TM and ETM+ provide thermal information based on a single long-wave infrared (LWIR) band, whereas it is captured with two LWIR bands by the Landsat OLI sensor [37]. Therefore, two separate methods were used to derive the LSTs from the three sensors. For deriving the LSTs from Landsat TM and ETM+ sensors, the Spectral Radiance Scaling Method was used [38]. The method first uses the following equation to convert digital numbers (*DNs*) into spectral radiance (L_λ):

$$L_\lambda = \frac{(LMAX_\lambda - LMIN_\lambda) \times (DN - QCALMIN)}{QCALMAX - QCALMIN} + LMIN_\lambda \tag{1}$$

where *DN* values range between 0 and 255; $LMIN_\lambda$ and $LMAX_\lambda$ are the minimum and maximum spectral radiances, respectively; and *QCALMIN* and *QCALMAX* are quantized minimum and maximum calibrated pixel values, respectively.

The second step involved utilizing Equation (2) to convert the spectral radiance to Brightness Temperature (in Kelvin) [39,40]. Temperatures were then converted to degrees Celsius.

$$T_K = \frac{K_2}{\ln(\frac{K_1}{L_\lambda} + 1)} \tag{2}$$

To derive the LSTs for the year 2014 from the Landsat OLI data, Equations (3) and (4) were used [41]:

$$TOA_r = mx + b \tag{3}$$

where TOA_r is the top of atmospheric radiance; *m* is the radiance multiplier (0.0003342); *x* is the raw band; and *b* is the radiance add (0.1) [42].

$$T_K = \frac{K_2}{\ln(\frac{K_1}{TOA_r} + 1)} \tag{4}$$

where T_K is the Temperature in degrees Kelvin; K_1 = 774.89 and K_2 = 1321.08. Temperatures were converted to degrees Celsius as well.

3.3. Modeling the Land Cover for 2026

Numerous urban growth prediction models (UGPMs) are available for modelling the future land use and land cover of an urban area [43]. In this study, modelling the LULC for the year 2026 was completed using IDRISI Selva GIS software package. To choose the most accurate modeling technique for forecasting the LULC of Dammam for 2026, we simulated the LULC of Dammam for 2014 using three separate but popular LULC modeling methods (Cellular Automata Markov (CAM), Multi-Layer Perception Markov (MLPM), and Stochastic Markov (SM)) and examined their accuracies. In all of the three models, the LULC for 2014 was simulated using the classification results from 1990 (earlier land cover) and 2002 (later land cover).

To model using the CAM method, a combination of Cellular Automata (CA) with Markov Chain Analysis was utilized. It first involved using the classified LULC data from 1990 and 2002 to produce the Markovian conditional probability areas. Then, Boolean images for each LULC type were generated using the 2002 classified results. Euclidean distances for each Boolean image were then calculated and suitability images using "FUZZY Factor Standardization" were generated. Finally, the CAM method utilized the Markovian conditional probability areas and suitability images to generate the LULC model for 2014. The CAM modeling technique has successfully modelled future LULC changes in several major cities in the world [44–47].

The MLPM modelling method utilizes and combines the concepts of Artificial Neural Network (ANN) and Markov Chain Analysis (MCA). To model, it first calculated the changes in the LULC between 1990 and 2002 and selected the driver variables (distance from bare soil; distance from vegetation; distance form water body; distance from built-up area; and empirical likelihood) to predict the transitions or changes. The strength of each variable was measured using Cramer's Value. The transition sub-model was created and modified until a maximum accuracy of 82.05% was achieved. The models showed that the LULC exhibited three potential transitional forms: bare soil to built-up areas, vegetation to built-up areas, and built-up areas to bare soil. It should be highlighted that conversion of built-up areas to bare soil/sand is usually rare. However, the study area experiences frequent sand-storms and strong winds and roads and highways are frequently found covered with sands due to these natural meteorological phenomena. It is possible for these areas to be classified as built-up areas in one year (when sands are not present) but as soil/sand at a later year when sands cover them. Hence conversion of built-up area to bare soil was found to be a valid potential transitional form. The transition potential map for each transition was produced and the LULCs for 2014 were modeled. Previous studies have found MLPM to produce better prediction accuracies in areas featuring stable and slow LULC changes [48]. Finally, in the SM model, Markovian conditional probability images from the years 1990 to 2002 were produced using the Markov module of IDRISI. The Stochastic choice was then applied to materialize the conditional probability images generated from Markov analysis.

Once the LULCs for 2014 were modeled using the three methods, the Map Comparison Kit (MCK) software was utilized to find the best-fit model. To select the best-fit model, the overall Kappa, Khisto, Klocation, and Fraction correct was calculated and examined [49]. The calculated results are shown in Table 3. Since the CAM model had the highest overall Kappa coefficients and Fraction correct, it was used to simulate the LULC for 2026.

Table 3. Comparison of overall Kappa statistics for three modeling results for 2014.

Kappa Components	CAM	SM	MLPM
Overall Kappa	0.56	0.30	0.45
Overall Klocation	0.67	0.36	0.62
Overall Khisto	0.83	0.83	0.73
Fraction Correct	0.75	0.60	0.69

3.4. Modeling of LST for 2026

After simulating the LULC for 2026, this study modified the methodology proposed by [28] to model and map the distribution of the LSTs across Dammam city. First, several land cover indices of each LULC type for the years 1990, 2002, and 2014 were derived. The indices were the Normalized Difference Bare Soil Index (NDBsI), NDVI, Soil Adjustment Vegetation Index (SAVI), Normalized Difference Water Body Index (NDWI), Modified Normalized Difference Water Body Index (MNDWI), and Normalized Difference Built-up Area Index (NDBI). The equations used to calculate the indices along with their references are given in Table 4.

Table 4. Indices used to model land surface temperature (LST) for 2026.

Index	Equation		Reference
	Landsat TM and ETM+	Landsat OLI	
NDBsI	$\frac{B5 - B6}{B5 + B6}$	$\frac{B6 - B10}{B6 + B10}$	[50]
NDVI	$\frac{B4 - B3}{B4 + B3}$	$\frac{B5 - B4}{B5 + B4}$	[50]
SAVI	$\frac{1.5\,(B4 - B3)}{B4 + B3 + 0.5}$	$\frac{1.5\,(B5 - B4)}{B5 + B4 + 0.5}$	[51]
NDWI	$\frac{B2 - B4}{B2 + B4}$	$\frac{B3 - B5}{B3 + B5}$	[52]
MNDWI	$\frac{B2 - B5}{B2 + B5}$	$\frac{B3 - B6}{B3 + B6}$	[53]
NDBI	$\frac{B5 - B4}{B5 + B4}$	$\frac{B6 - B5}{B6 + B5}$	[54]

Once the indices were derived, simple linear regression analysis was performed to examine their relations with LSTs to select the indices that contribute the most for LST modeling. In the regression analysis, indices were the independent variables and LST was the dependent variable. The NDBI,

NDBsI, and MNDWI were found to be the major significant indices ($p < 0.05$) contributing to LSTs. The three indices were normalized on a scale of 0 to 1 and reclassified into 20 equal classes to fit for the usage of "Markov Chain Analysis". We then conducted "Markov Chain Analysis" and combined Stochastic Choice with Markov Chain Analysis to simulate the selected indices for the year 2026. Using the three normalized indices data and the *LST* for 2014, the following equation (with $r^2 = 0.729$) was formulated to explain their relationships:

$$LST = 81.81 - (17.92 \times NDBsI) - (4.79 \times NDBI) - (60.39 \times MNDWI) \tag{5}$$

where *NDBsI*, *NDBI*, and *MNDWI* are the corresponding indices and *LST* is the land surface temperature. Finally, the equation and simulated 2026 indices were used to model and map the *LST* for 2026.

4. Results

4.1. Changes in LULC in Dammam

The classified LULC results for the years 1990, 2002, and 2014 are given in Figure 2 while Table 5 shows their areal statistics, and changes among classes between the years are provided in Tables 6 and 7. From 1990 to 2014, the amount of bare soil declined significantly (by 16,632 ha or 25.5%). During the same period, the built-up area increased by 18,899 ha (28.9% of the study area). The amount of vegetation declined from the years 1990 to 2002 (by 1415 ha or 2.16% of the study area). However, an increase in vegetation was observed between 2002 and 2014 (by 671 ha or 1% of the study area). Finally, the areas of water slightly increased by 380 ha (0.6% of the study area) between 1990 and 2002 and decreased by 1903 ha (3% of the study area) from 2002 to 2014.

Table 5. Areal statistics (in hectares) of classified land cover for 1990, 2002, and 2014.

Land Cover Class	1990		2002		2014	
	Area	%	Area	%	Area	%
Bare soil/sand	48,863	74.79	44,227	67.69	32,231	49.33
Built-up area	9368	14.34	15,039	23.02	28,267	43.27
Vegetation	1785	2.73	370	0.57	1041	1.59
Water body	5317	8.14	5697	8.72	3794	5.81
Total	65,333	100	65,333	100	65,333	100

Table 6. Change detection matrix showing the class changes (in hectares) between 1990 and 2002.

		2002				
		Bare Soil/Sand	Built-Up Area	Vegetation	Water Body	Total
	Bare soil/sand	42,803	5969	74	17	48,863
	Built-up area	883	8061	37	387	9368
1990	Vegetation	438	949	252	146	1785
	Water body	103	60	7	5147	5317
	Total	44,227	15,039	370	5697	65,333

Table 7. Change detection matrix showing the class changes (in hectares) between 2002 and 2014.

		2014				
		Bare Soil/Sand	Built-Up Area	Vegetation	Water Body	Total
	Bare soil/sand	30,131	13,824	250	22	44,227
	Built-up area	995	13,427	608	9	15,039
2002	Vegetation	49	147	174	0	370
	Water body	1056	869	9	3763	5697
	Total	32,231	28,267	1041	3794	65,333

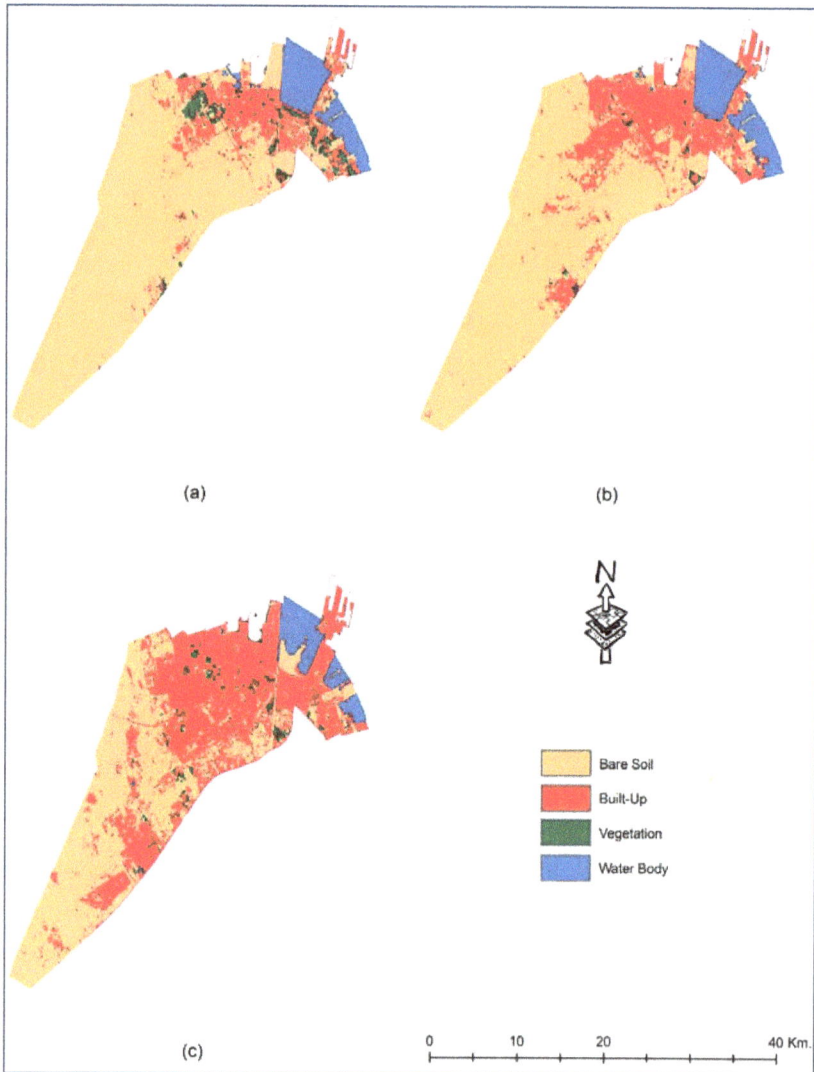

Figure 2. The study area's classification results for (**a**) 1990, (**b**) 2002, and (**c**) 2014.

4.2. Distribution and Changes of LST in LULC in Dammam

The average LSTs of individual LULC classes are provided in Table 8. It was observed that for each year, bare soil had the highest LST followed by built-up area, vegetation, and water bodies. The results also show that the average temperature for all the classes increased between 1990 and 2014. However, while the average temperature for water body increased slightly by 1.6 °C, the average temperature for bare soil, built-up area, and vegetation increased by an average of 7.5 °C. When we examined the changes in the decadal average temperature per class, significant increases (by 7 °C) were seen between 1990 and 2002 for bare soil and built-up area. The rate of increase reduced to 0.7 °C for these classes between 2002 and 2014. For vegetation class, the rates of increase were 3.14 °C (between 1990 and 2002) and 4.16 °C (between 2002 and 2014).

Table 8. Average LSTs in 1990, 2002, and 2014 for different land cover classes.

Land Cover Class	1990	2002	2014
Bare Soil	37.62	44.41	45.09
Built-Up Area	36.42	43.42	44.12
Vegetation	35.71	38.85	43.01
Water Body	29.04	30.69	30.64

The spatial distributions of the temperatures for 1990, 2002, and 2014 are given in Figure 3, while the relationship between temperature and areal coverage is provided in Table 9. In 1990, the entire study area had LSTs less than 40 °C, with the majority (87.07%) of the land cover having LSTs between 36 °C and 40 °C. In 2002, 77% of the land cover had LSTs between 41 °C and 50 °C. However, the percentage of LSTs between 41 °C and 50 °C increased to 91.26% in 2014, suggesting the study area's LSTs increased between 1990 and 2014, although it increased more in the last decade.

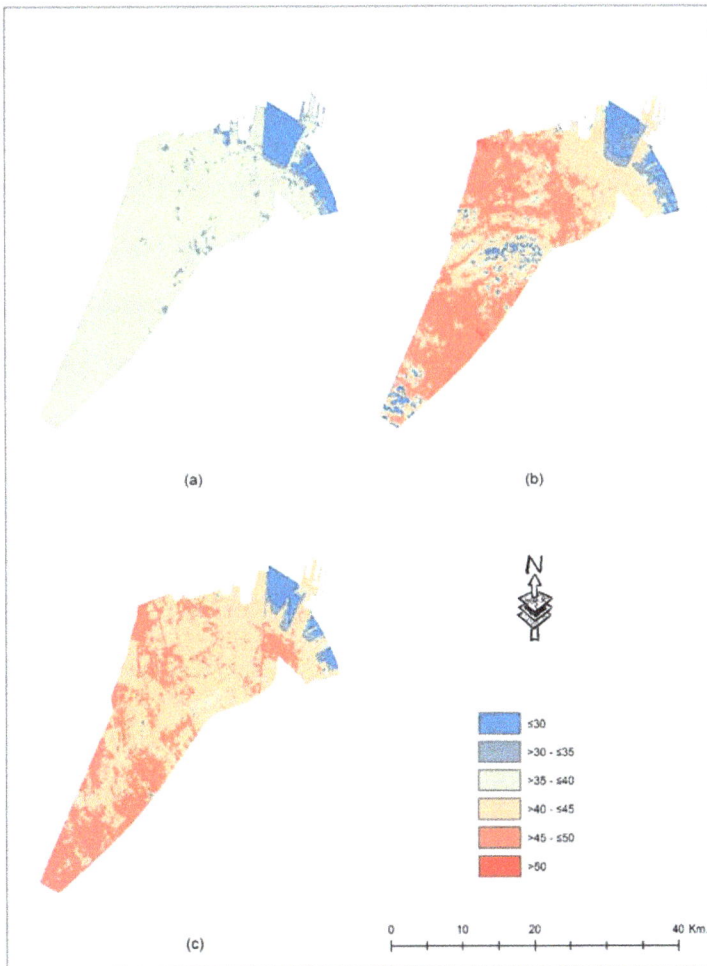

Figure 3. LST distributions for (a) 1990, (b) 2002, and (c) 2014.

Table 9. Distribution of areal coverage among different LST ranges for 1990, 2002, 2014, and 2026.

Ranges of LST (°C)	Areal Coverage (%)			
	1990	2002	2014	2026
≤30	7.86	6.14	3.81	0
31 to 35	5.07	6.44	2.3	0.29
36 to 40	87.07	9.9	2.58	1.38
41 to 45	0	36.11	58.69	2.35
46 to 50	0	40.9	32.57	35.06
>50	0	0.51	0.05	60.92

4.3. Modeling of LULC and LST for 2026

Figure 4 shows the simulated LULC for the year 2026 and Table 10 provides their areal statistics. The modeled LULC shows that 35,986 ha (55% of the study area) will change to built-up area, an increase of almost 27.3% from 2014. This increase will mostly occur in the city's northern and southeastern parts. Vegetation will also increase in the city from 1041 ha in 2014 to 3240 ha in 2026 (a gain of 211%) in central northern parts of Dammam and along the northern coastline. This increase in built-up area and vegetation will reduce the amount of bare soil by 8306 ha (a net loss of almost 26%) and water bodies by 1612 ha (loss of roughly 42.5%) from 2014 to 2026. Water bodies will be mostly lost due to new construction of residential and commercial zones along the eastern seashores of the city.

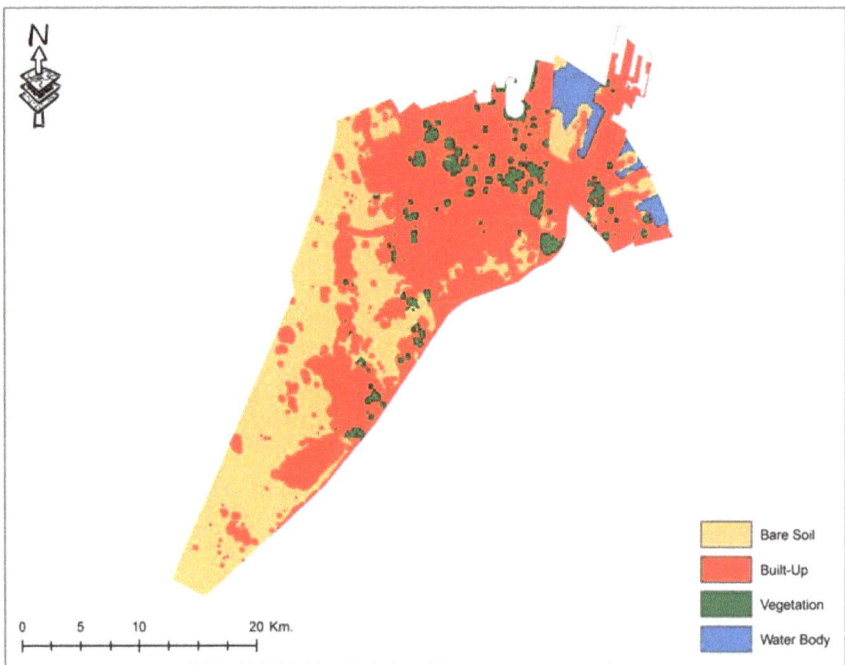

Figure 4. Modeled land use and land cover (LULC) for Dammam for 2026.

The spatial distribution of the modeled 2026 LST is provided in Figure 5 and the areal percentage statistics by LST range is shown in the last column of Table 9. Compared to the LST ranges of 1990, 2002, and 2014, most of the land coverage (98%) is predicted to have LSTs over 41 °C in 2026. The average LSTs for built-up area are forecasted to be 46 °C.

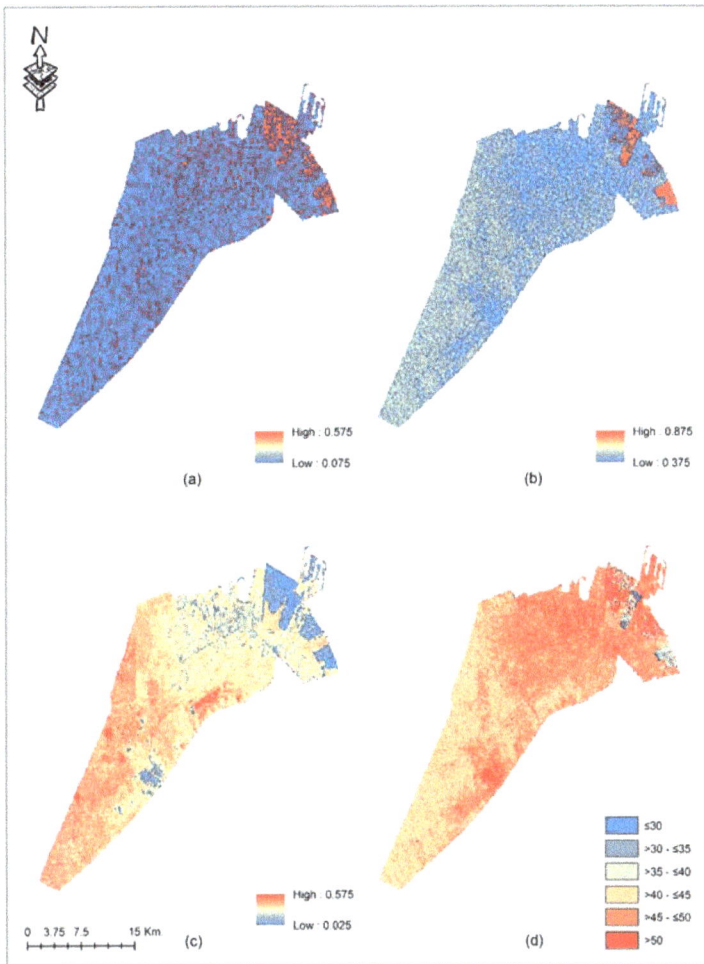

Figure 5. Distribution of the indices ((**a**) NDBI; (**b**) MNDWI; and (**c**) NDBsI) used to model the 2026 LST (**d**) for Dammam.

Table 10. Modelled areal statistics (in hectares) of land cover for 2026.

Land Cover Class	Area	%
Bare soil	23,925	36.62
Built-up area	35,986	55.08
Vegetation	3240	4.96
Water body	2182	3.34
Total	65,333	100

5. Discussions

5.1. Changes in LULC and LSTs

This paper first examined the changes in the LULC in the city of Dammam between 1990 and 2014. The city's urban areas expanded by 61% from 1990 to 2002 and 88% between 2002 and 2014.

Such decadal growth is higher than other cities in Saudi Arabia [19,55]. It is also higher than other developing cities in the world, including Kathmandu [13], Tripoli [56], and Dhaka [57]. Rapid population growth and economic prosperity are the primary reasons for such rapid urban growth [58]. Dammam's population increased from 127,844 (1974) to 260,048 (1986), and finally to 918,154 in 2010 [33,59] due to rapid migration of refugees during and after the 1991 Gulf war from neighboring countries of Iraq and Kuwait. The results also show that the vegetation in the study area decreased between 1990 and 2002 and increased between 2002 and 2014. Previous studies have also found similar patterns of decreasing vegetation coverage between the early 1990s and 2000 and an increase from the early 2000s to 2014 in other Saudi cities [55,60]. They suggested that increasing population resulted in vegetation increases in Saudi cities [60].

In tropical and sub-tropical urban environments, the LSTs are dependent upon the LULC, with urban built-up areas having the highest LSTs and significantly contributing to the formation of UHIs [24,28]. However, having a desert climate, the study area's bare soil (mostly sands) had the highest mean LSTs during the day followed by urban built-up areas in the all of the three years considered. The mean LSTs for vegetative areas were lower than urban built-up areas. Similar resulting patterns were also found in other neighboring desert cities of Abu Dhabi and Dubai, suggesting an inversion of UHIs where city centers are generally cooler than the outskirts of the city due to low vegetation coverage and sand being the main reflecting surface [31,61]. In both cities, a reduction of 5 °C (Abu Dhabi) to 12 °C (Dubai) of mean LSTs were due to the presence of green vegetative areas. For our study area, the vegetative areas lowered the mean LSTs by an average of 2 °C. We believe this slight lowering is due to the very low amount of vegetation present in Dammam (only 1042 ha in 2014), indicating that the reduction in LSTs is correlated to the amount of vegetation present in an area.

It was also found in this study that the mean LSTs of 2014 increased by an average of 7.5 °C when compared to the mean LSTs of 1990. Increases of mean LSTs were also found in Dubai and the semi-arid desert city of Santiago, Chile [61,62]. As highlighted by previous studies, the rapid population growth along with the urban expansion could be the contributing factors to such increases in the temperatures [63].

5.2. LULC and LST Modeling for 2026

The land use and land cover of Dammam are expected to change over the next decade, as 27% of the current LULC is modelled to be converted into urban built-up areas. This decadal growth rate is comparable to other cities in the world, including Setubal and Sesimbra (in Protugal with 25%), Asmara (in Eritrea with 25%), and Beijing (with 31%) [64–66]. Such growth will lead to urban sprawl and will have several benefits and consequences. The benefits include development of industrial infrastructures and facilities which can provide employment opportunities for the residents from small cities and rural areas of the kingdom as well as from other neighboring Arab countries and developing countries of South and Southeast Asia (i.e., Pakistan, Bangladesh, Indonesia, and the Philippines). The expansion can also provide better business, educational, and medical facilities for its residents.

The negative impacts are numerous and most often they outweigh the benefits of the urban expansions. With increasing employment opportunities, the population is projected to grow by almost 20% over the next decade (1,057,256 in 2015 to 1,264,227 in 2025) [33]. Due to increasing population and city expansions, travelling distances for the residents are expected to increase, resulting in more fuel consumption and traffic congestions. The high fuel consumptions will result in rising air pollution and cause various health problems for the elderly and children of the city. The cost for providing public utility services is also expected to rise. Urbanization has also been known to cause social disparities among the residents [67]. Finally, the LST of the city is expected to increase as well from the increasing built-up areas.

The LSTs modelled in this study for 2026 show that the majority of the city will have LSTs over 41 °C, with the average LSTs for built-up areas forecasted to be 46 °C. This is significantly higher than the modelled LSTs of cities in the tropical regions of the world [28]. Recently, Pal and

Eltahir [68] simulated the dry and wet-bulb temperatures for Middle Eastern cities between the years 2071–2100 using a regional climate model. Their study shows that by 2100, the regional average wet-bulb temperature will exceed 35 °C several times in the year and the average maximum dry-bulb temperature exceeding 45 °C in the low lying coastal cities of the region (i.e., Abu Dhabi, Dammam/Dhahran, and Dubai) will become the norm in the July, August, and September summer months. The LSTs modelled in this study are 1 °C higher than the dry-bulb temperature estimated by Pal and Eltahir [68]. This is to be expected, since LSTs and air-temperatures are highly correlated and as the air temperature increases, LST values will tend to be higher than the air temperature [69]. Such extreme high temperatures will be dangerous for human health as well as for animal and plant species. Instead of the end of the century as predicted by Pal and Eltahir [68], these results of this study suggest that Dammam city may experience very high temperatures that may be difficult for human inhabitability within the next one to two decades.

6. Conclusions

This study compared three separate Landsat images to evaluate LULC changes over the last two decades in Dammam, the capitol of Saudi Arabia's Eastern Province. It also examined the trends in the LSTs during these periods and their relationships with the four major LULC classes. Finally, based on the changes, this study projected the LULC and the LSTs for the year 2026. Since 1990, the urban area in Dammam has increased, resulting in decreasing bare soil. The results also show that the average LSTs have increased in the last two decades. If such a trend continues, built-up areas along with the LSTs will continue to increase over the next decade. Such increases in built up areas along with their temperatures will have numerous medical, environmental, and social impacts and consequences.

The study results will be beneficial for Dammam's government officials and planners, who can ensure that the city is growing in a restrictive manner by utilizing and comparing this study's maps with the city's future master plan. They can also create rules and regulations and create strategies that can reduce the LSTs in the city. Future studies should examine in detail the consequences and problems faced by the residents of Dammam due to urban expansions and LST increases as well as how to mitigate them. The growth and the distribution of LSTs of other cities in Saudi Arabia should also be examined and modeled to ensure they are growing in a sustainable manner.

Acknowledgments: The authors acknowledge and appreciate the financial support provided by King Abdulaziz City for Science and Technology (KACST) through the Science & Technology Unit at King Fahd University of Petroleum & Minerals (KFUPM) for funding this work through project number 13-ENE198-04 as part of the "National Science, Technology and Innovation Plan (NSTIP)" program. They also thank the King Fahd University of Petroleum and Minerals for providing the equipment and technical resources and funds to publish this study. Finally, we would also like to thank the two anonymous referees and the editors for their valuable comments for improving this manuscript.

Author Contributions: M.T. Rahman wrote Sections 1, 2 and 4–6 and edited the entire manuscript. A.S. Aldosary and M.G. Mortoja formulated the research design and edited the manuscript. M.G. Mortoja wrote Section 3 and collected and processed all the data for the study.

Conflicts of Interest: The authors declare no conflict of interest.

References

1. McKinney, M.L. Urbanization, Biodiversity, and Conservation. *Bioscience* **2002**, *52*, 883. [CrossRef]
2. Uttara, S.; Elliot, S. Impacts of Urbanization on Environment. *Int. J. Res. Eng. Appl. Sci.* **2012**, *2*, 1637–1645.
3. Maimaitiyiming, M.; Ghulam, A.; Tiyip, T.; Pla, F.; Latorre-Carmona, P.; Halik, Ü.; Sawut, M.; Caetano, M. Effects of green space spatial pattern on land surface temperature: Implications for sustainable urban planning and climate change adaptation. *ISPRS J. Photogramm. Remote Sens.* **2014**, *89*, 59–66. [CrossRef]
4. Gaffin, S.R.; Rosenzweig, C.; Khanbilvardi, R.; Parshall, L.; Mahani, S.; Glickman, H.; Goldberg, R.; Blake, R.; Slosberg, R.B.; Hillel, D. Variations in New York city's urban heat island strength over time and space. *Theor. Appl. Climatol.* **2008**, *94*, 1–11. [CrossRef]

5.   Wong, N.H.; Jusuf, S.K. GIS-based greenery evaluation on campus master plan. *Landsc. Urban Plan.* **2008**, *84*, 166–182. [CrossRef]

6.   Lai, L.W.; Cheng, W.L. Urban heat island and air pollution–an emerging role for hospital respiratory admissions in an urban area. *J. Environ. Health* **2010**, *72*, 32–35. [PubMed]

7.   Priyadarsini, R. Urban Heat Island and its Impact on Building Energy Consumption. *Adv. Build. Energy Res.* **2009**, *3*, 261–270. [CrossRef]

8.   Rahman, M.T. Detection of Land Use/Land Cover Changes and Urban Sprawl in Al-Khobar, Saudi Arabia: An Analysis of Multi-Temporal Remote Sensing Data. *ISPRS Int. J. Geo-Inf.* **2016**, *5*, 15. [CrossRef]

9.   Al-Ahmadi, F.; Hames, A. Comparison of Four Classification Methods to Extract Land Use and Land Cover from Raw Satellite Images for Some Remote Arid Areas, Kingdom of Saudi Arabia. *J. King Abdulaziz Univ. Sci.* **2009**, *20*, 167–191. [CrossRef]

10.   Madugundu, R.; Al-Gaadi, K.A.; Patil, V.C.; Tola, E. Detection of Land Use and Land Cover Changes in Dirab Region of Saudi Arabia Using Remotely Sensed Imageries. *Am. J. Environ. Sci.* **2014**, *10*, 8–18. [CrossRef]

11.   Al-Hathloul, S.; Rahman, M.A. Dynamism of metropolitan areas: The case of metropolitan Dammam, Saudi Arabia. *J. Gulf Arab. Penins. Stud.* **2003**, *29*, 11–43.

12.   Aguilar, A.G.; Ward, P. Globalization, regional development, and mega-city expansion in Latin America: Analyzing Mexico City's peri-urban hinterland. *Cities* **2003**, *20*, 3–21. [CrossRef]

13.   Thapa, R.B.; Murayama, Y. Examining Spatiotemporal Urbanization Patterns in Kathmandu Valley, Nepal: Remote Sensing and Spatial Metrics Approaches. *Remote Sens.* **2009**, *1*, 534–556. [CrossRef]

14.   Dewan, A.M.; Yamaguchi, Y. Land use and land cover change in Greater Dhaka, Bangladesh: Using remote sensing to promote sustainable urbanization. *Appl. Geogr.* **2009**, *29*, 390–401. [CrossRef]

15.   Jamali, N.A.; Rahman, M.T. Utilization of Remote Sensing and GIS to Examine Urban Growth in the City of Riyadh, Saudi Arabia. *J. Adv. Inf. Technol.* **2016**, *7*, 297–301.

16.   Rahman, M.T.; Rashed, T. Urban tree damage estimation using airborne laser scanner data and geographic information systems: An example from 2007 Oklahoma ice storm. *Urban For. Urban Green.* **2015**, *14*, 562–572. [CrossRef]

17.   Alwashe, M.A.; Bokhari, A.Y. Monitoring vegetation changes in Al Madinah, Saudi Arabia, using Thematic Mapper data. *Int. J. Remote Sens.* **1993**, *14*, 191–197. [CrossRef]

18.   Al-Rowili, M.S.; Fadda, E.H.; Vaughan, R.A. A Comparison of Data Fusion and Unsupervised Classification for Change Detection in Jeddah, Saudi Arabia. In Proceedings of the 22nd Symposium of the European Association of Remote Sensing Laboratories, Prague, Czech Republic, 4–6 June 2002.

19.   Aljoufie, M.; Zuidgeest, M.; Brussel, M.; Maarseveen, M.V. Spatial–temporal analysis of urban growth and transportation in Jeddah City, Saudi Arabia. *Cities* **2013**, *31*, 57–68. [CrossRef]

20.   Al-Harbi, K.M. Monitoring of agricultural area trend in Tabuk region—Saudi Arabia using Landsat TM and SPOT data. *Egypt. J. Remote Sens. Sp. Sci.* **2010**, *13*, 37–42. [CrossRef]

21.   Al-Gaadi, K.A.; Samdani, M.S.; Patil, V.C. Assessment of Temporal Land Cover Changes in Saudi Arabia Using Remotely Sensed Data Precision Agriculture Research Chair, College of Food and Agriculture Sciences. *Middle-East J. Sci. Res.* **2011**, *9*, 711–717.

22.   Rao, P.K. Remote sensing of urban "heat islands" from an environmental satellite. *Bull. Am. Meteorol. Soc.* **1972**, *53*, 647–648.

23.   Yuan, F.; Bauer, M.E. Comparison of impervious surface area and normalized difference vegetation index as indicators of surface urban heat island effects in Landsat imagery. *Remote Sens. Environ.* **2007**, *106*, 375–386. [CrossRef]

24.   Xiao, R.B.; Weng, Q.H.; Ouyang, Z.Y.; Li, W.F.; Schienke, E.W.; Zhang, Z.M. Land surface temperature variation and major factors in Beijing, China. *Photogramm. Eng. Remote Sens.* **2008**, *74*, 451–461. [CrossRef]

25.   Li, X.; Zhou, W.; Ouyang, Z.; Xu, W.; Zheng, H. Spatial pattern of greenspace affects land surface temperature: Evidence from the heavily urbanized Beijing metropolitan area, China. *Landsc. Ecol.* **2012**, *27*, 887–898. [CrossRef]

26.   Dousset, B.; Gourmelon, F. Satellite multi-sensor data analysis of urban surface temperatures and landcover. *J. Photogramm. Remote Sens.* **2003**, *58*, 43–54. [CrossRef]

27.   Chaudhuri, G.; Mishra, N.B. Spatio-temporal dynamics of land cover and land surface temperature in Ganges-Brahmaputra delta: A comparative analysis between India and Bangladesh. *Appl. Geogr.* **2016**, *68*, 68–83. [CrossRef]

28. Ahmed, B.; Kamruzzaman, M.; Zhu, X.; Rahman, M.S.; Choi, K. Simulating Land Cover Changes and Their Impacts on Land Surface Temperature in Dhaka, Bangladesh. *Remote Sens.* **2013**, *5*, 5969–5998. [CrossRef]
29. El Abidine, E.M.Z.; Mohieldeen, Y.E.; Mohamed, A.A.; Modawi, O.; AL-Sulaiti, M.H. Heat wave hazard modelling: Qatar case study. *QScience Connect* **2014**. [CrossRef]
30. Rasul, A.; Balzter, H.; Smith, C. Spatial variation of the daytime Surface Urban Cool Island during the dry season in Erbil, Iraqi Kurdistan, from Landsat 8. *Urban Clim.* **2015**, *14*, 176–186. [CrossRef]
31. Lazzarini, M.; Marpu, P.R.; Ghedira, H. Temperature-land cover interactions: The inversion of urban heat island phenomenon in desert city areas. *Remote Sens. Environ.* **2013**, *130*, 136–152. [CrossRef]
32. Radhi, H.; Sharples, S. Quantifying the domestic electricity consumption for air-conditioning due to urban heat islands in hot arid regions. *Appl. Energy* **2013**, *112*, 371–380. [CrossRef]
33. General Authority for Statistics Population Statistics. Available online: http://www.stats.gov.sa/en/ (accessed on 15 October 2014).
34. WeatherSpark Historical Weather for 2014 in Dammam, Saudi Arabia. Available online: https://weatherspark.com/history/32759/2014/Dammam-Eastern-Province-Saudi-Arabia (accessed on 1 January 2015).
35. NASA Landsat 4–7 Thermal Data to be Resampled to 30 Meters. Available online: https://landsat.gsfc.nasa.gov/landsat-4--7-thermal-data-to-be-resampled-to-30-meters/ (accessed on 30 April 2017).
36. Rozenstein, O.; Karnieli, A. Comparison of methods for land-use classification incorporating remote sensing and GIS inputs. *Appl. Geogr.* **2011**, *31*, 533–544. [CrossRef]
37. Rozenstein, O.; Qin, Z.; Derimian, Y.; Karnieli, A. Derivation of land surface temperature for landsat-8 TIRS using a split window algorithm. *Sensors (Switzerland)* **2014**, *14*, 5768–5780. [CrossRef] [PubMed]
38. Coll, C.; Galve, J.M.; Sánchez, J.M.; Caselles, V. Validation of Landsat-7/ETM+ Thermal-Band Calibration and Atmospheric Correction With Ground-Based Measurements. *IEEE Trans. Geosci. Remote Sens.* **2010**, *48*, 547–555. [CrossRef]
39. Ma, Y.; Kuang, Y.; Huang, N. Coupling urbanization analyses for studying urban thermal environment and its interplay with biophysical parameters based on TM/ETM+ imagery. *Int. J. Appl. Earth Obs. Geoinf.* **2010**, *12*, 110–118. [CrossRef]
40. Wukelic, G.E.; Gibbons, D.E.; Martucci, H.P.; Foote, H.P. Radiometric calibration of Landsat Thematic Mapper thermal band. *Remote Sens. Environ.* **1989**, *28*, 339–347. [CrossRef]
41. Butler, K. Deriving Temperature from Landsat 8 Thermal Bands (TIRS). Available online: http://blogs.esri.com/esri/arcgis/2014/01/06/deriving-temperature-from-landsat-8-thermal-bands-tirs/ (accessed on 3 January 2015).
42. Almutairi, M.K. Derivation of Urban Heat Island for Landsat-8 TIRS Riyadh City (KSA). *J. Geosci. Environ. Prot.* **2015**, *3*, 18–23. [CrossRef]
43. Triantakonstantis, D.; Mountrakis, G. Urban Growth Prediction: A Review of Computational Models and Human Perceptions. *J. Geogr. Inf. Syst.* **2012**, *4*, 555–587. [CrossRef]
44. Guan, D.G.; Li, H.F.; Inohae, T.; Su, W.; Nagaie, T.; Hokao, K. Modeling urban land use change by the integration of cellular automaton and Markov model. *Ecol. Modell.* **2011**, *222*, 3761–3772. [CrossRef]
45. Moghadam, H.S.; Helbich, M. Spatiotemporal urbanization processes in the megacity of Mumbai, India: A Markov chains-cellular automata urban growth model. *Appl. Geogr.* **2013**, *40*, 140–149. [CrossRef]
46. Gong, W.; Li, Y.; Fan, W.; Stott, P. Analysis and simulation of land use spatial pattern in Harbin prefecture based on trajectories and cellular automata-Markov modelling. *Int. J. Appl. Earth Obs. Geoinf.* **2015**, *34*, 207–216. [CrossRef]
47. Halmy, M.W.A.; Gessler, P.E.; Hicke, J.A.; Salem, B.B. Land use/land cover change detection and prediction in the north-western coastal desert of Egypt using Markov-CA. *Appl. Geogr.* **2015**, *63*, 101–112. [CrossRef]
48. Roy, H.G.; Fox, D.M.; Emsellem, K. Predicting Land Cover Change in a Mediterranean Catchment at Different Time Scales. In Proceedings of the 14th International Conference on Computational Science and Its Applications (ICCSA 2014), Guimarães, Portugal, 30 June–3 July 2014; Murgante, B., Misra, S., Rocha, A.M.A.C., Torre, C., Rocha, J.G., Falcao, M.I., Taniar, D., Apduhan, B.O., Gervasi, O., Eds.; Springer International: Cham, Switzerland, 2014.; Volume 8582, pp. 315–330.
49. Visser, H.; Najs, T.D. The Map Comparison Kit. *Environ. Model. Softw.* **2006**, *21*, 346–358. [CrossRef]
50. Chen, X.-L.; Zhao, H.-M.; Li, P.-X.; Yin, Z.-Y. Remote sensing image-based analysis of the relationship between urban heat island and land use/cover changes. *Remote Sens. Environ.* **2006**, *104*, 133–146. [CrossRef]

51. Huete, A.R. A soil-adjusted vegetation index (SAVI). *Remote Sens. Environ.* **1988**, *25*, 295–309. [CrossRef]

52. McFeeters, S.K. The use of the Normalized Difference Water Index (NDWI) in the delineation of open water features. *Int. J. Remote Sens.* **1996**, *17*, 1425–1432. [CrossRef]

53. Xu, H. A Study on Information Extraction of Water Body with the Modified Normalized Difference Water Index (MNDWI). *J. Remote Sens.* **2005**, *9*, 511–517.

54. Zha, Y.; Gao, J.; Ni, S. Use of normalized difference built-up index in automatically mapping urban areas from TM imagery. *Int. J. Remote Sens.* **2003**, *24*, 583–594. [CrossRef]

55. Alqurashi, A.F.; Kumar, L. Land Use and Land Cover Change Detection in the Saudi Arabian Desert Cities of Makkah and Al-Taif Using Satellite Data. *Adv. Remote Sens.* **2014**, *3*, 106–119. [CrossRef]

56. Al-sharif, A.A.A.; Pradhan, B.; Shafri, H.Z.M.; Mansor, S. Spatio-temporal analysis of urban and population growths in tripoli using remotely sensed data and GIS. *Indian J. Sci. Technol.* **2013**, *6*, 5134–5142.

57. Corner, R.J.; Dewan, A.M.; Chakma, S. Monitoring and Prediction of Land-Use and Land-Cover (LULC) Change. In *Dhaka Megacity: Geospatial Perspectives on Urbanisation, Environment and Health*; Dewan, A.M., Corner, R.J., Eds.; Springer: Dordrecht, The Netherlands, 2014; pp. 75–97.

58. Abou-Korin, A.A. Impacts of Rapid Urbanisation in the Arab World: the Case of Dammam Metropolitan Area, Saudi Arabia. In *5th Int'l Conference and Workshop on Built Environment in Developing Countries (ICBEDC 2011)*; University Sains Malaysia: Pulao Pinang, Malaysia, 2011; pp. 1–25.

59. Al-Hathloul, S.; Mughal, M.A. Urban growth management-the Saudi experience. *Habitat Int.* **2004**, *28*, 609–623. [CrossRef]

60. Aina, Y.A.; Merwe, J.H.V.; Alshuwaikhat, H.M. Urban Spatial Growth and Land Use Change in Riyadh: Comparing Spectral Angle Mapping and Band Ratioing Techniques. In Proceedings of the Academic Track of the 2008 Free and Open Source Software for Geospatial (FOSS4G) Conference, Incorporating the GISSA 2008 Conference, Cape Town, South Africa, 29 September–4 October 2008; pp. 51–57.

61. Abdi, A. Using Landsat ETM+ to Assess Land Cover and Land Surface Temperature Change in Dubai between 1999 and 2009. Available online: http://www.hakimabdi.com/20120729/using-landsat-etm-to-assess-land-cover-and-land-surface-temperature-change-in-dubai-between-1999-and-2009/ (accessed on 3 March 2016).

62. Peña, M.A. Examination of the Land Surface Temperature Response for Santiago, Chile. *Photogramm. Eng. Remote Sens.* **2009**, *75*, 1191–1200. [CrossRef]

63. Brazel, A.; Selover, N.; Vose, R.; Heisler, G. The tale of two climates—Baltimore and Phoenix urban LTER sites. *Clim. Res.* **2000**, *15*, 123–135. [CrossRef]

64. Tewolde, M.G.; Cabral, P. Urban Sprawl Analysis and Modeling in Asmara, Eritrea. *Remote Sens.* **2011**, *3*, 2148–2165. [CrossRef]

65. Araya, Y.H.; Cabral, P. Analysis and Modeling of Urban Land Cover Change in Setúbal and Sesimbra, Portugal. *Remote Sens.* **2010**, *2*, 1549–1563. [CrossRef]

66. Han, H.; Yang, C.; Song, J. Scenario Simulation and the Prediction of Land Use and Land Cover Change in Beijing, China. *Sustainability* **2015**, *7*, 4260–4279. [CrossRef]

67. Bhatta, B. Causes and Consequences of Urban Growth and Sprawl. In *Analysis of Urban Growth and Sprawl from Remote Sensing Data*; Advances in Geographic Information Science; Springer: Berlin/Heidelberg, Germany, 2010; pp. 17–36.

68. Pal, J.S.; Eltahir, E.A.B. Future temperature in southwest Asia projected to exceed a threshold for human adaptability. *Nat. Clim. Chang.* **2016**, *6*, 197–200. [CrossRef]

69. Li, Z.; Guo, X.; Dixon, P.; He, Y. Applicability of Land Surface Temperature (LST) estimates from AVHRR satellite image composites in northern Canada. *Prairie Perspect.* **2008**, *11*, 119–130.

MDPI

St. Alban-Anlage 66

4052 Basel

Switzerland

Tel. +41 61 683 77 34

Fax +41 61 302 89 18

www.mdpi.com

Land Editorial Office

E-mail: land@mdpi.com

www.mdpi.com/journal/land

www.ingramcontent.com/pod-product-compliance
Lightning Source LLC
Chambersburg PA
CBHW051852210326
41597CB00033B/5865